Plumbing:
Mechanical Services

We work with leading authors to develop
the strongest educational materials in Technology,
bringing cutting-edge thinking and best
learning practice to a global market.

Under a range of well-known imprints, including
Prentice Hall, we craft high quality print and
electronic publications which help readers to understand
and apply their content, whether studying or at work.

To find out more about the complete range of our
publishing, please visit us on the World Wide Web at:
www.pearsoneduc.com

Plumbing: Mechanical Services

Book 2

Fourth Edition

G.J. Blower

Eng Tech (CEI), MIP, LCGI, Technical Teachers Cert.

Formerly Senior Lecturer, Plumbing Mechanical Services Section, College of North East London.

Currently an NVQ Assessor and Training Advisor for J.T.L.

An imprint of **Pearson Education**

Harlow, England · London · New York · Reading, Massachusetts · San Francisco · Toronto · Don Mills, Ontario · Sydney
Tokyo · Singapore · Hong Kong · Seoul · Taipei · Cape Town · Madrid · Mexico City · Amsterdam · Munich · Paris · Milan

Pearson Education Limited
Edinburgh Gate
Harlow
Essex CM20 2JE

and Associated Companies throughout the world

Visit us on the World Wide Web at:
www.pearsoneduc.com

First published 1984
Second edition published 1989
Third edition published 1996
Fourth edition published 2002

ISBN 0582 43229 4

British Library Cataloguing-in-Publication Data
A catalogue record for this book is available from the British Library

Library of Congress Cataloging-in-Publication Data
Blower, G.J.
    Plumbing : mechanical services. Book 2 / G.J. Blower.—4th ed.
        p. cm.
    Includes bibliographical references and index.
    ISBN 0–582–43229–4 (pbk.)
    1. Plumbing. I. Title.

TH6123 .B58 2002
696′.1—dc21   2001059149

10 9 8 7 6 5 4 3
06 05 04 03 02

Typeset in 10/12 Times by 35
Printed and bound by in Malaysia, LSP

Contents

Foreword

It is said that every man is a builder by instinct and, while this may be true to a degree, the best way to become a craftsman is first to be an apprentice.

To be an apprentice or trainee means to learn a trade, to learn skills and have knowledge which will both be a means of earning a living and provide an invaluable expertise for life. But whereas this used to be the main purpose and advantage of serving an apprenticeship, nowadays many people in high positions in industry have made their way to the top after beginning with an apprenticeship.

The once leisurely pace of learning a trade has now been replaced by a much more concentrated period of learning because of the shortened length of training. In addition, there is an ever-increasing number of new materials and techniques being introduced which have to be understood and assimilated into the craftsman's daily workload. There is therefore a large and growing body of knowledge which will always be essential to the craftsman, and the aim of this craft series of books is to provide this fundamental knowledge in a manner which is simple, direct and easy to understand.

As most apprentices and trainees nowadays have the advantage of attending a college of further education to help them learn their craft, the publishers of these books have chosen their authors from experienced craftsmen who are also experienced teachers and who understand the requirements of craft training and education. The learning objectives and self-testing questions associated with each chapter will be most useful to students and also to college lecturers who may well wish to integrate the books into their teaching programme.

The needs have also to be kept in view of the increasing numbers of late entrants to the crafts who are entering a trade as adults, probably under a government-sponsored or other similar scheme. Such students will find the practical, down-to-earth style of these books to be an enormous help to them in reaching craftsman status.

L. Jaques
Formerly Head of Department
Leeds College of Technology

Preface

The contents of this volume complement those of Book 1 of this series in meeting the requirements of the National Vocational Qualification levels 2 and 3 in plumbing. It will also be helpful to students studying NVQ 3 in associated crafts such as domestic and industrial heating, and professional courses where a background knowledge of water services and sanitation is necessary. All those employed in the area of the building mechanical services industry must have a thorough practical and theoretical knowledge, not only of water services but also those relating to gas, oil and electrical systems, which in most modern installations are closely interwoven.

Some of the manual skills associated with plumbing and associated crafts may no longer be necessary, but their place has been taken by investigatory and diagnostic skills which are essential in fault finding and problem solving when installing and maintaining modern appliances. A good working knowledge of the Water Regulations and the requirements of Gas Regulations is essential for those working in these areas, as the responsibility for public health and unsafe installations is, in most cases, that of the installer. A lack of knowledge and poor workmanship can lead to potentially dangerous situations. A good understanding of basic sanitation principles is also essential to protect public health, and while the use of modern materials has simplified the methods of installation, design requires careful thought.

It is hoped the contents of this volume will help the reader to attain the standards necessary, thereby deriving pride and satisfaction from knowing a job has been well done and carried out in a professional manner. I would like to extend my thanks to the manufacturers of plumbing and heating materials and appliances and the staff of the Water Advisory Council who have furnished me with much of the specialist information necessary for this work. Thanks are also extended to my wife Vilma for her help and support in the production of this volume.

G.J. Blower.

Acknowledgements

We are grateful to the following for permission to reproduce copyright material: Table 2.2 based on Table 3 in the *Essential Gas Safety* manual, p. 73 (The Council for Registered Gas Installers, 1998); Figures 5.32, 5.34 and 5.36 reproduced with permission from the *Domestic Central Heating Calculator* (M.H. Mear & Company Limited, 1994). In some instances we have been unable to trace the owners of copyright material, and we would appreciate any information that would enable us to do so.

Extracts from British Standards are reproduced with the permission of BSI. Complete copies can be obtained by post from BSI Customer Services, 389 Chiswick High Road, London W4 4AL; Telephone: 020 8996 7000.

Copies of British Standards are also obtainable on loan from public libraries.

1 Welding and Brazing Processes

After completing this chapter the reader should be able to:

1. State the main safety precautions to be taken when working with compressed gases.
2. Identify the main causes of accidents in relation to gas heating equipment and describe how such accidents can be avoided.
3. Recognise the correct flame structure for various welding operations.
4. Select suitable methods of preparing sheet lead for welding.
5. Describe the principles and techniques of lead-welding processes.
6. Describe the techniques and underlying principles employed in bronze welding.
7. Identify the methods and techniques exploed with brazing processes for both sheet and pipes.

Introduction

Welding and brazing and a knowledge of the high-pressure gas equipment used for performing these operations are an essential part of a plumber's skill. Welding and brazing have been in use almost since the dawn of time, certainly since metals were first smelted. Long before oxy-acetylene equipment became commonly used, blacksmiths used a process of welding by heating iron or steel to white heat in a forge and hammering them together. The term *brazing* is derived from brass, when common brass known as spelter, consisting of approximately 50 per cent copper and 50 per cent zinc, was used with a flux called *borax* to join iron and copper without actually melting the parent metals. In modern terminology *parent metal* relates to the metals to be joined, and the metal used to join them is called the *filler rod*.

One of the first welding processes to be used was lead burning, more correctly known as lead welding. The term *lead burning* probably relates to a very early form of welding this metal by joining the edges with molten lead. In 1837 a French engineer invented the first welding blowpipe using compressed air with hydrogen as a fuel gas; it was used with reasonable success for lead welding. It was quickly realised that the small concentrated flame required for welding could only be achieved with pure oxygen, and when compressed oxygen became readily available, it was used extensively with hydrogen or manufactured gas for lead welding. The most common combination of gases now in use for all gas welding and brazing processes is oxygen and acetylene, mainly because of its flexibility and high flame temperature.

Welding safety

Compressed gases and fire are potentially dangerous and constant vigilance at all times is necessary when using welding equipment. There is an old saying 'familiarity breeds contempt', and it is a fact one can become so used to the equipment that one tends to forget the dangers involved. Lack of care and attention can be the cause of serious accidents. The following safety precautions must be rigidly adhered to in order to ensure the safety of the operator and those working around him. Compressed gas cylinders are subject to statutory regulations and British Standards which are listed at the end of this chapter.

Correct identification of gases is the first rule of safety — do not use any gas cylinder which is not clearly identified by colour or labels. Cylinders should always be secured while being transported and should not be allowed to project beyond the sides or ends of the vehicle. Cylinders should not be rolled along the ground and the valves must be closed when they are moved. When cranes are used to lift cylinders, chain slings or magnets must not be used, only approved webbing slings or cradles are permissible.

Storing cylinders safely

All compressed gases should be stored in well-ventilated apartments or compounds. Fuel gases must be stored separately from gases which are not combustible, and empty cylinders must be marked 'MT'. 'Highly flammable' and 'No smoking' notices must be prominently displayed in areas where cylinders are stored. Acetylene and propane cylinders should always be stored in the upright position — if horizontal storage is used for other than fuel gas cylinders, they must be securely wedged and not stacked more than three cylinders high. Any store used for gas cylinders must be used for that purpose only, be securely locked when not in use, and not allowed to become a dumping area for other materials. Oil, petrol or acids must never be stored with gas cylinders.

Avoiding dangers of fire or explosion

After setting up welding equipment (described in Book 1, Chapter 3) check all joints for leaks using soapy water. Pay special attention to the hoses and ensure, by bending them, that no cracks appear. Take care when 'snifting' cylinder valves in confined places where there may be naked lights. Do not allow cylinders to become overheated by storing or using them near a heat source, and keep the burning blowpipe well away from them. If for any reason an acetylene cylinder becomes hot, the valve should be closed immediately and

the cylinder removed to the open air where it should be cooled by being hosed down with cold water. The fire brigade should be contacted in all cases and the cylinder taken out of service. The supplier must be notified so that it can be tested to ensure it is safe for future use. The regulations governing acetylene cylinders are very strict indeed.

Take care not to allow sparks from welding or cutting operations to come into contact with cylinders or hoses as this can be a serious fire risk. If possible, do not allow hoses to trail over the floor where they are subject not only to falling fragments of red-hot metal but also to damage from machinery. A suitable fire extinguisher should always be at hand during any operation carried out using a flame. The dangers of allowing oil to come in contact with oxygen is mentioned in Book 1, Chapter 3, and should be carefully noted.

Cylinder valves

Valves should be opened slowly to avoid a sudden surge of pressure on the equipment. When the valve is closed, never apply more pressure than necessary and do not extend the cylinder key in any way. When the equipment is out of use, close the valves. Never use broken or damaged cylinder keys.

Personal safety

Do not wear overalls heavily contaminated with oil or grease. Where a welding operation causes sparks, leather aprons and gloves should be worn. Never use welding equipment without suitable eye protection. Many assume that because lead welding does not require high temperatures and subsequent glare, eye protection is not necessary; it is common sense always to protect the eyes, not only from glare but also from molten metal and sparks when welding or cutting steel. They are one of the most precious gifts man has and it is very foolish to take risks with them. When welding steel or copper specially tinted goggles conforming to BS 679:1959 must be used. Some welding processes employ the use of fluxes which, when heated, give off a bright glare causing severe eyestrain to the operator if the approved lenses are not used.

Environmental safety

Ensure there is adequate ventilation in any area in which a flame is used, as some oxygen is absorbed from the atmosphere and unventilated areas can be lacking in oxygen causing danger to the operator. Special care must be used when working in ducts or basements, and if possible the operator should not work in such conditions on his own. If he is out of sight of the equipment, a responsible person should be stationed near by to shut it down quickly in the event of an accident. Years ago when plumbers had to repair pumps in deep wells, they always took with them a lighted candle in a jar, the flame of which lengthened when the air became 'vitiated', meaning deficient in oxygen. Many plumbers do the same even today when working in confined spaces, as it serves as a warning when oxygen is lacking in the air.

Never be tempted to breathe in pure oxygen — it can result in pneumonia — and avoid the accidental enrichment of air by oxygen in a confined space which may lead to excessive fire risks. Material not normally combustible in the atmosphere will readily burn if it becomes enriched with oxygen.

When there are fume hazards, such as those encountered when working on painted surfaces or galvanised work, suitable respirators must be worn and the area must be well ventilated.

Do not use a flame on, or attempt to weld a tank or vessel which is suspected of containing (or having contained), flammable or explosive materials, unless one of the following treatments has been carried out:

(a) Boiling or steaming the vessel.
(b) Filling the vessel with water.
(c) Filling the vessel with a foam inert gas.

Treatments (b) and (c) ensure the exclusion of air, thus preventing the combustion of any traces of flammable materials which may be lingering in the vessel.

The flame

The type of flame to be used varies, depending on the process for which it is required. For soldering or brazing, a large spreading flame is employed so that the whole area to be joined is at a uniform temperature enabling the filler rod to flow freely into the joint. If fusion welding is employed the edges of the parent metal are melted and fused together with the filler rod which is made of similar material to that of the parent metal. The flame used for fusion welding must be very hot and concentrated, allowing full control of the relatively small area of molten metal.

One of the most important prerequisites of successful welding is the ability of the operator to set the flame, having the correct proportion of gases and being of sufficient size to enable the weld pool to be retained under control. The weld pool is the area of the parent metal which is melted prior to adding the filler metal. If the area is too large, it will collapse leaving a large hole.

A welding flame can assume one of three forms: neutral, oxidising and carburising. The characteristics of these flames are shown pictorially in Fig. 1.1.

The neutral flame is the most useful for fusion welding of steel and lead — too much oxygen will result in the formation of an oxide film over the molten pool and prevent the filler rod merging with it. An excess of acetylene will result in a carburising flame, giving rise to a weld containing impurities and leaving a sooty deposit on the surface of the finished job. An oxidising flame is necessary for bronze welding as will be seen later, and even a carburising flame has its uses, one example being hard surfacing low-carbon steel, an engineering process known as *stelliting*. A slightly carburising flame is also recommended for brazing aluminium. A neutral flame is recognised by the small blue rounded cone seen on the end of the nozzle. This small cone is known as the area of unburned gas — complete combustion takes place at a point about 3 mm in front of it and this is the hottest part of the flame. Throughout all welding processes, the aim should be to keep the cone at about this distance above the weld pool. If it is allowed to fall into the pool, a small explosion takes place blowing small particles of molten metal in all directions. This is one of the reasons why it is absolutely necessary to use goggles. It is also one of the causes of backfiring.

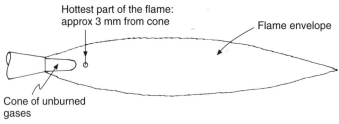

Hottest part of the flame: approx 3 mm from cone

Flame envelope

Cone of unburned gases

(a) Neutral flame
Used for most fusion-welding processes.

Sharp-pointed cone

(b) Oxidising flame
Used for bronze welding copper, copper zinc and copper tin alloys.

Cone of unburned gas

White acetylene feather

(c) Carburising or excess acetylene flame
Should not be used for welding copper, lead, iron or steel; mainly used in engineering for hard surfacing of steels.

Fig. 1.1 Flame characteristics.

(a) Butt weld
Note: gap between sheet or plate will vary depending on the material. For lead no gap is necessary.

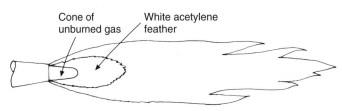

(b) Lap weld

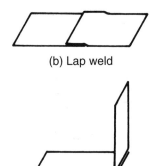

(c) Fillet weld (called 'angle' weld when on lead sheet)
May be 'set on' or 'set off' (see Fig. 1.22). Set on (no gap) is used for joining lead sheet.

Fig. 1.2 Set-up of welded joints.

The reducing flame

The reader may be puzzled in that no flux is normally required for fusion welding (there are one or two exceptions). The combustion of oxy-acetylene produces a 'reducing' flame which means the flame absorbs oxygen from the air surrounding it, causing the complete combustion of the hydrocarbons in the flame envelope which prevents the access of oxygen to the weld pool.

Set-up of welded joints

All types of welded joints can be broken down into three main types called butt, lap and fillet welds. The set-up of these joints is illustrated in Fig. 1.2. Although the following text relates to lead, the joints used are common to all welding processes. In some instances, especially where lead sheet is concerned, it is impossible to build up sufficient reinforcement in one run of welding and two or more depositions are laid to achieve a strong joint.

Lead welding

It will be seen that the area in which fusion takes place varies with the type of joint (see Fig. 1.3). Note that the penetration does not occur right through the thickness of the undercloak with the lap joint. For this reason this joint is suitable for positional work without danger of the flame causing a fire in the surrounding timber work.

Preparation for and welding a butt joint on lead
It is essential that the surfaces to be joined are thoroughly clean and this should be achieved using

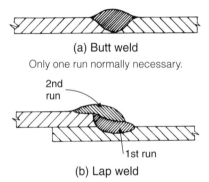

(a) Butt weld
Only one run normally necessary.

(b) Lap weld
Two runs required to build up and reinforce the joint.

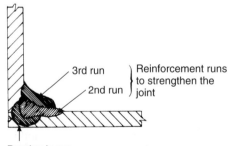

3rd run
2nd run } Reinforcement runs to strengthen the joint

Burning in run
the sheets are fused
together. Normally, if a good
fit is achieved no filler should be
necessary

(c) Angle weld
Reinforcement runs to strengthen the joint.

Fig. 1.3 Sections through completed lead welds showing number of loadings recommended.

a shave hook. Only the area covered by the weld should be shaved — a distance of about 6 mm on each edge is sufficient. If a larger area is shaved it not only spoils the appearance of the finished weld but also results in unnecessary thinning of the lead. Make sure that not only the top surface is clean but also the edges, and in the case of lap joints the underside too.

The lead should be assembled so that the edges to be joined are butted together and then tacked at 100–150 mm centres adding a little filler rod. This can be obtained in circular sections from lead manufacturers, although it is expensive to buy in this way if a lot of welding is to be done. It is usually more economical to cast one's own filler rods in specially prepared metal moulds of a type shown in Fig. 1.4. It is also possible to use strips of sheet lead cut from waste, but as these seldom exceed 2 mm in thickness (depending on the thickness of the lead from which they are cut) they are consumed too quickly. This results in a lot of stops and starts in a long run of welding. Do not forget to shave clean all filler rod before use to remove any oxide film from its surface.

Before actually commencing to weld, the operator should check that the correct nozzle is used, as if it is too small progress will be slow, and if too large it will be impossible to control the molten pool. Choice of nozzle size and correct adjustment of the flame are prerequisites to success in any welding operation; even an experienced welder will not be able to produce good work unless these conditions are right, much less one who is inexperienced.

The model 'O' blowpipe, illustrated in Book 1 Chapter 3, Fig. 3.5 has a range of five nozzle sizes most of which are suitable for lead welding. Table 1.1 shows suggested sizes for various thicknesses of lead, although this can vary slightly depending on the position of the work and the skill of the operator,

Table 1.1 Nozzle sizes for lead welding.

Nozzle no.	Thickness of lead (BS code no.)
2–3	4–5
3–4	5–6–7
5	Sand Cast lead

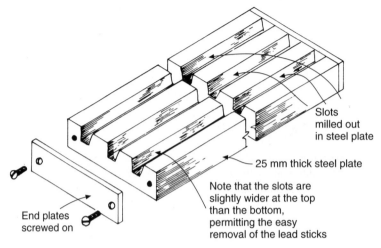

Fig. 1.4 Lead strip casting mould. Molten lead is poured into the grooves milled out in the steel where it quickly solidifies, producing a lead stick approximately 6 mm square which is ideal for most types of lead welding.

e.g. an experienced welder will use a larger nozzle size which will increase his welding speed. It will be noted that no mention is made of the number 1 nozzle in the table because it is seldom used for lead welding. One should be aware, however, that blowpipes used for lead welding are also suitable for light welding work on other metals, in which case a very small nozzle is often necessary.

Having selected the correct nozzle for the work, the next step is to turn on the blowpipe acetylene valve and ignite the gas, its volume being adjusted so that the flame burns on the end of the nozzle and no soot is given off. The presence of soot indicates that the quantity and velocity of the gas is insufficient to draw in oxygen from the surrounding air. If, on the other hand, the valve is turned on too much the gas will burn in the air away from the end of the nozzle, a condition referred to in the next chapter as 'lift off'. This must be corrected by reducing the volume of gas at the valve.

The oxygen valve is now turned on and adjusted until the white acetylene feather disappears. If it is found that the flame is too large and is overheating the work, it can be reduced by turning down the oxygen until the feather reappears, then shutting down the acetylene until the feather disappears into the area of the unburned gases.

The ability to adjust the blowpipe to give the correct flame for a specific work piece is something that will quickly be acquired by experience, but it is very important, and unless attention is paid to

it successful welding will never be accomplished. If the flame is too large the molten pool will be difficult to control, and if it is not large enough full fusion of the sheets to be joined will not be achieved.

Once one is in a comfortable position and the body is relaxed, the welding operation can be commenced. The leftward method is generally used for lead welding — that is to say, for right-handed operators the filler rod is held in the left hand, the blowpipe in the right, with the welding operation proceeding from right to left. This can be confusing to the left-handed operator who will normally hold the blowpipe in the left hand and the filler rod in the right, the weld proceeding in a rightward direction. The thing to remember, however, when using the leftward technique is that the *filler rod always precedes the blowpipe*. The angles of the filler rod and blowpipe in relation to the work are shown in Fig. 1.5. A little leeway is permissible in the recommended angles, depending on the position in which one is working, but it will be found that the best results are obtained using these angles. In short, if it is found that the weld is not going as it should and the flame is correct, check the angles.

In any welding operation a molten pool must be established before the filler is added, as failure to observe this will result in unsightly blobs of metal in the work and incomplete fusion. The blowpipe is now raised slightly to melt a piece of the filler rod which is held in close proximity to the molten pool.

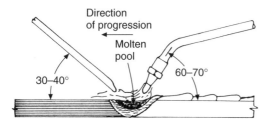

Fig. 1.5 Angles of rod and blowpipe for left-ward welding.

As it melts and merges into the pool, the flame is brought down in a stroking action until the outer edges of the pool reach the edges of the shaved area where they cool, causing the distinctive ripple effect common to all welding operations. The filler rod and blowpipe are now moved towards the left, when the whole action is repeated. Each time a small portion of the filler rod is deposited in the molten pool a ripple is formed, the speed of the operator and the size of the flame determining the shape of the ripple. Bear in mind the control of the molten pool is achieved not only by the correct flame, but by the amount of filler rod deposited into it and the consequent cooling effect brought about by the speed of the forward movement.

It is points such as this that must be learned from experience by the welder until they are as natural to him as walking and breathing. A large flame and high welding speeds will produce the herringbone pattern that looks so effective on a long run of welding. The use of a smaller flame and lower welding speeds will produce a weld that looks very similar to a weld on steel, sometimes called by lead welders the *thumbnail* effect. While the latter does not look quite so effective, it must be remembered that if an intricate piece of work is attempted in an awkward corner, a smaller flame and lower welding speed are much more likely to lead to good results. A fast flame can lead to loss of control of the weld pool in such circumstances. Both types of finished weld appearance are illustrated in Fig. 1.6. The advantages and limitations of each should be understood so that the best technique can be selected for a given set of circumstances.

The term *penetration* has been used earlier; the other important term is *reinforcement*. This is applied to the build-up of metal over and above the surface of the parent metal and, as its name

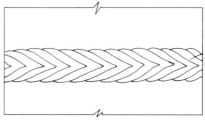

(a) Herringbone effect
Produced with a large flame and ideal for long flat runs.

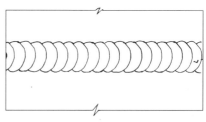

(b) Thumbnail effect
Produced with a smaller flame. Used in awkward corners and for intricate details.

Fig. 1.6 Lead welding patterns.

implies, its purpose is to strengthen or reinforce the area of the weld. It should not be excessive in height and the edge must merge into the parent metal smoothly.

Defects in welds

Although at this stage we are considering lead welding, any defects are common to all types of welding. These defects are illustrated in Fig. 1.7, the most common being insufficient penetration or lack of fusion which will obviously result in a weak job, which when subjected to any stress will fall apart.

The other main weakness is undercutting and, as the illustration shows, this reduces the thickness of the parent metal adjacent to the weld. This defect is very common in fillet or angle welds when the molten metal on the vertical side of the weld tends to run down into the molten pool. It can be corrected by careful application of the filler rod to the upper edge of the weld pool. Generally, when undercutting takes place it is caused by too great a welding speed, addition of insufficient filler metal, incorrect welding angles or the use of a flame that is too large.

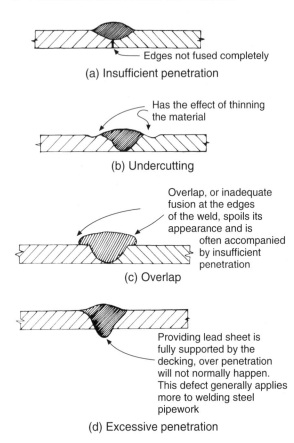

(a) Insufficient penetration
— Edges not fused completely

(b) Undercutting
Has the effect of thinning the material

(c) Overlap
Overlap, or inadequate fusion at the edges of the weld, spoils its appearance and is often accompanied by insufficient penetration

(d) Excessive penetration
Providing lead sheet is fully supported by the decking, over penetration will not normally happen. This defect generally applies more to welding steel pipework

Fig. 1.7 Common faults in welding.

Overlap results in incomplete fusion of the edges of the filler metal with the parent metal, and is caused by an excessive build-up of the reinforcement.

Excessive penetration is a defect which can happen when welding all types of materials. It is the result of using too big a flame which causes the weld pool to enlarge, resulting in its collapse. Some metals, notably steel, can be welded without support providing careful control is maintained over the weld pool, but in the case of lead adequate support of the underside is essential at all times. Figure 1.11 illustrates a typical example.

The word *inclusion* is often employed when discussing defects in welds. This relates to impurities in the deposited weld metal. While it is not common in lead welds, the practice of cleaning lead for welding with steel wool can lead to the inclusion of small particles of steel in the weld,

and for this reason it is not recommended. The other main point to bear in mind is the necessity to ensure that, when a weld is picked up or restarted after pausing for some reason, the weld pool merges into the metal already deposited. It is recommended that the restart is made 6–8 mm back into the previous deposit so that full fusion is maintained throughout the length of the weld. This also ensures that any oxides or scale contained in the weld pool, especially in the case of steel welding, are floated off to the surface and do not remain as an inclusion in the weld.

Lap welds on lead sheet

This process is very similar to butt welding when this type of joint is made in the downhand (flat) position. The main differences are that the weld must not penetrate right through the underside of the lap or undercloak, and more than one run of filler metal is necessary to give adequate strength to the joint. About 25 mm should be allowed for the lap, and remember to shave all the jointing surfaces. The underside of the overcloak is often forgotten, resulting in loss of fusion causing a defective weld. The overcloak should be tacked in position and the first run made. This will almost certainly cause undercutting, and to produce a weld of adequate strength a second run must be added to provide effective reinforcement. A careful study of the section of a lap weld shown in Fig. 1.3 will clarify this.

This second run should be made so that the edge of the first remains exposed by about 3 mm. This ensures complete fusion to the parent metal without the overlap referred to in the section on weld defects.

Lead welding in the vertical position

The lead should be shaved and prepared in a similar way to that of an ordinary lapped joint. Two main techniques of welding on a vertical face are employed, one for welding at 90° and referred to as an upright joint, the other where the joint lies at an angle across the vertical face and is called an inclined seam. These joints are illustrated in Figs 1.8 and 1.9, and as a rule they do not require the use of filler rod; the edge of the overcloak in each case is melted and fused to the undercloak.

To make these joints the blowpipe is held at about 90° to the face of the work, melting the undercloak to form a small pool while simultaneously describing an elliptical movement, melting the edge of the overcloak to flow into the molten pool. For control of the weld pool to be maintained when an upright joint is made, a step-by-step technique is employed. That is to say, as each weld bead is formed, the blowpipe is removed momentarily from the face of the work to allow the bead to congeal, forming a platform on which the next bead is deposited. This technique requires a lot of practice to perfect as if the molten edge of the overcloak is not carefully directed into the molten pool by the flame it will drop off. The only way to put this right is to make good the loss by using a very thin filler rod, preferably in this case cut from a strip of lead. This is easier said than done, and unless one is very skilled the appearance of the finished weld is spoiled. Careful selection of the welding nozzle and correct flame setting are essential in this method of welding.

Where possible, the inclined technique should be employed as the edge of the overcloak forms, to a greater degree, a support for the molten pool of lead and should control of the weld pool be lost, it is more easily rectified on an angled surface than on one at 90°. A stronger weld can be produced in this position by adapting the horizontal lap joint to the inclined position. The overcloak is turned out to a slight angle as shown in Fig. 1.10. This method

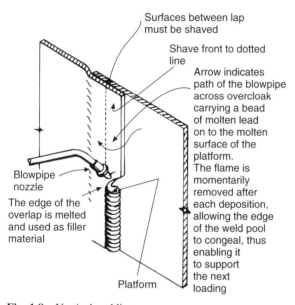

Fig. 1.8 Vertical welding.

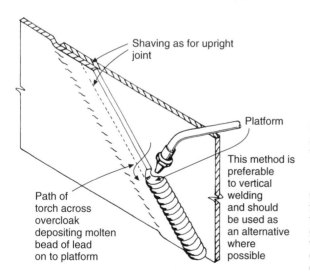

Fig. 1.9 Inclined vertical weld.

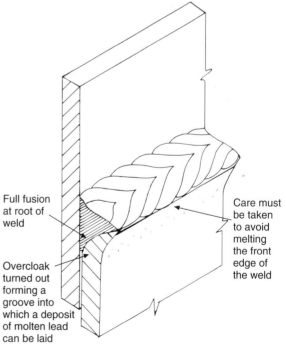

Fig. 1.10 Horizontal lapped weld.

requires the use of the filler rod, but as a greater area of lead is exposed to fusion it results in a much stronger joint. It does not, however, produce quite such a neat effect as the former technique.

Angle seams

If the work piece is relatively small and can be welded out of position, it can be set up on a wooden jig like that shown in Fig. 1.11 so that both faces are inclined. It is much easier to produce a good weld in this position than where one face is vertical.

The weld is accomplished by making three runs, the first simply fusing the two mating edges together, the second and third reinforcing the weld, the additional lead giving extra mechanical strength. It is on the first or 'burning in' run that accidents are likely to happen, such as burning away the edges, and it must be stressed that careful control of both the flame and the weld pool is essential.

It is more difficult to make an angle seam where one face is vertical: this is called *fillet welding* and the main difficulty encountered is undercutting the vertical sheet. The actual welding technique employed on lead is different from that for steel, but the problem of undercut is the same. It is

essential to become skilled in welding angles in this position, however, as large pieces of lead are not easily handled and must often be welded *in situ*. Three runs are recommended as in the case of angle welds made in the inclined position, the last run being made taking great care to avoid undercutting the lead on the vertical surface. A thinner filler rod may be used with advantage here and should be held as close as possible to the vertical surface so that it melts at the same time as the surface of the lead, merging with it before it falls. Manipulation of the blowpipe from side to side as each bead is deposited, and avoiding the use of a fast flame, will also prevent undercutting.

The application of these basic techniques of lead welding are illustrated in Chapter 10 and with practice can show considerable savings in time over traditional bossing techniques.

Lead-welded pipe joints

The use of lead pipe is now very limited and it is a fact that many plumbers will never come into contact with it. For this reason no specific information regarding this subject is given here. Most remarks relating to the welding of sheet, however, also apply to pipework, and with a little ingenuity the techniques previously described can be adapted for welding pipes should an occasion arise. In cases where help or advice is required on this subject it is recommended that contact is made with the Lead Sheet Association.

Bronze welding

The flame
One of the most important factors to bear in mind when bronze welding is that, unlike fusion welding of lead and steel, a slightly oxidising flame must be used. The reason for this is that the zinc in the alloy melts at a temperature of 410 °C and volatilises (gasifies) at higher temperatures. If the flame is incorrectly adjusted the gas will bubble through the molten filler metal as welding proceeds, leaving a series of blowholes which can result in a defective joint. The excess oxygen in the flame, however, combines with the zinc gas to form zinc oxide which melts at around 1,800 °C, a much higher

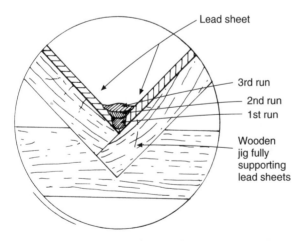

Lead sheet

3rd run
2nd run
1st run

Wooden jig fully supporting lead sheets

Fig. 1.11 Welding angle seam on a work piece which is out of position. Suitable for bench fabricating small components such as lead outlets for flat roofs and catchpits.

temperature than either the copper or the filler rod, and in this way no loss of zinc occurs. Experience and practice soon enables one to identify the correct flame, but the following information will be useful. If the finished weld has a matt, yellow appearance, with some evidence of blowholes and poor weld ripple formation, the flame lacks oxygen. If, however, the flame is excessively oxidising it will cause the joint to blacken due to the oxide film and result in poor adhesion of the filler metal. This condition can be identified further by the exceptionally bright, almost golden colour of the deposited metal.

Bronze-welded joints

The basic concept in preparing joints for bronze welding is to allow sufficient space for a body of the filler rod to be built up, as this and its adhesion by 'wetting' the parent metal are the sources of its strength. Figure 1.12 illustrates a typical profile for bronze welding on metal thicknesses upwards of approximately 2.5 mm. This process was used extensively for jointing copper tubes for both water supply and sanitary pipework. Due to the fact that the filler metals used are subject to dezincification, jointing by this method is no longer acceptable and has been superseded by brazing techniques. It still has its uses in mechanical engineering services, especially for repair work on iron castings. Figures 1.13 and 1.14 illustrate typical repairs using the method. It is also suitable for light construction work such as hand railing or other light support structures as shown in Fig. 1.15 especially where galvanised steel tubes are used. A more satisfactory joint will be obtained using bronze-welding

techniques on this material, causing less damage to the protective zinc coating, than by attempting to use fusion welds. It is very important to remember that welding galvanized work is very dangerous due to the fumes given off. Any such work must be undertaken in the open air or in a properly ventilated workshop, and a suitable respirator should be used.

It is essential that, prior to welding, the work is thoroughly clean. All joint surfaces should be scoured with card wire or a wire brush or prepared with a suitable mechanical grinder.

Any residue must be removed after welding because, like all fluxes, they are corrosive. One should be careful to use the right amount of flux as the residues set as a hard glass-like structure which is difficult to remove. Chipping with a chisel or screwdriver can cause damage to the work. The simplest way is to wash the joint with a weak solution of phosphoric acid. It is possible to obtain flux-impregnated filler rods at little extra cost (see

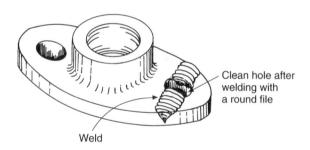

Fig. 1.13 Repairing a broken cast iron flange.

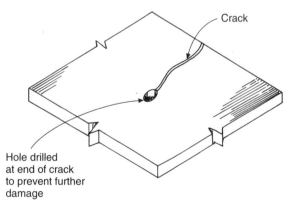

Fig. 1.14 Bronze welding a cracked cast iron plate.

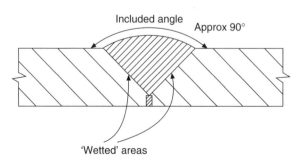

Fig. 1.12 Preparation of steel or cast iron plate for bronze welding.

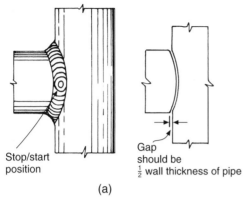

Stop/start position

Gap should be $\frac{1}{2}$ wall thickness of pipe

(a)

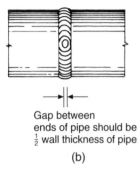

Gap between ends of pipe should be $\frac{1}{2}$ wall thickness of pipe

(b)

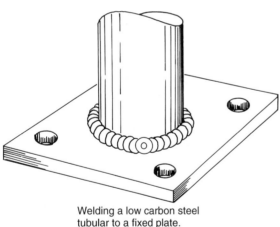

Welding a low carbon steel tubular to a fixed plate.

(c)

Fig. 1.15 Butt-type bronze-welded joints for steel pipes. Mainly used for steel fabrications and general-purpose work where the use of high welding temperatures is undesirable, e.g. galvanised work. Note that although the finished appearance of these welds is identical to fusion welds on steel pipe the gap for bronze welding is much less.

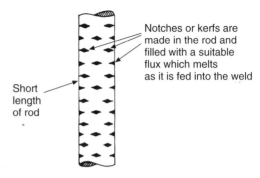

Notches or kerfs are made in the rod and filled with a suitable flux which melts as it is fed into the weld

Short length of rod

Fig. 1.16 A flux-impregnated filler rod.

Fig. 1.16). The use of these ensures that only the right amount of flux is used and little trouble will be experienced in its subsequent removal. The rods must be kept dry to prevent deterioration of the flux content.

Fluxes
Special fluxes consisting mainly of borax and silicon are used for bronze welding. They may be applied to the joint as a paste by mixing the powder with water or by dipping the heated filler rod into the flux causing it to adhere to the rod. Table 1.2 shows the commonly used filler metals and their suitability for various types of parent metals.

The welding process
Most bronze-welded joints are normally made in the downhand (flat) position. Positional welding can be done using similar techniques to those used for lead, but it requires a lot of practice and is seldom necessary. After fluxing, the joint should be positioned and tacked if necessary. Heat the joint, keeping the flame moving from side to side to ensure the whole area is evenly heated. The filler rod should now be applied by means of a stroking movement until it is seen to run, wetting the edges of the joint and flowing into the root of the weld. The blowpipe should then be moved from side to side, or on smaller joints rotated so that the nozzle describes a series of circles as the weld proceeds in a forward direction round the joint, forming the characteristic weld ripples. Ensure that the filler rod at the stop position of the weld merges thoroughly with that at the start by remelting and overlapping it by at least 6 mm. To avoid a hollow at the stop/start

Table 1.2 Bronze welding filler metals.

BS 1485 classification	Copper %	Nickel %	Silicon %	Manganese %	Iron %	Zinc	Melting range °C	Suitable parent metals
C2	60	—	0.3	—	—	Remainder	Above 850 °C	Brass bronze copper and low carbon steel
C4	60	—	0.2	0.2	0.3	Remainder	Above 850 °C	Galvanised low carbon steel
C5	50	10	0.3	0.5	0.5	Remainder	Above 850 °C	Cast and malleable iron
C6	43	12	0.3	0.2	0.3	Remainder	Above 850 °C	Combination of any of the above

position of the weld due to contraction of the filler metal, the flame should be momentarily removed at this point. This allows the weld pool to cool slightly, permitting the edges to congeal before a little more filler metal is applied. The stop/start would then appear as in Fig. 1.15.

Brazing processes

This is a method of making joints on metals using similar techniques to those employed with soft soldering; however, as the filler metals used have a higher tensile strength than soft solder, the depth of the sockets may be reduced. As with soft soldering, the filler metals used for brazing penetrate small gaps between the surfaces of the metals to be joined by capillary attraction. This means that some care is required to ensure the correct tolerances of the gap which will enable the filler metal to penetrate fully the surfaces of the joint. The considerable permissible variations in the gap depend on the type of filler alloy used. A branch joint, for instance,

made with a hammer and bent bolt, might not be so accurately formed as one made with a purpose-made tool, and in such circumstances a gap-filling alloy would be used to make the joint.

Brazing alloys

Those most commonly used in plumbing and associated crafts are based on silver–copper and copper–phosphorus alloys. Both these alloys melt at temperatures well below 800 °C, which is lower than those used for bronze welding. Copper–silver alloys nearly always contain zinc and cadmium in varying proportions, which have the effect of lowering the melting point and increasing their fluidity. The less silver the alloy contains the greater the temperature difference between the completely solid and the liquid state of the alloy. This gives it a long pasty range like that of plumbers' solder, and filler rods having this characteristic are capable of bridging over and filling larger gaps than those having a higher silver content. Table 1.3 indicates the whole range of

Table 1.3 Silver brazing alloys (reference BS 1845).

Type	Nominal composition					Melting range (°C)		Characteristics
	% Silver content	Copper	Zinc	Cadmium	Tin	Solidus	Liquidus	
Ag 1	50	X	X	X	—	620	640	Free flowing
Ag 2	42	X	X	X	—	610	620	Free flowing
Ag 3	38	X	X	X	—	605	650	Gap filling
Ag 4	61	X	X	—	—	690	735	Gap filling
Ag 5	43	X	X	—	—	700	775	Gap filling
Ag 6	60	X	—	—	X	600	720	Gap filling
Ag 7	72	X	—	—	—	780	780	Free flowing

Table 1.4 Copper–phosphorus brazing alloys (reference BS 1854).

Type	Nominal composition			Melting range (°C)		Characteristics
	% Silver	% Phosphorus	Copper	Solidus	Liquidus	
CP1	15	5.0	X	645	700	All types gap filling
CP2	—	6.5	X	645	740	CP1 only suitable for resisting torsional
CP3	2	7.5	X	705	800	stresses, shock loads or flexing.
CP4	5	6.0	X	645	825	For copper tube/brazing

silver–copper alloys in general use. The letters Ag are the chemical symbol for silver and will indicate that the alloy contains this metal.

It should be noted that brazing alloys containing zinc and cadmium must be used in well-ventilated areas due to the dangerous fumes they give off. Copper–phosphorus alloys are more commonly used in plumbing as in most cases they contain little or no silver and are much cheaper than those having a high silver content. A further advantage of these alloys is that when brazing copper no flux is required as phosphorus is a deoxidising agent. While copper–silver alloys can be used to join both ferrous and most non-ferrous metals, copper–phosphorus alloys are more limited in their use, being confined to copper and copper–zinc or copper–tin alloys. When used on metals other than copper, a flux recommended by the manufacturer of the filler rod must be used. Table 1.4 lists the four main copper–phosphorus alloys in common use.

Joint design for brazing

As a brazed joint is stronger than those made using soft solder as a filler, less surface contact between the metals to be joined is necessary. Where the depth of socket on a soft-soldered capillary joint is about 15–18 mm, the same joint using a brazing alloy will need a socket depth of only 6–8 mm. For making brazed socket joints on pipes, socket-forming mandrels can be used (see Fig. 1.17) or the special tools described in Book 1 Chapter 3. Branch joints can be formed with a hammer and bent bolt, but attention must be paid to ensuring an accurate fit between the mating surfaces due to the limitation of capillary attraction. When forming sockets or

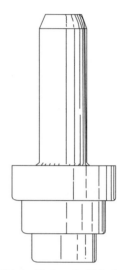

This tool is inserted into the annealed (softened) end of the tube and hammered home to form a socket

Fig. 1.17 Multi-diameter steel mandrel for forming brazed socket joints.

branch openings in copper pipes, they must always be worked in the annealed or softened state, the one exception being when using the special branch opening tools described in Book 1 Chapter 3. Similar methods of branch opening can be accomplished using power tools which enable branch holes to be opened very quickly. All mechanical methods have two disadvantages: they will only form 90° branches, and due to the size of the hole necessary to allow the entry of the forming tool, the size of the branch is lower than the crotch. The branch must be profiled as shown in Fig. 1.18. Where oblique branches are required traditional

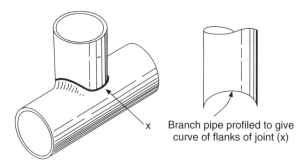

Branch pipe profiled to give curve of flanks of joint (x)

Fig. 1.18   Profile of a branch formed with patent tools.

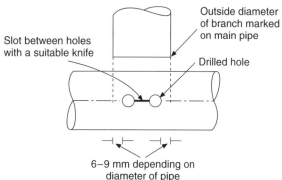

Outside diameter of branch marked on main pipe

Slot between holes with a suitable knife

Drilled hole

6–9 mm depending on diameter of pipe

Nominal pipe diameter (mm)	Suggested drill diameter(mm)
25–40	6–8
50–100	8–10

(a)

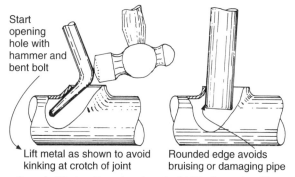

Start opening hole with hammer and bent bolt

Lift metal as shown to avoid kinking at crotch of joint

Rounded edge avoids bruising or damaging pipe

Round off hole with a short length of LCS pipe or bar. The final shape of the hole should be made using pipe or bar fractionally smaller than the branch diameter.

(b)

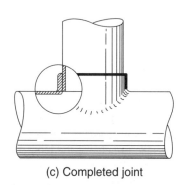

(c) Completed joint

Fig. 1.19   Setting out and preparation of bronze welded branch joints.

methods using a hammer and bent bolt are normally used. Figure 1.19 illustrates the techniques employed.

Working up branch joints in copper tubes using hand tools

To prepare a branch joint the first step is to mark its position on the main pipe. Two holes are then drilled, the edges of which (not the centres) should be 6–8 mm inside the confines of the marks indicating the branch position. After annealing, a slit is made between the holes with a strong knife such as a hacking knife. Take care not to nick the outer edges of the holes as this will cause them to split when the branch is opened. The branch can then be worked up with a bent bolt and hammer. The hole should then be rounded up with a piece of steel pipe or bar having as near as possible the same diameter as the copper branch. Finally, the top is filed level and the joint cleaned prior to welding.

Two outstanding points should be remembered:

(a) Always work the copper hot.
(b) Take care to 'lift' the area that will form the crotches of the branch as this will prevent the formation of a crease which would obstruct the waterway.

Brazed joints for sheet metal

Figure 1.20 shows typical types of joint design for sheet metal fabrications which the plumber may encounter during the course of his work. Notice that they are made in such a way that a suitable area of mating surfaces is achieved and, due to the turns made, distortion is minimised as this can present problems on light sheet metal work.

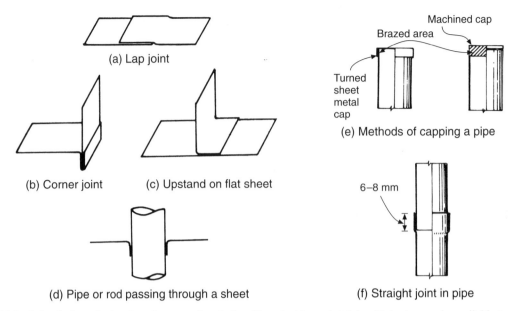

(a) Lap joint

(b) Corner joint   (c) Upstand on flat sheet

(d) Pipe or rod passing through a sheet

(e) Methods of capping a pipe

(f) Straight joint in pipe

Fig. 1.20 Joint designs for brazing sheet metal and pipe. Note that in each joint sufficient space is available to permit penetration of the filler material.

Methods of heat application for brazing

Welding requires a small concentrated flame, but a large spreading flame is necessary for brazing to ensure that the filler metal penetrates the full depth of the joint. Small-diameter pipes can be both fabricated into desired designs and brazed successfully on site using propane or butane blowlamps. For workshop use, natural gas/air brazing blowpipes are very effective, the air being pressurised by a small compressor. Oxy-acetylene equipment is also very useful — and essential where large-diameter pipes are to be brazed on site. Special nozzles having a series of holes similar to those in the top of a pepperpot are available which provide a more suitable flame structure for brazing than a normal welding nozzle.

One of the problems encountered with both welding and brazing copper is its high conductivity rate. In some cases heat losses due to conduction make it impossible to achieve the temperatures required and a secondary heat source, such as another blowlamp, is necessary to keep the work hot while the joint is made. While it is not often suitable for pipework, a brazing hearth can be constructed on site by placing bricks round the work. Heat absorbed by the bricks is reflected

back by radiation enabling quite large pieces of work to be brazed with a relatively small heat source. A very important point to watch when brazing with a welding blowpipe is that one becomes so used to holding the flame close to the work that there is a tendency to do the same with brazing. This can lead to local overheating on one part of the joint and often leads to unnecessary spillage of the brazing rod over the edges of the joint, resulting in a very poor external appearance. The thing to remember is that brazing is a form of soldering, not welding, and if oxy-acetylene equipment is used with a welding nozzle the flame should be employed as illustrated in Fig. 1.21.

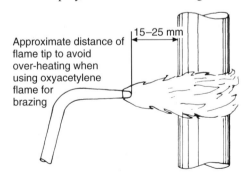

Approximate distance of flame tip to avoid over-heating when using oxyacetylene flame for brazing

15–25 mm

Fig. 1.21 Using oxyacetylene flame for brazing.

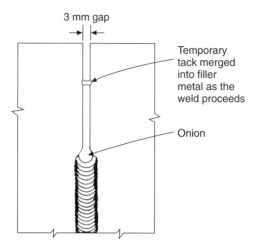

Fig. 1.22 Fusion welding a low-carbon steel butt joint on plate.

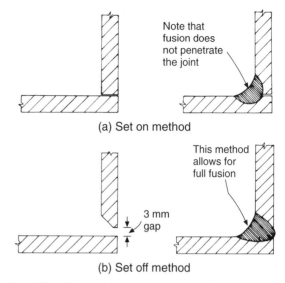

Fig. 1.23 Fillet welds on low-carbon steel sheet or plate.

Fusion welding low-carbon steel

At this level the trainee should be capable of effectively using the fusion-welding process for fabricating brackets and fixings for pipe clips and hangers. Most of the remarks about fusion welding of lead also apply to steel, especially those relating to penetration and full fusion of the metal. The most important point to bear in mind when butt welding two plates together is that a gap of about 3 mm must be left between them. This enables the edges of the metal to be melted, leaving what welders call an *onion*, or hole, which is filled, as the weld progresses, with a filler rod having roughly the same composition as the parent metal. Providing the onion is maintained full fusion of the edges is assured (see Fig. 1.22).

Fillet welds can be made in two ways (see Fig. 1.23). For making brackets and hangers the 'set on' method is acceptable. For pipework joints the 'set off' method is necessary to achieve full fusion and is mandatory in any pipe-welding examinations.

A detailed study of either bronze welding or steel welding is outside the scope of this book and is indeed the subject of welding NVQs and special examinations set by the Plumbing Joint Industrial Board and the Heating and Ventilating National Joint Council. Plumbers wishing to qualify in this subject should contact these bodies or their local technical college for further information.

Further reading

Much useful information can be obtained from the following sources.

'T' Drill copper pipe branch forming tools/'T' Drill Portable Tools, Baiford, 29 Thornycroft Lane, Downhead Park Miltun Keynes MK15 9BR, Tel. 01908 667667.

REMS, 1a Greenleaf Rd Walthamstowe, London E17 6QQ Tel. 020 8521 9168.

Welding and Brazing Filler Metals and Fluxes/ Johnson Matthey Metal Joining Products, Unit C, Arundel Gate Court, 1, Frogget Lane, Sheffield S1 2NL Tel. 0114 241 9400.

Joining of Copper and ITS Alloys, Copper Development Association Publication No. 98. CDA Verulam Industrial Estate 224 London Rd ST Albans Herts AL1 1AQ Tel. 01727 731200.

Safe Under Pressure. British Oxygen Company.

British Standards:

BS 120:1975 Rubber hose for gas welding and allied processes.

BS 341:1991 Part 1 Valve fittings for compressed gas cylinders.

ES EN 1089 PT3 1997 Identification of industrial cylinder containers.

BS 5045:1976 Low-carbon steel cylinders for the storage and transport of gases.

Self-testing questions

1. (a) Describe what action should be taken if it is discovered that an acetylene cylinder is hot to the touch.
   (b) List the main considerations for storage of compressed gas cylinders.
2. (a) Explain the safety hazards likely to be encountered when using oxy-acetylene welding equipment in basements and ducts.
   (b) What precautions must be taken prior to welding containers suspected of having contained flammable substances?
3. (a) Sketch a neutral flame and describe how it differs from oxidising and carburising flames.
   (b) State the type of flame that should be used for steel, lead and bronze welding.
4. State why no flux is necessary when fusion welding lead and low-carbon steel.

5. (a) Explain the two factors that influence control of the molten pool when welding sheet lead.
   (b) Describe two methods of welding lead on a 90° face.
6. List the common defects that can be found in welds produced by inexperienced operators.
7. Make a simple sketch of the set-up prior to the welding of a butt joint, a lap joint and a fillet joint in lead sheet.
8. Explain the differences between bronze welding, brazing and fusion welding.
9. (a) Sketch and describe the method used to open a branch hole for a brazed joint in copper pipe using hand tools.
   (b) State the constituents of two common brazing alloys used in plumbing.
10. (a) Explain why no flux is necessary when brazing copper pipe joints with copper–phosphorus filler alloys.
    (b) State why the socket depth for a brazed joint is less than that for a soft-soldered joint.

2 Gas Installations

After completing this chapter the reader should be able to:

1. State the main characteristics of natural gas.
2. Outline the main regulations relating to gas installations and appliances.
3. Explain the need for pressure testing gas installations and equipment.
4. Identify the potential dangers of gas/air mixtures and explain the procedure for purging pipework and appliances. List the appliances that do not normally require a flue and explain the reasons for this.
5. Understand the necessity for the provision of an adequate air supply for the combustion of gas and identify the dangers of incomplete combustion.
6. State the working principles of a flue and sketch and describe the precautions to be taken when siting flue terminals. Explain the precautions to be taken where flues pass through combustible materials.
7. Indentify the causes of condensation in gas appliance flues and list the measures taken to minimise its effects.
8. Describe the main principles of balanced flue appliances and their advantages.
9. Explain the working principles of simple gas controls.
10. Identify the main methods of gas ignition on gas fires, water heaters and boilers.
11. Explain the working principles of simple flame failure devices.
12. Identify types of gas fires, and their installation and flueing requirements.

Legislation

Due to the potential dangers of gas, suppliers, installers and users are subject to certain regulations. All competent gas installers must have a good working knowledge of regulations affecting their work. The following gives a broad outline of their requirements. Further details will be found in the gas (Installations and Use) Regulations.

Certification

All installers of gas appliances must have a certificate of competence for each area of gas work they undertake. The validating certificates must be issued under one of the following schemes:

(a) Approved Code of Practice (ACOP).
(b) Nationally Accredited Certification Scheme (NACS).

NACS has superseded ACOP.

The Council of Registered Gas Installers (CORGI)

Although originally membership was voluntary, as a result of gas-related accidents it became mandatory in 1991 for all companies undertaking gas work using natural gas to be registered with a body approved by The Health and Safety Executive (HSE). CORGI is currently the organisation which maintains a register of all approved operators. It also carries out regular inspections of gas installations fitted by its members, runs a public enquiry and complaints service and is responsible for publishing the importance of gas safety. Members of CORGI are kept up to date with changes in the law, technology and safe working practices.

The Gas Safety (Installation and Use) Regulations
These are mandatory and deal with safe installation and maintenance practices in most types of building. The Regulations place responsibilities on installers, maintenance engineers, and suppliers and users of gas, including landlords of rented property. The Regulations are followed by guidance notes which (although not mandatory) will normally be sufficient to comply with the law if observed by an installer. This is an important document and all operators carrying out gas work should be familiar with it.

The Gas Act 1995
This updates previous legislation to include new licensing arrangements for Public Gas Transporters, permitting competition in the domestic gas market and allowing consumers to purchase gas from any supplier they wish to use. It also includes provision for safety regulation to be made in (a) The Gas Safety and (b) The Gas Safety (Rights of Entry) Regulations.

Health and Safety at Work Act
This has been dealt with in Book 1 of this series. It applies to everyone concerned with work activities, both employer and employees. It also includes provisions to protect the public from exposure to risks to health and safety. Failure to comply with the general requirements of the Act and those of other documents relating to this subject may result in legal proceedings.

The Gas Safety (Management) Regulations
These are designed to protect the public against dangers caused by failure to observe safe working practices when transporting or supplying gas. The HSE (Health and Safety Executive) is the enforcing body for gas safety and has legislation with which gas suppliers·and transporters must comply. The main points of the legislation are as follows:

(a) Under these Regulations a gas supplier *must* produce a case showing systems and procedures

that will be adopted to ensure a safe supply of gas. Subject to this being approved by HSE, permission to supply gas will be granted.
(b) The supplier must operate a full gas emergency service.
(c) It must provide a gas incident service which reports gas explosions and cases of carbon monoxide poisoning.
(d) It must operate an emergency telephone number for customers reporting gas leaks.

The Gas Safety (Rights of Entry) Regulations
These give authorised officials of gas transporters the right to enter a property and inspect any equipment connected to the gas supply. They have the power to disconnect any appliance they consider to pose an immediate danger. British Gas Transco is currently the main transporter and is responsible for dealing with all emergency calls. Lack of ventilation, seriously defective flues and gas leaks exceeding the permissible limits all constitute an immediate danger. In cases where the responsible person, e.g. owner or landlord, refuses to allow an appliance or supply of gas to be isolated, the National Gas Emergency Service Provider or the gas supplier must be notified immediately. Only they have the necessary powers of enforcement.

The Building Regulations
Compliance with these regulations is mandatory in England, Wales and Northern Ireland. Building Standards in Scotland have similar requirements. Approved Documents 'B' (fire safety), 'F' (ventilation) and 'J' (heat-producing appliances) give guidance on some of the specifications relating to the Gas Safety Regulations. In Northern Ireland the Approved Document part L, (heat-producing appliances) applies. In Scotland, Technical Standards part F and K of Building Standards are applicable. They relate to heat-producing installations, storage of liquid, gaseous fuel, and ventilation.

Gas Appliance Safety Regulations (GASE)
These implement an EC directive which requires appliances and fittings used for gas to conform with specified essential requirements, not the least being safety in use. Both the supply and installation of

any gas appliance are prohibited unless it bears the CE mark. The object of this legislation is to ensure all gas appliances for sale in the European market meet agreed safety standards and are designed to protect the consumer.

Reporting of Injuries, Diseases and Dangerous Occurrence Regulations (RIDDOR)

Reference should be made to Book 1 of this series where this legislation has been explained more fully. It is designed to allow the HSE to investigate dangerous situatations and accidents. The responsibility for reporting dangerous gas fittings and installations lies with the service fitter, whether self-employed or of employed status. In the latter case it should be reported to the company for whom the fitter works. Only after notifying the National Gas Emergency Services or the gas supplier should a report be made to the HSE. Some examples of what should be reported are listed as follows:

(a) Gas escapes outside the tolerances of soundness tests due to poor or unsatisfactory workmanship.
(b) Open-ended uncapped pipes, which may or may not be connected to a gas supply.
(c) Any evidence of 'spillage', the cause of which has not been rectified.
(d) Defective flues or chimneys not clearing the products of combustion. This includes appliances that should be flued but are not, and flues discharging into a roof space or excessively outside the parameters of the recommendations of BS 5440-1:2000.
(e) Defective or insufficient ventilation of areas in which an appliance is fitted.
(f) Use of unsuitable materials for gas pipes.
(g) Faulty servicing making an appliance unsafe for use.

Having regard to the foregoing, if the installation/appliance is immediately dangerous it must be disconnected and the supply capped off. If it is at 'risk' but not immediately dangerous it can simply be turned off. In both cases the responsible person, e.g. householder, landlord, etc., must be given an approved warning notice and a DO NOT USE notice attached to the appliance. The permission of the responsible person must be obtained prior

to carrying out these procedures, but in the event of non-cooperation the National Gas Emergency Service must be contacted immediately. In all cases where an appliance or installation is considered **immediately** dangerous the HSE must be notified. If any appliance or installation does not comply with current regulations, standards or specifications but does not fall into the preceding categories, the responsible person should be informed orally. However, a record should be kept for reference at a later date.

The Office of Gas Supply (OFGAS)

This is a regulating body having powers under the Gas act to:

(a) Issue licences to gas transporting and supply companies.
(b) Ensure the quality and calorific value of the gas.
(c) Appraise and certificate gas meters.
(d) Protect the interests of the consumer.

Properties of natural gas

Natural gas is predominantly a mixture of hydrocarbons but, unlike the town gas used previously, contains no hydrogen. Its composition varies slightly according to its source, but a typical example of its constituents is shown in Table 2.1. Note that all these gases, except nitrogen, are combustible. Natural gas is non-toxic (which means it is not poisonous) as it contains no lethal gases such as carbon monoxide. It will, however, produce carbon monoxide if it is not completely burned, and for this reason careful attention must be given to flues and ventilation. In its natural state it has no smell, and to avoid possible dangers from leaks a chemical is added to give it a distinctive odour.

Table 2.1 Constituents of natural gas.

Constituents	%
Methane	93
Ethane	3
Propane	2
Butane	1
Nitrogen	1

Calorific values

The calorific value of a fuel may be defined as the heat units it contains per unit volume. The volume by which gas is measured is the cubic metre and the unit of heat is the joule. One cubic metre of natural gas contains approximately 38 (megajoules), but there are sometimes slight differences in this figure due to variations in the source of supply. The prefix *mega* means 1 million and enables one to deal with fewer digits, as 38,000,000 J (joules) is rather an unmanageable number when used for calculations.

Specific gravity

Natural gas is lighter than air. If air is taken to have a specific gravity of 1, then natural gas has a specific gravity of about 0.58. This is one of the reasons why higher pressures are required on the main and service pipes.

Stoichiometric mixture

This term relates to the quantity of air required to burn a fuel to achieve complete combustion. In the case of natural gas the figure is 10.57. In simple terms, 10.57 volumes of air are required to achieve the combustion of 1 volume of gas. Thus it will be appreciated how important it is to make sure that there is adequate ventilation to provide sufficient air for combustion. *Incomplete combustion will result in the production of carbon monoxide, a very dangerous toxic gas.*

Combustion and gas burners

Products of combustion

When natural gas is completely consumed the products of combustion are harmless, being water vapour, carbon dioxide and the nitrogen originally contained in the air. Some gas appliances are flueless, typical examples being cookers and small water heaters, the products of combustion producing no ill-effects on the occupants of the rooms in which the appliances are installed providing they are correctly fitted and maintained. It is important, however, to ensure that there is adequate ventilation in such rooms, as although these products are not toxic, it is possible that they could cause vitiated air (air from which the oxygen has been used) to recirculate in the appliance, thus producing carbon monoxide.

Ignition temperature

Natural gas requires a temperature of 700 °C to cause it to ignite; this is slightly higher than for town gas.

Limits of flammability

This relates to the amount of gas in air required to produce a flammable mixture, and is usually expressed as a percentage. In the case of natural gas it is between 5 and 15 per cent. Anything less than 5 per cent will be too weak; if it is more than 15 per cent the mixture will be too rich.

Burning velocity

This term relates to the speed at which, on ignition, the flame spreads through the gas/air mixture. This speed is affected by the pressure of the gas which can accelerate its movement to such a degree that the gas/air mixture is unable to burn quickly enough. The result is an unstable flame burning in the atmosphere some distance away from the end of the burner as illustrated in Fig. 2.1. This is often referred to as *lift-off* and must not be tolerated, as the flame can easily be extinguished, allowing gas to enter the room with consequent danger of explosion. Burners are designed in such

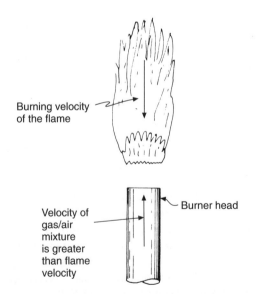

Burning velocity of the flame

Velocity of gas/air mixture is greater than flame velocity

Burner head

Fig. 2.1 Lift-off. This very undesirable situation takes place when the velocity of the gas flow is greater than the burning velocity of the flame. Because the flame is not stabilised on the burner it can very easily be extinguished.

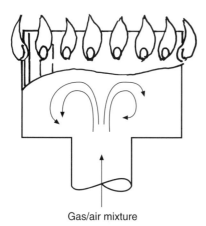

Gas/air mixture

(a) Turbulence caused by the sudden enlargement of the burner head reduces the velocity of the gas/air mixture

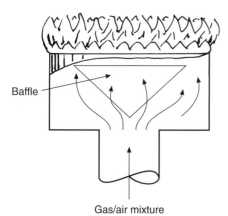

Baffle

Gas/air mixture

(b) Recirculating hot gases using a baffle to increase the effect of turbulence

Fig. 2.2 Stabilisation of gas flames. Both types of burner head are designed to create turbulence.

a way that the flame is stabilised as shown in Fig. 2.2.

The opposite of lift-off is *light back* which occurs when the velocity of the gas/air mixture is so low that the flame speed is greater and passes back through the burner to light on the injector as shown in Fig. 2.3. This is also dangerous as complete combustion is not achieved, giving rise to the production of carbon monoxide.

Combustion air

When a fuel is burned it combines with the oxygen in the air to produce heat. From a purely academic view, a chemical change takes place during which oxygen in the air joins with the hydrocarbons in the gas, producing heat as it does so. The products of this chemical change are carbon dioxide and water vapour, which in this case are called products of combustion. The remainder of the air, consisting mainly of nitrogen, has no effect on the chemical change and simply passes out of the appliance with the products of combustion. Neither nitrogen, carbon dioxide nor water vapour are toxic, but it is essential that they are removed from the room in which the appliance is situated, as they can cause acute discomfort to the occupants. As stated previously, a build-up of these products will deter the entry of fresh air and may lead to an oxygen deficiency causing incomplete combustion in the appliance. This is very serious, because if the hydrocarbons in the gas are unable to combine with sufficient oxygen, carbon monoxide is produced. If this happens with a flueless appliance

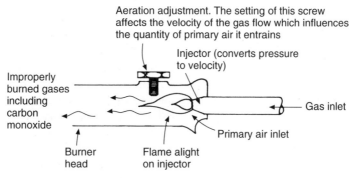

Aeration adjustment. The setting of this screw affects the velocity of the gas flow which influences the quantity of primary air it entrains

Injector (converts pressure to velocity)

Improperly burned gases including carbon monoxide

Gas inlet

Primary air inlet

Burner head

Flame alight on injector

Fig. 2.3 Light back. This occurs when the speed of gas passing through the injector is reduced to such a degree that its velocity is less than the flame speed. This situation is potentially dangerous as it can result in the production of carbon monoxide.

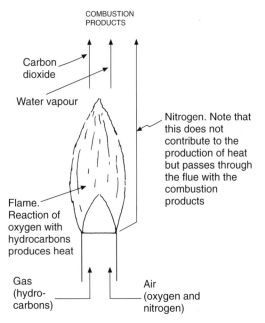

Fig. 2.4   Combustion of gas.

(a) Ragged yellow flame indicates an insuffient air supply

(b) Short poorly defined inner cone indicates too much air

The appearance of a flame can be a guide to combustion defects. Some flames are designed to be softer or quieter E.G. Grills and boilers. Others such as cooker hubs and water heaters burn very fiercely.

Fig. 2.5   Examples of incomplete combustion.

such as a cooker or small water heater or an appliance with a blocked or defective flue, it can result in the death of the occupants of the room.

All types of burners are classified as (a) pre-aerated and (b) post-aerated. Most modern appliances use pre-aerated burners where some air is mixed with the gas before it is burned. Some employ atmospheric or natural draught, others forced draught where the air for pre-mixing is supplied under pressure by a fan. Figure 2.4 illustrates a typical atmospheric burner correctly adjusted to give complete combustion, while Fig. 2.5(a) and (b) illustrate examples of incomplete combustion. The physical appearance of the flame is a good guide to combustion fault finding. Defects due to inadequate supply of air are usually due to causes (a) to (c):

(a)   Blocked or undersized flues.
(b)   Inadequate ventilation causing 'vitiation' or lack of oxygen.
(c)   Under-aeration (insufficient air). The most common cause of this is blockage of the primary air ports of the burner by 'lint'; furnishing fibres and pet hairs are typical

examples. Open-flued appliances are more prone to this than those that are room sealed.
(d)   Over gassing — this is usually due to an excessive gas rate or pressure.

Installation of pipework

This is defined as any pipework fitted to the outlet side of the primary meter. To comply with the Gas Safety Regulations, all pipework and appliances must be installed using approved materials and recommended procedures. All open-ended pipes must be capped if left unattended. This avoids the risk of others, being unaware of any open ends, connecting or reconnecting the gas supply to the meter. If a blowlamp is used to repair or extend an existing supply, the meter outlet must be disconnected and both ends sealed to prevent a flashback. When using flux to make soldered joints, it is important that it is used sparingly to avoid corrosion inside the pipe. Although active fluxes are permissible, those of the non-active type are recommended for gas services. Pipework under screeded concrete floors and in wall chases should be protected against corrosion. Pipes having a factory-finished plastic sheath are recommended but grease-impregnated tape or yellow-coloured wrapping tape are suitable. Compression fittings of any type must not be concealed or fitted in such a way as to preclude access to them. The main gas cock, filter, governor and meter must be installed in an accessible position to afford the consumer, installers and service engineers easy access to them. All domestic gas installations must have main equipotential bonding conforming to the IEE wiring regulations. It must be connected to the pipework within 600 mm of the meter outlet. Reference should be made to Chapter 11 for more detailed information on this subject. If a piece of pipe or a meter is to be removed, the electrical continuity of the installation must be maintained. If this is broken, even only temporarily, it could result in a fatal shock or an explosion due to a spark igniting a gas–air mixture.

To conform to the Gas Regulations a temporary bonding wire should be used as shown in Fig. 2.6.

All pipes and fittings used for a gas supply must be of suitable strength and comply to BS 6891. Materials used for making joints on threaded pipes and fittings must be suitable — not all jointing pastes meet this requirement and it is wise to check the label on the container before use. Polytetrafluoroethylene (PTFE) tape and most types of jointing media for oil pipelines are also suitable for natural gas. No pipework may be fitted in a cavity wall, and where it passes through such a wall it must be sleeved. The sleeve must have sealed ends to ensure that a leak in the pipe does not result in the cavity becoming filled with gas. Gas pipes must not be fitted in such a way that they are subjected to a compressive load. This means that they must not be installed under footings or loadbearing walls. When any new or additional work on an existing supply is carried out, the pipework and appliance (if applicable) should be air tested, using methods described in the section on testing.

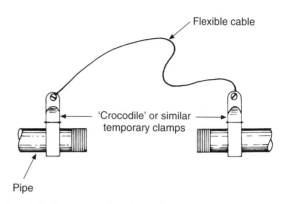

Fig. 2.6 Temporary bonding wire. This must be used when a section of pipe or a meter is temporarily removed to ensure the continuity of electrical bonding.

Before fitting any gas appliance it is essential to ascertain that sufficient air is available for combustion, and should the appliance be fitted with a flue pipe it must be installed in such a way that the products of combustion are carried away to the external air.

All floor-standing appliances must be adjusted so that they are level, firm and stable. This is very important in the case of free-standing gas cookers or refrigerators. A very effective way of checking the level is to stand a pan of water on top of the appliance when any inaccuracy in level will be easily seen. In the case of free-standing boilers the floor or hearth must be of adequate strength and conform to the building regulations regarding its combustibility.

Gas meters

The fixing of gas meters and the running of gas service pipes from the main is normally undertaken by the supplier's employees or one of their recognised subcontractors. The siting of meters is the subject of strict control relating to fire precautions and escape routes in the event of an emergency. If it is necessary to disconnect a meter in the course of one's work, the connections must be sealed to prevent any gas it contains becoming a fire or explosion risk, and when it is refitted any connection must be tested for leaks and purged (as described later) before any appliances are used.

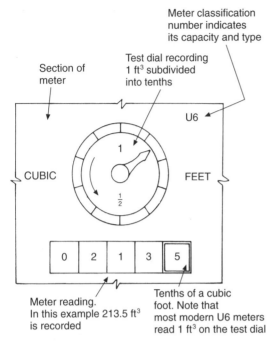

Meter classification
number indicates
its capacity and type

Test dial recording
1 ft³ subdivided
into tenths

Section of
meter

U6

CUBIC

FEET

Meter reading.
In this example 213.5 ft³
is recorded

Tenths of a cubic
foot. Note that
most modern U6 meters
read 1 ft³ on the test dial

Fig. 2.7 U6 gas meter dial.

Figure 2.7 illustrates the readings on a typical U6 meter.

Conversion of meter reading to kilowatts
U6 meters are calibrated to give readings in cubic feet (ft³), and it may be necessary in some cases to convert this to cubic metres (m³). In rounded-off figures it can be assumed that 1 m³ = 35.336 ft³. To convert cubic feet directly to kilowatts the following formula is used:

$$1 \text{ unit of gas } (100 \text{ ft}^3) = 2.83 \text{ m}^3$$

This is multiplied by the calorific value of the gas which may vary slightly from area to area. The calorific value used here is 38.5 — the total thus produced is then divided by 3.6 which converts the original reading to kilowatts. Example: Assume a meter has passed 33,000 cubic feet — convert this to kilowatts:

$$33,000 \div 100 = 330 \text{ units}$$
$$330 \times 2.83 = 933.9 \text{ ft}^3 \text{ (cubic feet)}$$
$$933.9 \times 38.5 = 35,955.15 \text{ (total calorific value)}$$
$$35,955.15 \div 3.6 = 9,987.542 \text{ kW (kilowatts)}$$

E6 meters
These meters do not have a test dial but a liquid crystal display. They are calibrated in m³ and the method of calculating gas flow rates is slightly different to that used with U6 meters.

Testing appliance pressures and gas rates

To check the working pressure at a meter, connect a pressure gauge to the meter test point. It should read 21 mbar plus or minus 1 mbar.

After an appliance has been fixed it must be checked to ensure it is working at the pressure recommended by the manufacturer and adjusted so that it consumes no more than the recommended quantity of gas. 'Over-gassing' may result in the production of carbon monoxide, flame lift-off, and loss of efficiency. The procedures are as follows.

Setting burner pressure
A manometer should be fitted to the test point on the appliance. The pressure shown on the gauge should be as specified in the manufacturer's instructions. Any alteration necessary is made at the appropriate control, usually the governor.

Setting the gas rate
This is required by the Gas Safety Regulations. The object is to check the volume of gas used by the appliance, which will be found both on the badge affixed to it and in the manufacturer's instructions. As an example a gas central heating boiler with an input rating of 52,000 Btu/h is to be checked. The procedure using a U6 meter which measures gas in cubic feet is as follows:

1. Check that the burner pressure is correctly set.
2. Turn off all other gas appliances in the premises.
3. Assume the calorific value of the gas to be 38 MJ/m (1,040 Btu/ft).
4. Turn the boiler on and ensure the burner is alight. Check the flame appearance for signs of poor combustion and allow approximately 10 minutes for it to reach operating temperature.
5. Watching the cubic foot dial on the meter, record the time in seconds for the pointer to rotate once; accurate timing is important.

Assuming it takes 72 seconds to burn 1 cubic foot, the gas rate can be determined using the following simple formula:

$$\frac{\text{seconds in 1 hour} \times \text{calorific value Btu/h}}{\text{time in seconds to burn 1ft}^3}$$

$$\therefore \frac{3,600 \times 1,040}{72} = 52,000 \text{ input rates in Btu/h}$$

$$3,421 \text{ Btu/h} = 1 \text{ kW}$$

$$\therefore \frac{52,000}{3,421} = 15.2 \text{ kW input}$$

E6 Ultrasonic gas meters   This type of meter give a direct reading in m^3. To check the gas rate of an appliance, assuming the pressure is correct and all other appliances are turned off, the procedures are as follows:

1. Note the reading and time for 2 minutes.
2. Record the second reading.
3. Add the number of seconds until the next digit is shown.
4. Subtract the first reading from the second.
5. This gives the gas rate in m^3 over 2 minutes.
6. Convert these figures to kW using an E6 meter chart, or using the following formula:

$$kW = \frac{\dfrac{\text{Number of seconds}}{\text{in 1 hour}} \times \dfrac{\text{recorded}}{\text{volume in m}^3}}{\dfrac{120 \text{ seconds}}{(2 \text{ mins})} + \dfrac{\text{number of seconds until}}{\text{next digit on test dial}}}$$

Example: 1st reading 00836.322
2nd reading 00836.374 + 10 seconds
00836.374 − 00836.322 = 0.052
(recorded volume)

$$\frac{3,600 \times 0.052}{120 \text{ seconds} + 10} = 1.44 \text{ m}^3/\text{h}$$

Depending on the calorific value of the gas, 1 m^3 gives approximately 10.8 kWh
∴ input = 1.44 × 10.8 = 15.552, approximately 15.6 kWh

Pressure

In practical terms the earth can be said to be surrounded by air to a height of approximately 7 miles and as a consequence its surface is subjected to a pressure equivalent to its mass, which is 101.3 k P/m or 1.013 bar. It will be seen that for all practical purposes atmospheric pressure can be expressed as 1 bar. Normally this can be ignored when pressure readings are taken because atmospheric pressure affects both sides of the gauge in the same way, any pressure recorded being technically called *gauge* pressure. When the pressure shown on a gauge is added to that of the atmosphere the combined pressures are called *absolute*. To give an example of absolute pressure, if 2 bar are recorded on a pressure gauge and added to that of the atmosphere, the result will be an absolute pressure of 3.013 bar (atmospheric pressure = 1.013 bar). The pressure in an installation when no appliances are in use is known as the standing pressure. Working pressures are those recorded when an appliance is in use. The causes of pressure loss are considered in the following text and must be taken into account when sizing pipes for both gas and water.

Gas pipe sizing

Great care must be taken to ensure the pipe sizes for gas installations are adequate. Failure to do this may result in unsatisfactory performance of an appliance, e.g. low hot water temperature supplied from water heating equipment and poor performance of cooking appliances. In extreme cases an insufficient volume and pressure of gas at an appliance could cause the pilot and even the main burner flame to be extinguished. This is a dangerous situation which could lead to an explosion if the gas is re-ignited, and in some circumstances the production of carbon monoxide. The design of a gas piping installation should take into account the following. It should be capable of supplying sufficient gas at all the appliances connected to it and consideration should be given to possible future extensions. As with water supplies, allowances must be made for pressure loss caused by the frictional resistance of the pipe walls. The longer the pipe the greater this will be, and losses due to turbulence in tees and elbows must also be taken into account. It should be noted that pipes of various materials have differing carrying capacity and it is important that reference is made to the

Table 2.2 Discharge in a straight horizontal copper or stainless steel tube with 1.0 mbar differential pressure between the ends, for gas of relative density 0.6 (air = 1). (Courtesy of CORGI)

Copper piping in accordance with BS EN 1057 or corrugated stainless steel tube to BS 7838

Size of pipe	Length of pipe (m)							
mm	3	6	9	12	15	20	25	30
				Discharge m³/h				
10	0.86	0.57	0.50	0.37	0.30	0.22	0.18	0.15
12	1.5	1.0	0.85	0.82	0.69	0.52	0.41	0.34
15	2.9	1.9	1.5	1.3	1.1	0.95	0.92	0.88
22	8.7	5.8	4.6	3.9	3.4	2.9	2.5	2.3
28	18	12	9.4	8.0	7.0	5.9	5.2	4.7

Note: When using this table to estimate the gas flow rate in pipework of a known length, this length should be increased by 0.5 m for each elbow and tee fitted, and by 0.3 m for each 90° bend fitted.

pipe sizing table which applies to the material being used. To give an example, a steel pipe having a nominal diameter of 15 mm and a length of 3 m will discharge 4.3 m³ per hour. A copper pipe of the same nominal diameter will discharge only 2.9 m³ per hour. Gas appliance manufacturers specify the recommended pipe diameter for their appliance only. Where more than one appliance is used it is necessary to determine a pipe diameter that will satisfy the total possible demand. **The permissible pressure drop on a gas pipeline during periods of maximum demand must not exceed 1 mbar.**

Example of pipe sizing
Figure 2.8 shows a gas carcass supplying a variety of gas appliances. Copper tube as shown in Table 2.2 is used as the pipework material. Typical ratings of gas appliances are shown in Table 2.3. Bear in mind the two principle factors involved are the gas rate of each appliance and the frictional resistance of the pipework. A tabulation chart similar to that shown in Table 2.4 will be useful. The procedure is as follows:

(a) On the tabulation chart enter the appropriate section of pipe from the drawing.
(b) Enter the gas rate from Table 2.3 (in practice the gas rate will be taken from the manufacturer's instructions).

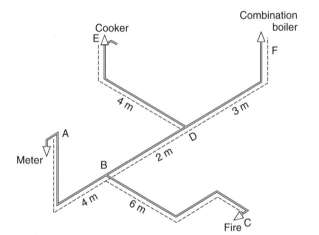

Fig. 2.8 Pipe sizing for gas appliances.

Table 2.3 Typical gas ratings.

Appliance	Gas rate m³/hour
Warm air heater	1.0
Multipoint water heater	2.5
Cooker	1.0
Gas fire	0.5
Central heating boiler	1.5
Combination boiler	2.5

Table 2.4 Tabulation of data.

Pipe section	Gas rate	Measured pipe length	Allowances for fittings		Total length	Pipe diameter
ref fig 2	m^3/hr	m	type	equivalent length	m	mm
A–B	4	4	3 elbows	1.5	5.5	22
B–D	3.5	2	–	–	2	22
B–C	0.5	6	3 elbows 1 tee	2.0	8.0	12
D–E	1.0	4	3 elbows 1 tee	2.0	6	15
D–F	2.5	3	1 elbow	0.5	3.5	22*

Note: A 3 m length of 15 mm pipe will carry 2.9 m³/hour. 3 m is just short of 3.5 m so only a 0.500 m length would need to be 22 m diameter.

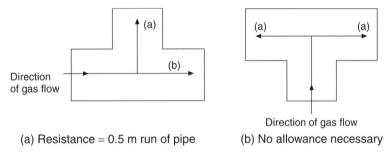

(a) Resistance = 0.5 m run of pipe   (b) No allowance necessary

Fig. 2.9 Tees showing where allowances for resistance to flow must be allowed for.

(c) Add the measured length of pipe to the allowances made for fittings. In this system of sizing a standard length of 0.5 m is allowed for both elbows and tees of all sizes (see Fig. 2.9).

Referring to Fig. 2.8 it will be seen that pipe A–B is supplying all the appliances in the installation, so the gas rate will be the sum of the following:

gas rate for fire = 0.5
gas rate for cooker = 1.0
gas rate for combination boiler = 2.5

Figure 2.8 shows the effective length of pipe A–B to be 4 m. Referring to Table 2.2 it will be seen that the pipe lengths are increased in multiples of 3 m. The nearest to 5.5 m is 6 m (column 3 from left): reading down this column it will be seen that a 6 m length of 22 mm pipe will pass 5.8 m³/h which is well within the limits required. The remaining pipes are sized in a similar way.

In situations where it is necessary to convert cubic feet to cubic metres the factor 0.0283 is used.
Example: convert 60 cubic feet to m³

60 × 0.0283 = 1.698, aproximately 1.7 m³

Testing for soundness

The U gauge, or manometer
This instrument is used both for measuring the pressure and testing for leaks in gas installations. It is a very simple device consisting of a glass tube bent in the form of a letter U. A clear plastic tube is sometimes used and is less liable to breakage, but it has the disadvantage that it discolours after a period of time making it difficult to read. The tube is contained in a plastic or metal case which offers some degree of protection.

A graduated scale is attached to the two legs of the U, each large division on the scale representing 1 mbar, which is one-thousandth part of a full bar of pressure. It is important to note that low gas pressures are usually quoted in millibars, 1 mbar being in effect the pressure exerted by a head of water 9.8 mm high. For practical purposes a millibar is taken as 10 mm or 1 cm. Prior to metrication U gauges were graduated in inches and tenths of an inch, and where these gauges are still in use it is useful to remember that 1 inch water gauge is equivalent to 2.5 mbar.

U gauges may be of the direct reading or indirect reading type: both are shown in Fig. 2.10. In the case of Fig. 2.10(a), the direct reading type, it will be seen that although the scale is measured in 1 mbar divisions, each division is numbered to represent two. Providing the gauge has been zeroed properly a direct reading can be obtained from one leg only. A reading on the indirect type is taken by adding together the number of millibars each side of the zero. Figure 2.10(b) shows how the same pressure is read on each type of gauge.

Careful scrutiny of Fig. 2.10(c) shows that the surface of the water in each leg of the tube is curved due to the adhesion of the water to the sides of the tube. The name given to the curve is the *meniscus* and it is important to make sure when taking a reading it is taken from the bottom of each curve as shown.

Whenever a new gas installation is fitted, or additions are made to existing systems they must be checked for leaks on completion of the work. This should be done before any pipework is covered over or any protective coating is applied. A soundness test must also be carried out when an appliance is serviced, if it has been fitted with new components or has been altered in any way.

Testing a new installation

Many gas pipes have to be buried in walls and floors and must be fitted before the walls are plastered or any floor finish is applied. This pipework is often referred to as the *gas carcass* and the procedure for testing is as follows. All the open ends must be properly capped or plugged except one, to which is fitted a tee piece having a small test cock on one branch to allow air to be admitted, and a test nipple on the other end to which the U gauge, carefully zeroed, is attached by means of a rubber tube.

(a) Direct reading type

Sometimes known as a half-scale manometer. Each large division measures 1 mbar but is numbered in 2 mbar. When this gauge has been correctly zeroed it can be read directly. The reading shown here is 6 mbar.

(b) Indirect reading type

With this type the number of millibars on either side of zero must be added. In this instance the water levels read 3 millibars on each leg. Thus 3 + 3 = 6 mbar. Note that in both cases the reading is taken from the lowest point of the meniscus.

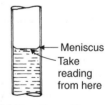

(c) Reading a U gauge

Fig. 2.10   The U gauge (manometer).

The set-up for testing is shown in Fig. 2.11. The U gauge must always be in an upright position when tests are made to ensure a correct reading. The air in the system is pressurised by being pumped or blown through the test cock until a

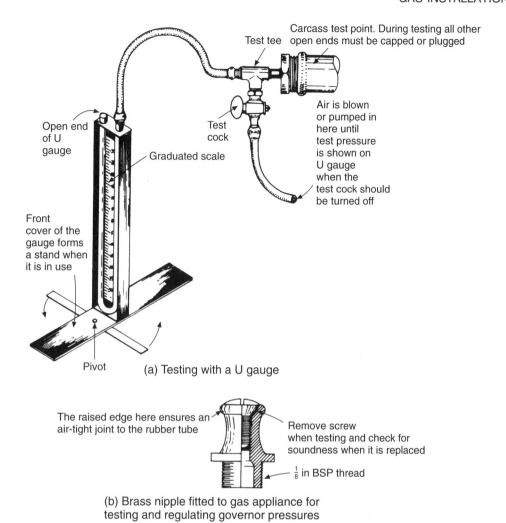

Carcass test point. During testing all other open ends must be capped or plugged

Test tee

Open end of U gauge

Test cock

Graduated scale

Air is blown or pumped in here until test pressure is shown on U gauge when the test cock should be turned off

Front cover of the gauge forms a stand when it is in use

Pivot

(a) Testing with a U gauge

The raised edge here ensures an air-tight joint to the rubber tube

Remove screw when testing and check for soundness when it is replaced

$\frac{1}{8}$ in BSP thread

(b) Brass nipple fitted to gas appliance for testing and regulating governor pressures

Fig. 2.11 Use of a U gauge (manometer) for testing soundness of gas installations.

reading on the U gauge of 20 mbar is shown. The test cock is then shut and a period of time (usually 1 minute) is allowed for the air temperature to stabilise. This is important as the moisture content of warm air will condense when admitted to cold pipework and cause a reduction in its volume, resulting in a drop in the pressure showing on the U gauge. After a 1 minute stabilisation period has elapsed, the pressure should hold, without any further drop showing on the gauge, for a period of 2 minutes. If the distance between the water levels in each leg of the gauge lessens, a leak is indicated and must be found by painting the joints with leak detection fluid, any leaks being detected by the formation of bubbles at the source of leakage. Soap solutions are no longer recommended, as it has been found they have a slight corrosive effect on metals. It should also be noted that where possible, gas should be used for testing in preference to air.

Testing extensions to existing pipework in domestic properties
Before the extension is fitted or connected, the existing installation should be tested in the following way. The main gas valve on the inlet side of the meter and all appliance valves and pilot lights should be turned off, and the U gauge connected to the main test nipple, which is usually fitted on or

adjacent to the meter. On most installations U gauge connections are fitted on the outlet side. The system should then be pressurised by turning on the main gas cock until a pressure of 20 mbar is recorded on the U gauge — the gas should now be turned off. After a 1 minute period for the temperature to stabilise, no loss of pressure should be recorded on the gauge for a period of 2 minutes. It is an accepted fact that many existing installations are not completely gas tight, e.g. minute leaks on appliance valves. A slight leakage may be acceptable on an appliance, providing it does not exceed that shown in Table 2.5 — but under no circumstances is it acceptable if the pipework is not sound. The procedure here is to isolate all appliances on the system and conduct a test on the pipework only. If it is defective, the leak must be found. However, if the test proves the pipework to be sound and no complaints are made about a smell of gas, there is no undue cause for alarm. Of course, if gas can be smelt the leak must be found and rectified. The differing pressure drops shown in the table relate to the size of the installation and the number of appliances used, which is indicated by the type of meter. A U6 (DO7) meter is used for normal domestic installations whereas a P4 (D4) would be used with a larger pipe diameter. If it is understood that the pressure loss in a small pipe over a given period of time will be greater than that in a pipe having a larger diameter, it will be seen why there is a difference in the permissible drop in various installations. Although this difference exists, the ratio of the pressure loss has been calculated to be approximately the same for all types of installations. The rating of a meter relates to its capacity, i.e. the quantity of gas it will pass, and is indicated by a stamp on the front of the case.

One other very important point must be made here in relation to testing existing installations.

Table 2.5 Permissible gas pressure drops.

Meter	Pressure drop (mbar)	Capacity in cubic feet per revolution
E6	8.0	N/A
U6 (D07)	4.0	0.071
P1 (D1)	2.5	0.100
P2 (D2)	1.5	0.200
P4 (D4)	0.5	0.400

Never search for leaks with a match or naked light — it could result in a one-way ticket to the hereafter!

On existing installations any pressure test on existing pipework may be false if the main gas valve 'lets by'. After a period of time the lubricant used to enable it to turn freely and maintain a positive on/off action, may dry out, and a small quantity of gas can seep past the valve. This can be checked by using a similar testing procedure to that used for existing installations, turning off the valve when a pressure reading of only 10 mbar is shown on the gauge. If the gauge indicates an increase in pressure the valve is defective and the fact must be reported to the gas supplier. This test must always be applied to existing installations after soundness tests are conducted.

Purging domestic installations

One of the most important things that must be done when commissioning a new gas installation is to purge it of air, as if a mixture of gas and air occurs in the pipes, a blow back when lighting the appliance could cause an explosion. It is therefore essential to 'air test' the installation before purging is carried out.

The procedure for purging new installations is as follows. The main gas cock should be turned off before any of the appliances are turned on, preferably the one at the end of the main run of pipe. If the appliance is fitted with a flame failure device no gas will be able to pass to the burner, and the usual practice in such a case is to disconnect the burner union or remove the screw in the appliance test nipple. During the purging operation, it is essential that the area into which any gas/air mixture is discharged is well ventilated by opening any windows or doors. Avoid using any electrical switch which may cause a spark, ensure that there are no naked lights and, of course, smoking during the operation is not permitted. The main gas valve should be turned on until the meter is purged by passing a volume of gas not less than five times the capacity per revolution of the meter mechanism. This capacity is shown in the window housing the meter dials, and on a U6 meter is shown as 0.071 ft^3 per revolution, thus $5 \times 0.071 = 0.355$, just over a third of a cubic foot. Note that the volume of gas shown on the test dial of the meter varies. That shown in Fig. 3.5 will pass 1 ft^3 per revolution, thus to purge this meter a movement of the pointer through just under 3.5 divisions is required. This will ensure that no air remains in the meter or pipework. (Note that some U6 meters have a 2 ft^3 test dial.) The same volume of gas must pass an E6 meter when purging is carried out.

The main cock is now turned off and any connections that were broken for the purpose of purging should be tightened and retested for soundness. This is done in the usual way using leak detection fluid, after having re-established the gas supply by turning on the main cock. Each of the other appliances connected to the system should then be turned on until gas is smelt from the burner. As already stated, some appliances are equipped with flame failure devices and air in the pipes supplying them can only be purged as previously described. Only when all the air in the system has been removed should any attempt be made to light the appliances. If a new appliance or pipe run only has been fitted there is no need to purge the meter or the main pipe runs. It should only be necessary to disconnect the new appliance and proceed as previously described.

No appliance should be permanently connected to the gas supply until it has been commissioned, tested and purged.

Ventilation

A good supply of air is necessary when all types of fuel are burnt and if an appliance is fitted in a confined space, e.g. a cupboard, a flow of air may be necessary to keep it cool. A supply of air is usually introduced into the building via vents built into the structure. The following relates to the air supply for open-flued appliances which take air for combustion from the area in which they are fitted. It is not exhaustive and for more detailed information reference should be made to CP 5440 Pt2.

Adventitious ventilation is the term applied to the natural ingress of air into a building via skirting boards and around windows and doorways, which occurs despite all efforts made to reduce it. For this reason an air vent is not normally necessary for appliances rated up to 7 kW. This does not apply to Northern Ireland where a permanent ventilation opening of at least 4.5 cm^3 is required for any open-flued appliance, plus 4.5 cm^3 for every kW of input rating in excess of 8 kW. It must be noted that adventitious air is not taken into account when calculating the air requirements for certain fires, flueless appliances, and those fitted in compartments.

Care must be taken when fitting air vents. They should not be located where they would become blocked, flooded or allow the ingress of contaminated air. In the past it has been the practice to take air for combustion from the ventilated area under a suspended floor. However, there are risks due to radon gas, which is produced naturally from the decay of uranium in rocks and soils. The level of concentration varies nationally, but in high risk areas any ventilation should be sited in, or ducted to, an external wall. If in doubt the local building control officer's advice should be sought.

Figure 2.12 illustrates the method of providing air

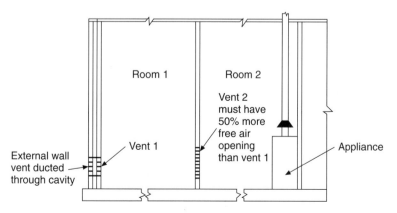

Fig. 2.12 Air vents in series.

Table 2.6 Minimum effective air vents serving gas appliances up to 60 kW input.

Open-flued appliances	Position of vent levels	Free area of vent cm^2 per kW input
In a room ventilated directly to the external air	High or Low	4.5 in excess of 7 kW
In a compartment ventilated to the external air	High	4.5
	Low	9.0
In a compartment ventilated via an internal room or space	High	9.0
	Low	18.0
Room-sealed appliances		
In a room	—	No vent necessary
In a compartment ventilated to the external air	High	4.5
	Low	4.5
In a compartment ventilated via an internal room or space	High	9.5
	Low	9.5

Notes: A compartment is an enclosure designed or adapted to house a gas appliance. High and low level vents are necessary for cooling. Air for ventilation must always be taken from the same source, either from outside or inside the building. Vents must be fitted on the same wall of the compartment. All combustion air must be provided via low level vents. The top of low level vents should not be higher than 450 mm to avoid the passage of smoke in the event of a fire.

for combustion to an open-flued appliance which is in a room with no access to an outside wall. It will be seen that air both for combustion and for ventilation of the room must pass through an adjoining room in which a vent can be fitted to the outside air. The important point here is the increase in size of the internal ventilation grills. Room-sealed appliances take air for combustion externally to the building but in certain instances ventilation is necessary for cooling. Table 2.6 gives a guide based on the requirements of BS 5440 Pt 2 for providing air for combustion and ventilation in both room-sealed and open-flued appliances.

Air bricks

These are the usual method of providing ventilation in a building. They are usually made of terracotta, although cast iron may be used. A section of a terracotta air brick is shown in Fig. 2.13. Their size varies but is normally measured in millimetres as follows: 75 × 225, 150 × 225, and 225 × 225. A duct must always be provided through the wall connecting the ventilating grills. It will be seen that the holes taper and it is important that the free area is calculated by multiplying the area of one hole by the number of holes in the vent. Their sizes must

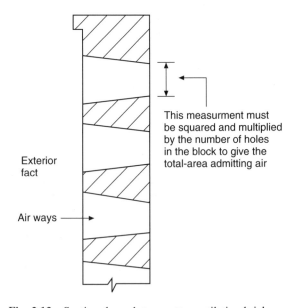

This measurment must be squared and multiplied by the number of holes in the block to give the total-area admitting air

Exterior fact

Air ways

Fig. 2.13 Section through terracotta ventilating brick.

be taken from the inside surface of the ventilator, which will be the effective area of each hole. Assuming the area of one hole is 64 mm and the vent contains 36 holes, the free area will be 2,304 mm or 23 cm.

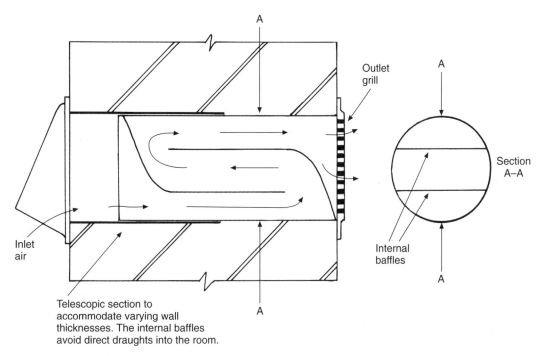

Fig. 2.14 Draught-proof wall ventilator.

Figure 2.14 illustrates an alternative to the bricks described. Because these ventilators are draught-proof the possibility of the occupant of the building blocking a draughty vent is avoided. They are circular in shape so that a round hole can be drilled using a core drill on the structure.

Note that the installer or service engineer is responsible for determining and checking ventilator sizes in both new and existing buildings. All manufacturers of gas appliances provide detailed instructions for their installations which include recommendations relating to the supply of fresh air. These instructions must be carefully studied and complied with to prevent the occurrence of accidents.

Flues

Function of a flue

The flue is an important part of the installation and must be sound, safe and efficient. Its main function is to remove the products of combustion from the appliance. In addition to this, an open flue acts as a ventilator, whether or not the appliance is in use, and this contributes to the physical comfort of the building's occupants and minimises the formation of condensation, especially in kitchens.

Flueless appliances

Whether or not a gas appliance is fitted with a flue depends mainly on (a) the period of time during which it is in use and (b) the quantity of gas it consumes. Flueless appliances are permissible in rooms or spaces under the following conditions, providing adequate fixed ventilation is fitted. In most cases an openable window is also required.

(a) With low-rating appliances, e.g. refrigerators.
(b) When the period during which they are in use is of relatively short duration, e.g. a water heater serving a sink.
(c) When conditions associated with their operation render it likely that the user will increase the ventilation of the room by opening the windows, e.g. cookers and hobs causing cooking smells and steam.

Gas appliances which fall into the categories listed above are: gas cookers; instantaneous water

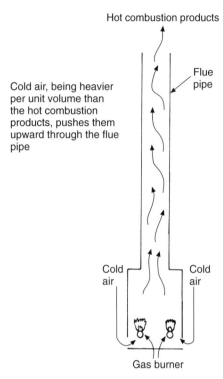

Hot combustion products

Flue
pipe

Cold air, being heavier
per unit volume than
the hot combustion
products, pushes them
upward through the flue
pipe

Cold
air

Cold
air

Gas burner

Fig. 2.15   Operation of a flue.

heaters up to 12 kW, providing they are not liable
to prolonged use or fitted in a confined space
(not less than 5 m³); storage water heaters where
the heat input is less than 3 kW, or 4.5 kW if
the storage capacity is less than 45 litres; gas
circulators, having a heat input of less than 3 kW,
providing they are not installed in a bathroom,
airing cupboard or badly ventilated area.

Open flues (natural draught)

A flue works on the principle of convection
(see Fig. 2.15), a method of heat transfer which
applies to fluids, i.e. liquids and gases. Convection
has many applications in plumbers' work including
the circulation of hot water and drain ventilation,
the principle relating to the latter being exactly the
same as that for a flue. In a natural draught flue the
upward movement of the combustion products is
brought about by the difference between their
temperature and that of the air surrounding the
flue.

Factors affecting the performance of flues

Height   An increase in the height of a short flue
will raise its performance within certain limits.
However, if it is too high the frictional resistance
of the pipe walls will slow down the updraught and,
more important, as the products of combustion cool,
the convective current will become weaker. These
factors do not apply to flues of less than 6–9 m
in height, and longer flues are seldom encountered
in domestic dwellings. However, if they are,
a mechanical extractor may be necessary and
specialist advice should be sought. It is important
to note that any appliance in which the products
of combustion are removed by mechanical
extractors must be made in such a way that the
valve admitting gas to the burner will not open
unless the extractor is operating.

Flue gas temperatures   Generally, the updraught
in a flue is improved as the temperature of the
combustion products increases, but this tends to
become proportionally less at higher ambient air
temperatures.

Flue runs   Where possible, flues should be
vertical. If bends are necessary those having an
angle of 135° are preferred, as those having a more
acute angle offer more resistance to the flow of
combustion products. Bends of 90° and horizontal
flue runs for natural draught appliances are no
longer permissible. The diameter of the flue is
dependent on the flue outlet spigot on the boiler
and must never be reduced. The flue must always
be fitted to the appliance in such a way that it can
easily be disconnected; this can be accomplished
with a sliding or split collar as shown in Fig. 2.16.
The flue above the point of disconnection must be
securely fixed to prevent it dropping down when
the appliance is removed. Flue pipes should always
be fitted with the socket upward, as this avoids any
condensation running down the outside of the flue
pipe leaving an unsightly stain.

Flue materials

Pipes   Lightweight asbestos cement was for many
years used almost exclusively for flue pipes but is
now not permissible.

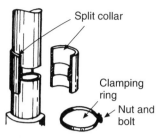

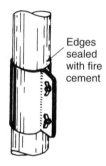

(a) The two halves of the split collar are secured with a clamping ring prior to sealing joints with fire cement
These must be fitted as closely as possible to the appliance.

(b) A sheet metal clamp fixed round the pipe and secured by wing nuts

Fig. 2.16 Methods of flue pipe disconnection for cleaning and maintenance.

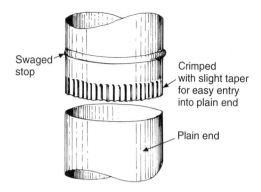

Fig. 2.17 Push-fit joint for single-wall stainless-steel flue pipe.

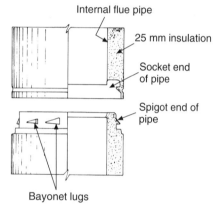

To make a joint the two ends are brought together and given a twist to engage the bayonet lugs in the socket and spigot ends of the pipe.

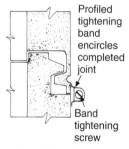

Detail of completed joint showing bayonet lugs engaged.

Fig. 2.18 Steel twin-wall flue pipes.

Similar pipes made of sisal fibres bound in cement, trade name 'Nurastone' are available but they tend to be very brittle. Care must be taken to prevent damaging them when handling and jointing. Joints are made in the traditional way using fireproof string and fire cement. Another suitable material for gas flues is stainless steel. It has been used as a flexible flue liner for many years, but it is now made in rigid form. A typical joint for this material is shown in Fig. 2.17. One end of each pipe is crimped and slightly tapered enabling a tight push-fit joint to be made into a plain end of pipe. Where necessary to give added support to the joints, self-tapping screws or pop rivets may be used to form a permanent fixing. Twin-wall insulated flue pipes may be used in situations where there is no existing brick flue. Unlike the previous materials discussed, which are normally only used in short lengths for connecting an appliance to an existing flue, twin-wall pipes provide a purpose-made

insulated chimney, see Fig. 2.18. Their construction and the materials used vary slightly depending on the manufacturers, but the basic principles are the same. In some cases both the inner and the outer

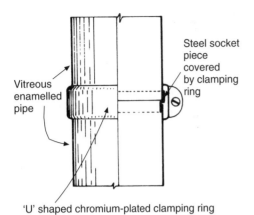

'U' shaped chromium-plated clamping ring

Fig. 2.19 Vitreous enamelled sheet steel flue pipes. These pipes are obtainable in black or white and have a very pleasing appearance. They are socketless, being joined with a pressed steel socket piece. This is hidden underneath the chromium-plated steel clamping ring which is secured with a screw.

pipes are made of stainless steel, in others the outer casing may be galvanised steel. Some have push-fit joints, others employ a bayonet-type joint which locks the pipes together when given a slight twist. In both cases an external locking band is used to ensure rigidity of the joint. All manufacturers of this type of flue produce a complete range of components such as bends, appliance adaptors, varying pipe lengths and brackets. When ordering insulated flues of this type, remember it is not possible to cut and joint it on site, therefore the exact lengths should be obtained.

Enamelled steel socketless pipes are sometimes used to connect appliances to a flue, the joints being made with shaped rings and clamps as shown in Fig. 2.19. These pipes are made in a variety of sizes and lengths and can be cut with a hacksaw, but this is best avoided due to possible damage to the enamel. If careful measurements are taken the flue can, in most cases, be constructed using stock sizes of pipe. This particular material is normally only used to connect an appliance to the main flue or chimney. Flue lining is dealt with later in this chapter.

Pre-cast flue blocks These are blocks made of Portland or high-alumina cement conforming to BS 1289:Parts 1 & 2. They may be built into the building structure by the bricklayer or in a slightly different form for external use to construct what is best described as a traditional chimney. They incorporate an insulated ceramic flue liner, and although more expensive than a flue pipe, they do have a more aesthetic appearance.

Flues for multi-storey buildings
The two systems illustrated in Fig. 2.20 were developed by the gas industry and are extensively used in multi-storey buildings for the discharge of combustion products into a common flue. It must be stressed that while under normal circumstances it is not permissible to discharge more than one appliance into a flue, these systems are designed for this purpose. Special room-sealed appliances are required for use with the system shown, and advice should be sought from the appliance manufacturer prior to its installation in such schemes.

The SE duct system This is basically a single open-ended duct running from the bottom to the top of the building into which is fitted the terminal of appliances on each floor. Each appliance draws its combustion air from, and discharges its combustion products into, this common duct. The degree of air dilution ensures no vitiated air is drawn into any appliance connected to the duct.

The 'U' duct system This is very similar to the SE duct and can be used where it might be difficult to arrange suitable horizontal ducts at the base of the building or where there might be a danger of the inlets becoming obstructed in any way. A study of Fig. 2.20 shows that, as its name implies, two ducts are employed forming the legs of a 'U'. Both air for combustion and the products of combustion are drawn from, and discharged into, one leg of the 'U', air for combustion and dilution air being drawn down through the other. A good flow of air through the duct is maintained by convection due to the discharge of the hotter products of combustion into one leg of the 'U' only.

Flues passing through combustible materials Although the flue temperature of most gas appliances is normally very low, accidents can happen, and if the risk of fire is to be avoided the

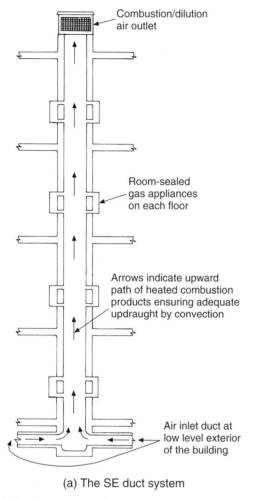

(a) The SE duct system

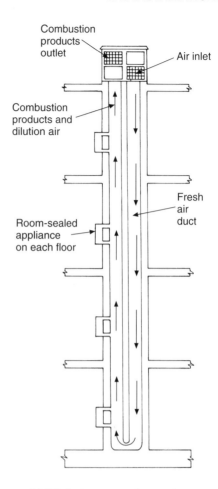

(b) 'U' duct common flue system

Fig. 2.20 Common gas flues.

flue must not be in direct contact with any combustible materials, e.g. wood floor or rafters. If it is within 50 mm of any such material it must pass through a sleeve of non-combustible material of sufficient width to form a space of 25 mm minimum between the sleeve and the flue pipe. Where the flue passes through a floor or ceiling, some method of preventing smoke or flame (in the event of a fire) from passing through the space around the pipe into the space above is necessary. The air space in these circumstances can be filled with a non-combustible insulation material such as fibreglass.

To comply rigidly with these fire regulations is not an easy task, but they are important.

Figure 2.21 gives some idea of what is required, and it should be appreciated that as only a few special components are purpose-made for sleeving flues, in most cases it will be left to the plumber's ingenuity to devise something suitable. Where a flue pipe passes through a roof, it must be weathered with a pipe flashing as shown in Fig. 2.22.

Fitting appliances to existing flues It is often convenient to use an existing brick flue for a gas appliance such as a boiler or fire. If the flue has been out of use for a period of time, or if it has been used for an appliance burning solid

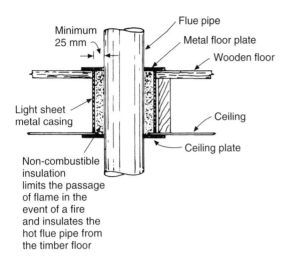

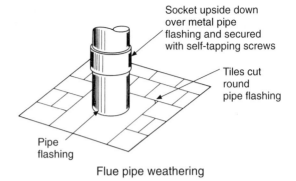

Flue pipe weathering

Fig. 2.22 Weathering of flue pipes passing through a roof.

Fig. 2.21 Passing a gas flue through combustible materials. This illustrates a flue passing through a wooden floor, but similar arrangements must be made where the flue passes through the roof if it is within 50 mm of any combustible material.

fuel, it must first be properly swept. This will ensure that it is clean and is clear of birds' nests and other obstructions. The brickwork must be checked for soundness, and in old buildings it is recommended that a check is made to ascertain that the flue is serving only one appliance. It has been known for two fireplaces to be connected to one flue or chimney, and in the case of a gas appliance being fitted this could mean that combustion products are discharged into another room. Even where the flue is to be lined this could mean that the insulation the lining provides is not as effective as it might be. To test an existing flue for soundness, cap the top and warm it up by placing a blow lamp in the opening for about 10 minutes. Remove the lamp and light a smoke pellet prior to sealing the opening. This will verify whether the flue is sound and not interconnected with another. Do not confuse this with testing flues for updraught.

If an appliance is likely to be out of action for long periods of time, say during the summer months when no space heating is required, it is essential that some means is used to prevent birds from nesting in the chimney. A terminal will prevent this, but as it is not usual for these to be fitted in the

case of gas fires it is recommended that the top of the chimney is covered with a corrosion-resistant mesh or a suitable terminal.

Testing flues for updraught
Before an appliance is fitted the flue must be tested to ensure it has a positive updraught. This is carried out using a smoke match enabling any sign of insufficient draught to be seen. Unless these tests are satisfactory, the appliance must not be fitted until the cause of any defect has been ascertained and corrected. Typical causes for adverse draught conditions are shown in Fig. 2.23.

Smoke tests should be conducted under the worst possible conditions; in other words, all doors and windows through which air can be admitted should be shut. If the room contains an extractor fan, this must be working during the test. In the event of the draught being insufficient, under no circumstances should the appliance be fitted until the draught problem is solved. Any form of ventilation must be of a type which cannot be closed, to deter anyone, who may not be aware of the dangers, from sealing them. Ventilators of a type which avoids direct draughts into the room should be fitted; Fig. 2.14 illustrates a ventilator of this type. Spillage tests must be conducted when an appliance is commisioned or serviced and carried out as shown by the examples given in Fig. 2.24(a), (b), (c) and (d). The term *spillage* relates to the products of combustion being discharged into a room or compartment due to downdraught. Spillage tests are conducted in a similar way

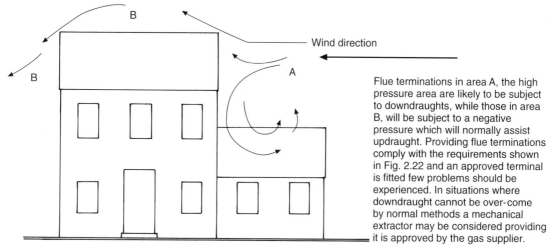

Wind direction

Flue terminations in area A, the high pressure area are likely to be subject to downdraughts, while those in area B, will be subject to a negative pressure which will normally assist updraught. Providing flue terminations comply with the requirements shown in Fig. 2.22 and an approved terminal is fitted few problems should be experienced. In situations where downdraught cannot be over-come by normal methods a mechanical extractor may be considered providing it is approved by the gas supplier.

Fig. 2.23   Wind effect on natural draught (conventional) flues.

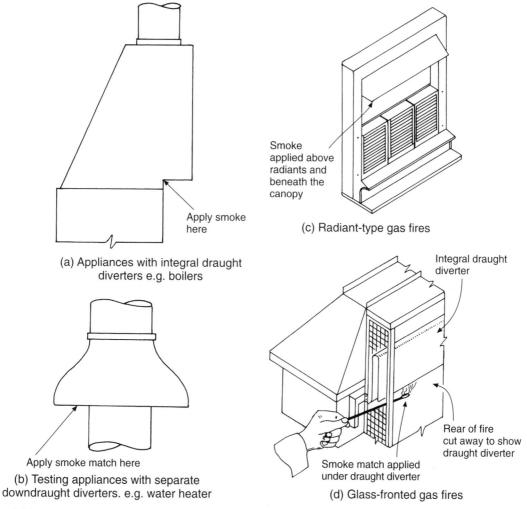

Apply smoke here

(a) Appliances with integral draught diverters e.g. boilers

Smoke applied above radiants and beneath the canopy

(c) Radiant-type gas fires

Apply smoke match here

(b) Testing appliances with separate downdraught diverters. e.g. water heater

Integral draught diverter

Rear of fire cut away to show draught diverter

Smoke match applied under draught diverter

(d) Glass-fronted gas fires

Fig. 2.24   Testing gas appliances for spillage.

to that described for initially testing for updraught, e.g. when the worst possible conditions are prevailing.

Condensation in flues

It has already been stated that one of the principal products of combustion is water vapour and the reader will be aware that the effect of warm, moisture-laden air coming into contact with a cold glass window is to produce condensation. The same situation will occur if flue gases containing water vapour are discharged into a cold flue.

An explanation of condensation may be helpful here. The amount of vapour contained in the air varies according to its temperature, e.g. the higher the temperature, the greater the quantity of water vapour the air can absorb. There is a limit, however, and should a body of air at a specified temperature reach saturation point (i.e. it cannot absorb any more vapour at that temperature), any temperature drop will cause some of the water vapour to be precipitated, the temperature at which this takes place being called dew point. In the case of a gas flue two steps are taken to minimise the incidence of condensation; one is to keep it warm by various methods of insulation, the other is to make provision for what is called *dilution* of the combustion products. Dilution of the flue gases is achieved by allowing more air into the flue to absorb some of the water vapour. All products of combustion contain a certain amount of moisture, but its effect is aggravated in the case of gas or oil equipment as their high degree of efficiency means that the flue gas temperature is comparatively low in comparison with, for example, that of solid fuel, and therefore cannot absorb so much water vapour. Dilution air is admitted to the flue via the downdraught diverter or an open end in the boiler flue as shown in Fig. 2.25.

Insulation of flues

See Fig. 2.26. The usual method of avoiding excessive condensation in existing flues is to line them with a stainless steel flexible liner. This is marketed in coils in nominal sizes of 100 mm, 125 mm and 150 mm diameter. It can be cut with a fine tooth hacksaw, but beware — it is both springy and sharp and has been the cause of many badly cut

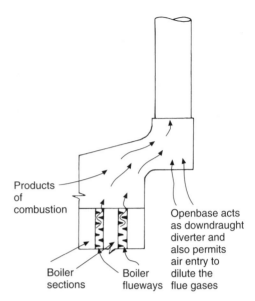

Fig. 2.25 Admitting dilution air to appliance flues. If air is permitted to mix with the products of combustion it has the effect of absorbing some of the water vapour, thus reducing condensation problems.

fingers and hands. A good idea is to tape the edges when it is cut as shown in Fig. 2.25(d). Figure 2.25 also illustrates the details relating to this method of lining. In some instances it will be necessary to remove an existing chimney pot and the cement mortar securing it. Extreme care must be taken if the flue is an old one with cracks and poorly jointed brickwork, as it may collapse. If in doubt, a proper scaffold must be used and the stack rebuilt. The fact that both the top and bottom of the flue are sealed prevents any movement of the air surrounding the liner, still air being an excellent insulating agent.

Flue terminals

Typical examples are shown in Fig. 2.27 (part 1) and are constructed of metal or terracotta. A wide variety of types is available. The 'O H' terminal shown in Chapter 6 is also suitable for gas. All gas appliance flues must be provided with a terminal, except for certain types of open fire, and even here they are recommended. Their purpose is to prevent blockage by material such as leaves and birds' nests. Some types are made in such a way that they increase the flue updraught.

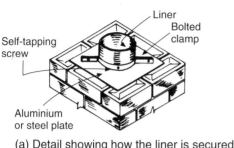

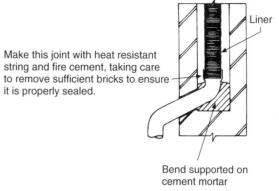

Make this joint with heat resistant string and fire cement, taking care to remove sufficient bricks to ensure it is properly sealed.

(a) Detail showing how the liner is secured at the top of the flue

(c) Alternative method of connecting an appliance flue to the liner

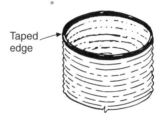

(d) To avoid accidents, seal the cut edges of the flue liner with suitable tape

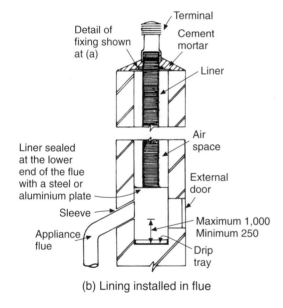

(b) Lining installed in flue

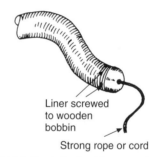

(e) Method of attaching rope to lining

The rounded bobbin, which can be hired or bought from the supplier of the lining, enables it to negotiate any bends in the brick flue.

Fig. 2.26 Flue lining details with flexible stainless steel flue linings. Flexible stainless steel flue linings are the most convenient to use, especially in older buildings with previously unlined flues.

The illustrations shown in Fig. 2.27 (part 2) are based on the recommendations of BS 5400 Part 1 2000. They show most of the configurations likely to be encountered during the normal course of work when dealing with open-flued appliances. It will be seen the general requirements are such that (a) the terminal is high enough to avoid pressure zones and adverse draught conditions and (b) vitiated air is unlikely to enter the structure via windows and ventilation apertures. In some cases it may be necessary to provide for a fan-induced draught, but this should be avoided in domestic properties where possible.

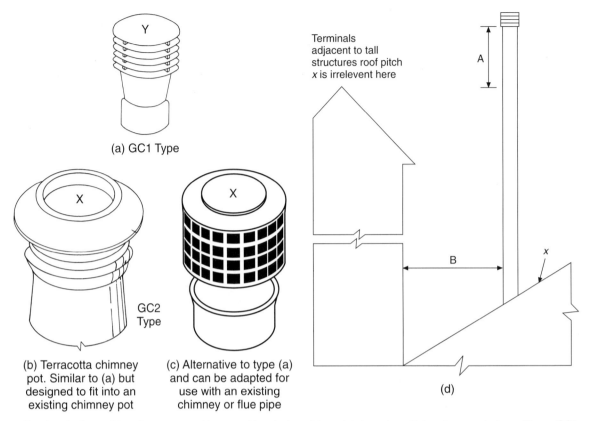

(a) GC1 Type

GC2 Type

(b) Terracotta chimney pot. Similar to (a) but designed to fit into an existing chimney pot

(c) Alternative to type (a) and can be adapted for use with an existing chimney or flue pipe

Terminals adjacent to tall structures roof pitch x is irrelevent here

(d)

Flue terminals must be of an approved type and be designed to resist downdraugth in adverse wind conditions. GC1 terminals are made of Nurastone or stainless steel. GC2 terminals are similar but made of terracotta and designed for terminating a gas flue passing through traditionally built brick chimney stacks.

Fig. 2.27 (part 1)   Flue terminals.

Downdraught diverters

Most gas appliances having a conventional flue are fitted with a downdraught diverter. The main object of this component is to prevent any vitiated air from affecting the proper combustion of the gas at the burner. (Remember that the term *vitiated air* means air lacking in oxygen.) Downdraught diverters also prevent the flames lifting off the burner if the updraught is excessive. Figure 2.28 shows how a downdraught diverter functions. Providing the downdraught is not persistent, it will have no harmful effects because the products of complete combustion are harmless. They are, however, undesirable, and if the appliance is not burning correctly and is producing carbon monoxide the results could be fatal. Not all open flued appliances are provided with downdraught diverters.

Room-sealed appliances

These do not require a conventional flue or chimney for the discharge of the combustion products, the flue being an integral part of the appliance. Those having open or conventional flues draw air for combustion through ventilators and openings in windows and doors, often causing uncomfortable draughts in the building, this does not apply to room-sealed appliances. Air for combustion is drawn in from outside the building via a wall-mounted terminal, the products of combustion being discharged through the same terminal in a separate duct. All the terminals for balanced flue appliances are made in such a way that there is no danger of vitiated air re-entering the appliance providing they are correctly installed.

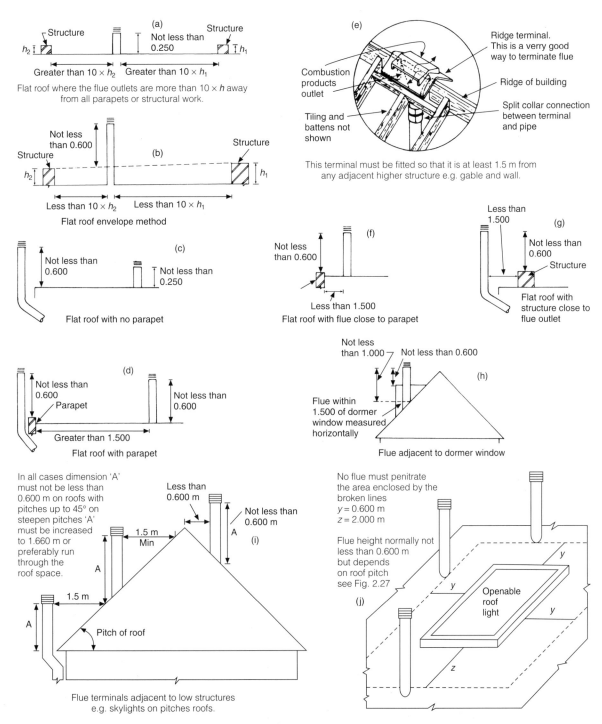

Fig. 2.27 (part 2) Approved positions for flue terminals. All the illustrations included in this group comply to BS 5400:Part 1:2000 and show the recommendations for fixing terminals to minimise the effects of downdraught in flues where they terminate in exposed positions. All dimensions are in metres. It should be noted that in most cases where the roof or a structure on the roof can affect the free discharge of combustion products, the minimum height of the flue above the structure is 0.600 m.

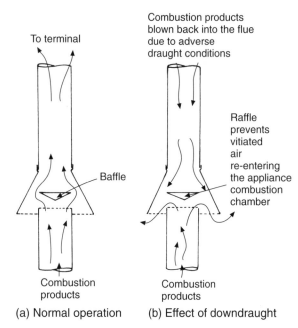

To terminal

Combustion products blown back into the flue due to adverse draught conditions

Baffle

Raffle prevents vitiated air re-entering the appliance combustion chamber

Combustion products

Combustion products

(a) Normal operation

(b) Effect of downdraught

Fig. 2.28 Downdraught diverter: its purpose is to prevent downdraught affecting complete combustion of the gas.

Figure 2.29(a) illustrates the basic principles of a natural draught room-sealed appliance. One of its limitations is that it must be fixed on an outside wall. Many appliances of this type have fan-assisted draught which enables flues and terminals to be made much smaller and longer, giving more flexibility to their installation. Other advantages of balanced flues are that there are no condensation problems and there is no possibility of combustion products entering the room. One disadvantage is that they are slightly less efficient than appliances with conventional flues due to the colder combustion air which is drawn from outside the building. This is, however, minimal with modern equipment having smaller flues and where the incoming air is warmed prior to entry into the combustion chamber. In a new building where the hot water and heating system is gas fired, a chimney or specially constructed flue will be unnecessary so there is a considerable saving. The flue terminals of any type of appliance must be carefully sited to avoid the possibility of combustion products entering the building through fresh air inlets or open windows. Table 2.7 and Fig. 2.30 show what are

generally acceptable positions for room-sealed appliance terminals (the corresponding minimum distances are given).

These terminals get very hot, however, and if touched could inflict serious burns, therefore a properly constructed guard must always be fitted where the terminal is less than 2 m from the ground level. High-level terminals can cause damage to plastic gutters and the approved methods of overcoming this are shown in Fig. 2.31. The terminal positions shown comply with the relevant British Standards and those of appliance manufacturers. Prior to fitting an appliance however, the installer should conduct a survey to ensure that any products of combustion do not cause a nuisance to neighbours of adjacent properties. Thought should also be given to the possibility of future extensions to neighbouring property which may prevent the efficient function of the flue leading to a dangerous or high-risk situation.

Gas controls

Modern gas appliances employ a wide diversity of control systems, most of which require the use of electricity. Originally all gas appliances, including gas boilers, had control systems operated by the gas pressure, and while they were very effective and did not require an undue amount of maintenance, their flexibility had some limitations. By today's standards some would not meet the requirements of the Gas Regulations as they did not shut down in a fail-safe position.

The most important controls from the viewpoint of the consumer, are those which provide means of varying the temperature of the appliance and the periods when it is on or off. Controls must also be safe in use, and before approval, they are examined and tested very thoroughly. It should be understood that although the controls dealt with in the following text are treated in isolation, it will be found that on most modern equipment they are fitted into what is commonly called a multifunctional, or composite control. If a defect occurs in a control, they can sometimes be repaired by someone who is qualified to do so, but it is often cheaper to replace it and many specialist suppliers

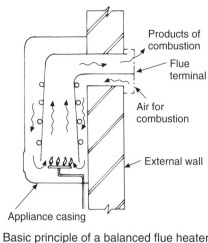

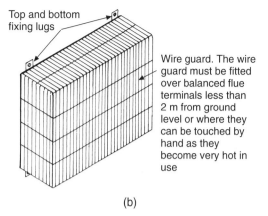

Basic principle of a balanced flue heater

(b)

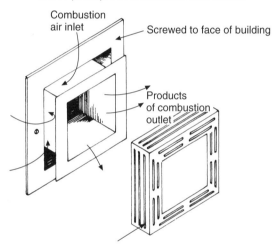

(a) Terminal for natural draught balanced flue
appliance

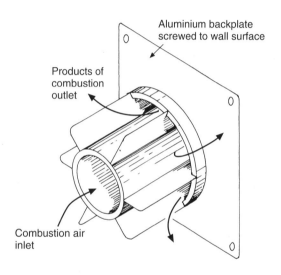

(c) Typical fanned flue terminal for balanced flue
appliance

Fig. 2.29   Principle of balanced flue room-sealed appliances and details of terminal with guard removed to show ducts. Room-sealed appliances draw no air for combustion from inside the room. Combustion air enters the appliance via a duct in the flue terminal, the products of combustion being discharged through a separate duct in the same terminal.

offer replacements on a part-exchange basis. Should it be necessary to order new parts, do make sure the reference number of the control, and the appliance to which it is fitted, is quoted to the supplier as there are many variations.

Main burner controls
The controls employed will vary according to the appliance, but the following types are common to water-heating equipment. Instantaneous heaters rely on the pressure of the water to lift the main gas valve, which ensures that the heater is charged with water before the main burner will fire. This type of valve is dealt with in Chapter 4.

Solenoid valves
As explained later in the section on thermoelectric valves, a solenoid, when energised, acts in the same way as an electromagnet and lifts a valve off its seating, permitting the passage of gas.

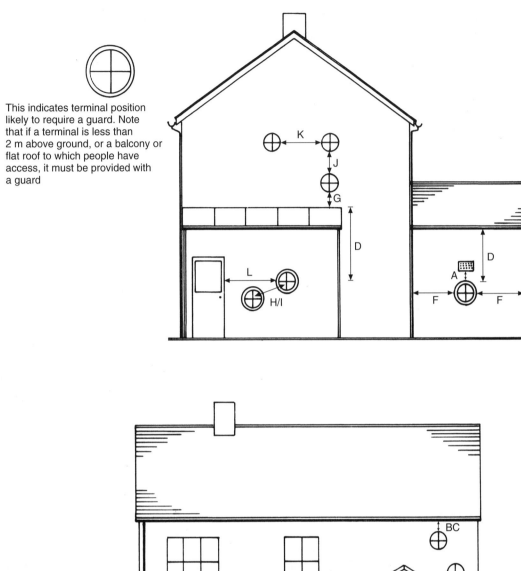

This indicates terminal position likely to require a guard. Note that if a terminal is less than 2 m above ground, or a balcony or flat roof to which people have access, it must be provided with a guard

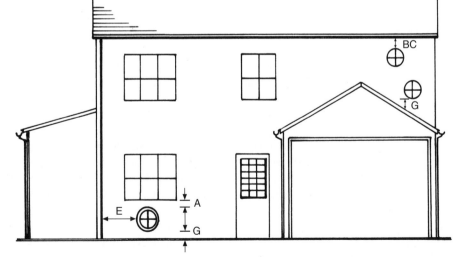

Fig. 2.30   Specifications for room-sealed appliance terminals.

Table 2.7 Suitable positions for room-sealed appliance terminals. Figures are permissible distances in mm.

Terminal position (see Fig. 2.29)	Natural draught type	Fanned draught type
A Directly below an openable window or other opening, e.g. an air brick	300	300
B Below gutters, soil pipes or drain pipes	300	75
C Below eaves	300	200
D Below balconies or car port roofs	600	200
E From vertical drain pipes and soil pipes	75	75
F From internal or external corners	600	300
G Above ground, roof or balcony level	300	300
H From a surface facing a terminal	600	600
I From a terminal facing a terminal	600	1,200
J Vertically from a terminal on the same wall	1,500	1,500
K Horizontally from a terminal on the same wall	300	300
L For an opening in a car port (e.g. door, window) into a dwelling	1,200	1,200

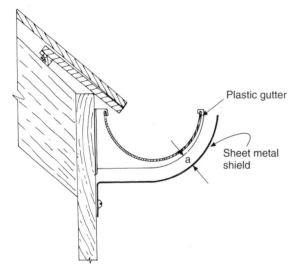

Fig. 2.31 Protection of plastic gutters from hot products of combustion ref BS 5440 Part 1. Where a natural draught balanced flue terminal is fitted less than 1 m below a plastic gutter or less than 0.5 m below a painted surface, a sheet metal shield at least 1 m in length must be fitted as protection against the hot combustion products. The requirements for fanned flues may differ and the manufacturer's instructions must be complied with.

Most solenoid valves are operated by 240 V so it is very important that any electrical work conforms with the wiring regulations. Figure 2.32 shows a solenoid valve illustrating its main features.

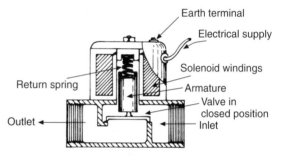

Fig. 2.32 Solenoid (magnetic) valve. This illustrates the basic principle of magnetic valves. When the solenoid is energised it overcomes the pressure of the return spring and pulls the valve off its seating.

Thermostats
The function of a thermostat is to cause the gas supply to the main burner to shut down when the appliance reaches a preset temperature. One exception to this is a cooker thermostat which allows a small quantity of gas to bypass the valve. Unlike most water-heating equipment, cookers have no permanent pilot, and should the thermostat shut down the gas completely, the oven burner would not be relit when the thermostat calls for heat.

One of the most common is the rod-type thermostat which has been fitted on cooking appliances for many years. The principle employed is the differential expansion rate of invar steel and brass, invar steel having a relatively low rate of expansion, while that of brass is comparatively high.

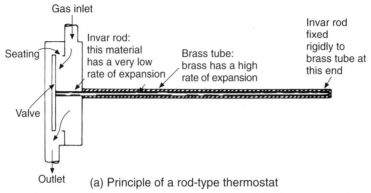

(a) Principle of a rod-type thermostat

When the brass tube is heated it expands at a higher rate than the invar rod, thus pulling the valve on to the seating and closing off the gas supply. It will be seen, however, that no adjustment of temperature is possible.

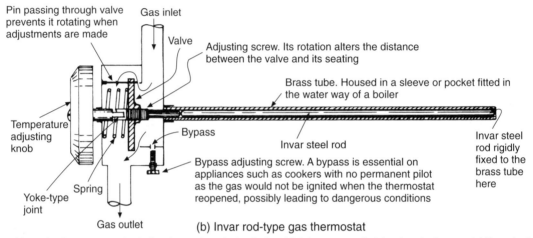

(b) Invar rod-type gas thermostat

Here, the thermostat valve is fitted on a threaded spindle which is not actually joined to the invar rod. When the brass tube expands, carrying with it the invar rod, pressure exerted by the spring pushes the valve on to its seating, temperature variation is achieved by turning the adjusting knob which rotates the threaded spindle, moving the valve closer to or further away from its seating. The further away it is, the higher will be the temperature before the brass tube expands sufficiently for it to close.

Fig. 2.33 Rod-type thermostats.

Figure 2.33(a) shows the action of this simple valve. It will quickly be seen that it will close at one temperature only, the valve having no provision for adjustment.

In practice these thermostats are made as shown in Fig. 2.33(b). Their use for domestic appliances is very limited as most now have electrical control. They are still commonly used for gas cookers in large commercial establishments.

Fluid expansion thermostats   These employ a sealed bulb or sensor and a bellows filled with a heat-sensitive fluid. The sensor is situated in a

pocket in the appliance waterways, and on a rise in water temperature the fluid expands, extending the bellows. The illustration in Fig. 2.34(a) operates directly on a gas valve, causing it to open or close, depending on the temperature. The same principle can be employed to operate a microswitch, enabling it to be used for gas appliances controlled by electricity as shown in Fig. 2.34(b).

Gas governors
As the name implies, these are devices for controlling the pressure and flow of gas to an installation or an individual appliance.

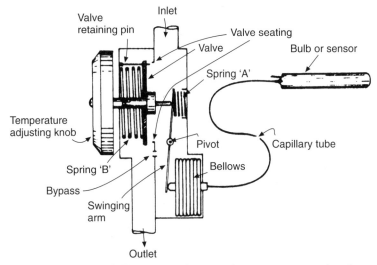

(a) Fluid expansion type thermostat – non electric

When the sensor is subjected to heat, the heat-sensitive fluid it contains expands, causing the bellows to distend and apply pressure via the swinging arm on spring 'A', simultaneously allowing spring 'B' to close the valve. When the sensor cools and the fluid it contains contracts, the bellows returns to its normal position and pressure on spring A overcomes that of B, opening the valve.

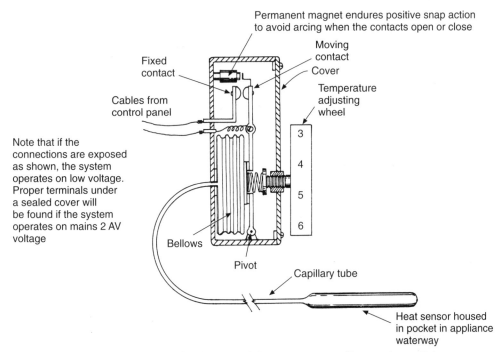

(b) Typical fluid-operated thermostat controlling a microswitch

Diagrammatic illustration. Operation: When the thermostat is at the correct temperature the contacts will be closed, but as the watch temperature increases the heat-sensitive fluid in the sensor expands causing the bellows to push the moving contact away from the fixed contact which has the effect of causing the main gas supply to the burner to be closed.

Fig. 2.34 Fluid expansion thermostats.

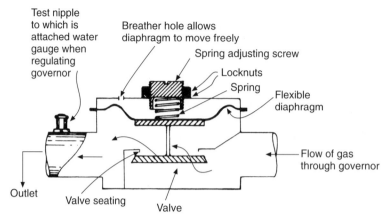

Test nipple to which is attached water gauge when regulating governor

Breather hole allows diaphragm to move freely

Spring adjusting screw

Locknuts

Spring

Flexible diaphragm

Flow of gas through governor

Outlet

Valve seating

Valve

Fig. 2.35 Simple gas appliance governor. A governor is designed to even out any variations of pressure that might occur in a gas installation, and to ensure a constant pressure at the appliance to which it is fitted.

Service governor Gas pressure in the main distribution system is normally about 2 bar. This is reduced by the service governor fitted to the inlet side of the meter to give a working pressure at the consumer's appliance of 20 mbar. In the event of malfunction causing high pressure gas to enter the gas services in the building with possible dangerous effects, it will automatically shut off the gas. It can be manually reset when the cause of the problem has been rectified. Adjustments and repair to these regulators must not be undertaken by an installer and any defects must be reported to the gas transporter or supplier.

Constant pressure governors Individual appliance governors are not now considered necessary except for gas boilers, enabling fine adjustments to be made to ensure maximum economy. Constant pressure governors are provided for this purpose. A simple governor of this type is illustrated in Fig. 2.35. It is really a variable restrictor in the gas supply. An increase of the inlet pressure will exert a greater force on the flexible diaphragm causing it to lift, carrying with it the valve. This has the effect of reducing the aperture through which the gas can pass to the outlet, causing a pressure drop. A lowering of the inlet pressure will result in less pressure on the diaphragm, thus allowing the valve to drop and thereby increasing the size of the outlet aperture. The movement of the valve enables a balance to be maintained between the inlet and outlet gas pressures.

Pressure can be increased by compressing the spring with the adjusting screw. The outlet pressure should be checked with a U gauge at the test nipple on the outlet side of the governor, and if it is found to be insufficient the diaphragm can be loaded by compressing the spring until the pressure recommended by the manufacturer is achieved. It should be noticed that the top of the governor has a small vent orifice which is open to the atmosphere, and if this becomes obstructed the governor will not function properly.

When a gas appliance is fitted it is important that it operates at the pressure recommended by the manufacturer. Most governors on domestic boilers are now integral with the multifunctional control. The proceedures for setting gas pressures are described on page 26.

Multifunctional controls

The method of gas controls employing components as separate units, is obsolete for domestic appliances. They are now housed in a unit called a multifunctional control, and are produced by various manufacturers, varying only slightly in detail. Some types employed with modern boilers, i.e. combis, incorporate a modulating valve, which enables the gas rate to meet the varying demands made on the heat-producing equipment. A typical example of the type used for domestic boilers is shown in Fig. 2.36. Their main advantages are the saving of space in boiler compartments; they are also more convenient for the full electrical control

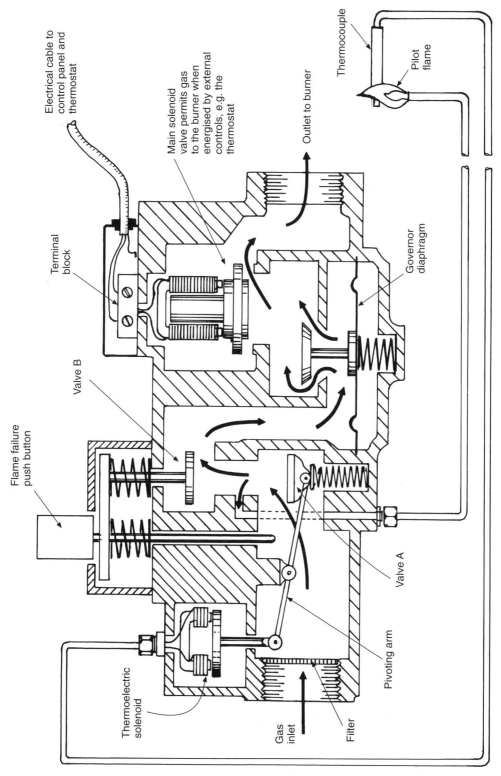

Electrical cable to control panel and thermostat

Main solenoid valve permits gas to the burner when energised by external controls, e.g. the thermostat

Thermocouple

Pilot flame

Outlet to burner

Terminal block

Governor diaphragm

Valve B

Flame failure push button

Valve A

Pivoting arm

Thermoelectric solenoid

Gas inlet

Filter

Fig. 2.36 Simplified diagram of a typical multifunctional gas control.

systems employed on modern appliances. A further advantage is that their use enables user controls to be conveniently situated on the front panel of the boiler. The flow of gas through the control should be carefully noted, and it will be seen how the individual components previously described are now combined into one integrated unit.

Ignition devices

Some gas appliances employ a pilot flame, its function being to ignite the gas in the main burner when the controls are calling for heat. Others, such as gas fires and cookers, use a system whereby the main burner is lit directly by a spark or filament coil. Some pilot flames are permanent, irrespective as to whether the main burner is on or off. Non-permanent pilots only ignite when a thermostat calls for heat, and it is claimed that some economy may be achieved using this system.

Manual ignition

This system is mainly confined to older appliances where the pilot jet or main burner is easily accessible and can be lit with a match or taper. Manual ignition applies mainly to older types of gas appliances.

Spark ignition

Most readers will be aware that in an internal combustion engine the fuel/air mixture is fired by the sparking plug. A similar arrangement is used to light either the pilot jet, or, in some cases, the main gas burner, by means of a spark caused by a high voltage electric current arcing across a gap. The two main methods employed are (a) mains spark, or (b) piezoelectric ignition, the latter being illustrated in Fig. 2.37. The crystals are made of lead zirconate–titanate, which, when subjected to pressure, produce an electromotive force of approximately 6,000 V, which is transmitted via the metal pressure pad to the spark electrode. The illustration shows both a cam-operated ignitor, commonly employed with gas fires, and an impact type which works on a similar principle. Pressure, in this case, is applied by a blow from the hammer, which incorporates a trip mechanism and is activated by the operating knob.

The spark generator illustrated in Fig. 2.38, operates on the main electricity supply. Unlike the piezoelectric system, where the ignitor is operated manually and used with an appliance having a permanent pilot, spark generation is used with those employing non-permanent pilots. They are automatically operated by the control system when a signal from the thermostat indicates the appliance is required to fire. Some economy may be achieved using this system, as the pilot is only alight while the burner is firing. With all spark ignition devices one or two points should be noted. The electrodes must be clean and the gap between them must be maintained to the manufacturer's recommendations as the spark will not bridge it if it is too wide. The high-tension leads must be well insulated and preferably not be in contact with metal parts on the boiler if short-circuiting is to be avoided, and all earth wires must be effectively connected to the appliance.

Flame failure devices

The object of a 'fail-safe' device is to prevent gas reaching the main burner until a pilot flame has been established. If, for instance, the main burner is 'gassed' and some delay takes place before ignition, the ratio of gas to air on ignition could result in an explosion. The following devices are those most commonly found in domestic water heaters and boilers.

The bimetallic strip

This device will only be found on a very old appliance and has been superseded by more modern controls. The main purpose for retaining the illustrations is to demonstrate a principle which has many other applications in the gas and electrical industry. It has been used on gas water heaters and gas boilers since they were first marketed. It works on the principle that metals have differing expansion rates (see Fig. 2.39). Two common metals which have widely varying coefficients of linear expansion are brass and invar steel and these metals are the most commonly used in bimetallic controls. Figure 2.39(a) indicates how the bimetallic strip

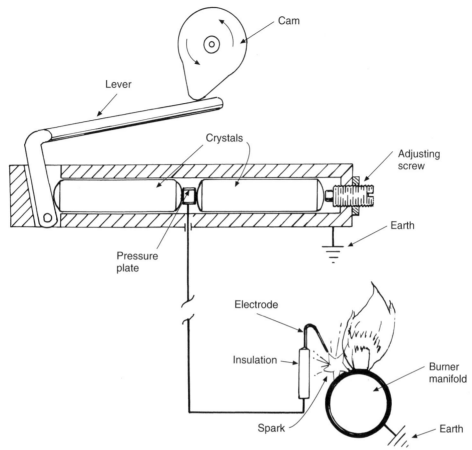

(a) The cam is operated when the gas control knob is rotated, simultaneously turning on the gas and exerting pressure on the crystals. This causes a voltage build-up causing a spark to arc across the gap between the electrode and burner igniting the gas. This arrangement is used to light the main burner directly and is commonly used in gas fires

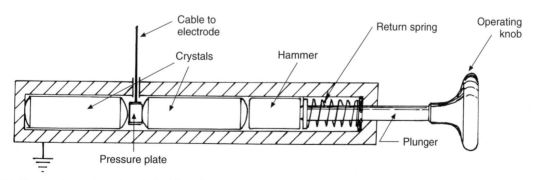

(b) Plunger type piezoelectric ignition. Pressure on the plunger exerts pressure on the hammer, causes it to strike a blow on the crystals which, as with (a) produces a spark

Fig. 2.37 Piezoelectric ignition.

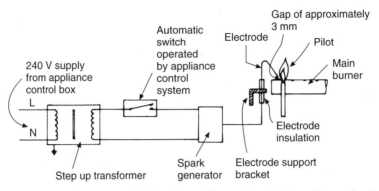

Fig. 2.38 Mains spark ignition. This system of ignition is usually employed with appliances having non-permanent pilots. Its action is normally fully automatic, being part of the control system of the appliance.

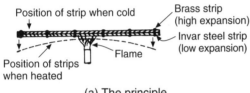

(a) The principle

Shows the effect on two strips of metal riveted together when they are subjected to heat. The metal having the higher rate of expansion will cause the other to bow.

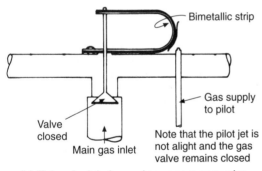

(b) This principle is used to open a gas valve

Fig. 2.39 Bimetallic strip flame failure device.

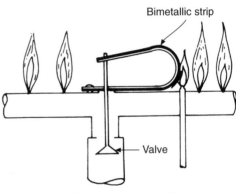

(c) The valve in operation

When the pilot flame is alight it causes the bimetallic strip to bend and open the gas valve, simultaneously lighting the gas on the main burner. Should the pilot flame be extinguished for any reason the bimetallic strip will open and pull the gas valve upward on to its seating, thus shutting off the gas supply. It is therefore not possible for the burner to be gassed unless the pilot flame is alight to ignite it.

bends when it is subjected to heat, causing the gas valve to open. If the pilot light is either turned off or blown out, quite a common occurrence if the appliance is in a draughty environment, the bimetallic strip returns to its normal position closing the valve.

Thermoelectric valve

This valve has been used on gas boilers for many years and is now commonly employed in many other gas appliances such as cookers and water heaters. As the name implies, it is an electrically operated valve, electricity being generated by the

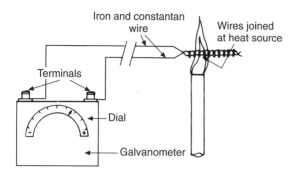

Fig. 2.40 Illustration showing the principle of a thermoelectric valve. When heat is applied at the junction of the two wires an electromotive force is produced and is registered on the galvanometer.

hot contact of the thermocouple when the pilot flame is ignited.

Figure 2.40 shows diagrammatically the principle upon which this valve works. Two wires, one made of iron and the other of constantan, are joined together at one end and heated. If the free ends of the wires are connected to a galvanometer it will show that a small flow of electricity is generated due to the differential movement of the molecules of the two metals when subjected to heat. The thermoelectric valve utilises this small current of electricity to energise a solenoid in the thermoelectric valve to hold the gas valve open.

Figure 2.41 illustrates diagrammatically a thermoelectric valve. Figure 2.41(a) shows the valve in the closed position and it will be seen that gas cannot pass to either the main burner or the pilot. Figure 2.41(b) shows the position of the valves when the reset button is depressed. This has the effect of allowing gas to be admitted to the pilot jet while maintaining the outlet valve to the main gas burner in the closed position. At this stage it is possible to light the pilot flame, which impinges on the thermocouple, causing the generation of an electromotive force. This energises the solenoid and holds the main gas valve open when the reset button is released. The release of the reset button, shown in Fig. 2.41(c), due to the action of the integral springs, allows the lower valve to open

and admit gas to the main burner. This will remain open until, for some reason, the pilot flame is extinguished, causing the solenoid to be de-energised and the main valve to be pushed back on to its seating, closing off supplies to the main burner and the pilot. It will be seen that, should the gas supply fail for any reason, both the pilot and the main burner are completely isolated. It will also be seen that a permanent pilot flame must be established before the main burner will function.

Electricity is conducted from the thermocouple to the solenoid via a mineral-insulated lead which looks like a small-diameter copper pipe. A section through the lead is shown in Fig. 2.42.

The cause of defects is the subject of more advanced study, but the most common faults affecting thermoelectric valves are:

(a) Pilot jet partly obstructed, preventing sufficient heat from reaching the thermocouple.
(b) Loose connection of the thermocouple lead to the valve.
(c) Damage to the lead due to careless handling.

Mercury vapour flame safety valve

Unlike the thermoelectric valve which requires a permanent pilot flame, the mercury vapour valve can be used where the pilot flame is ignited only when the appliance is in use, resulting in small savings on gas consumption. It is generally used with a solenoid-type main gas control and this, when electrically energised, allows gas to pass to the inlet of the mercury vapour valve and to the pilot jet. When the pilot is lit the flame heats the phial causing the mercury vapour which it contains to expand and distend the circular bellows, thus causing the lever to open the main gas valve to the burner outlet, overcoming the downward pressure of the spring. When the appliance thermostat is satisfied, it de-energises the solenoid in the control box, closing off the gas supply to both the main burner and the pilot jet. The mercury vapour then cools and contracts, pressure on the spring forcing the valve back on to its seating. These valves are quicker acting than thermoelectric valves but more prone

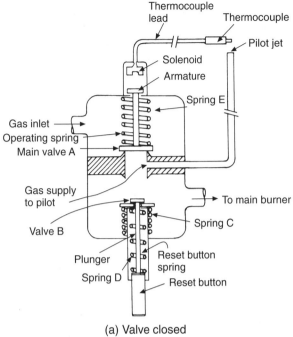

(a) Valve closed

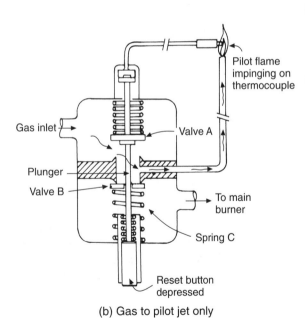

(b) Gas to pilot jet only

By depressing the reset button the plunger, passing through valve B, pushes valve A off its seating, permitting the passage of gas to the pilot jet which is ignited and heats the thermocouple. As valve B is retained in the closed position due to pressure from spring C, no gas can yet pass to the main burner.

Fig. 2.41 Operation of a thermoelectric valve.

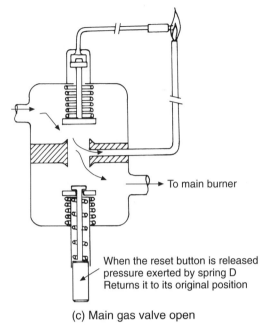

(c) Main gas valve open

After approximately 30 seconds sufficient electrical energy is developed by the action of the pilot flame on the thermocouple to energise the solenoid which holds the main gas valve A open. Both the reset button and valve B are now in their original positions due to the release of the reset button and pressure from spring D. Gas can pass to the main burner. If the pilot is extinguished the thermocouple will cease to produce electrical energy and the main valve will close due to pressure exerted by the operating spring E.

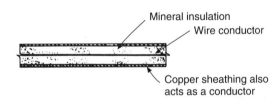

Fig. 2.42 Section through thermocouple lead. This material is used extensively as an electrical conductor in situations where other conductors are unsuitable due to the relatively high temperatures, e.g. in boiler or water heater casings. When handling avoid very sharp bends and do not kink as this can result in short-circuiting the conductors.

to breakdown. If the valve fails to open it is usually due to a damaged bellows unit which has allowed the vapour to escape, and a replacement valve will be required. Figure 2.43 illustrates the working principles of these valves.

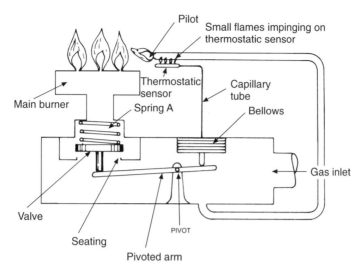

Fig. 2.43 Mercury vapour flame safety device. These valves are used with a different type of control system from that used with a thermoelectric valve, the main gas control being a solenoid operated by an electrical thermostat. Upon a supply of gas reaching the mercury vapour valve the pilot is lit simultaneously by spark ignition and heats the thermostatic sensor. Expansion of the fluid opens the bellows, pushing down the pivot arm, thus opening the main valve admitting gas to the burner. When the thermostat is satisfied the main solenoid will be de-energised, closing off the gas supply to both the pilot and the burner. On contraction of the bellows, spring A will push the valve back on to its seating. It is impossible for this valve to open unless a pilot flame has first been established.

Flame conduction and rectification

This is an electronic flame protection device that was designed originally for commercial appliances, but due to the use of microelectronics on modern appliances, the components used with this type of flame-failure equipment can now be produced small enough to be used with domestic appliances such as cookers and boilers.

The basic principle of flame rectification is as shown in Fig. 2.44, based on the fact that when a substance burns, in this case gas, a chemical reaction or change is taking place. The flame we see when this happens also produces minute electrically charged particles called ions which can be made to pass between two conductors through the flame. In effect the flame is acting as a conductor to the flow of electricity. If a flame is not established, obviously there will be no flow of electrons and the control system of the appliance will not pass gas. A basic knowledge of electrical principles is required to fully understand the process, but the following information will be helpful. The flow of electrons through the flame produces an alternating current

which means the flow is constantly reversing backwards and forwards, and to be effective for the purposes being considered, it must be changed or 'rectified' to 'direct' current. This type of current flows in one direction only. Rectification is achieved, in this case, by the electrodes which must be of a suitable type. The direct current thus produced is amplified and operates a relay which in turn operates the electrical controls on the main gas supply. An amplifier is an electronic device which makes it possible for a small current to operate a relay. A relay is basically a solenoid or electromagnet which enables one source of electricity to control another, in this case the solenoid valve controlling the gas inlet to the appliance.

Photo-electric flame failure components

The use of an 'electronic eye' as a fail-safe device is well established in oil-firing practice and the development of similar burners for gas combustion has led to its use for the same purpose in some types of industrial gas appliances. The electronic eye, more correctly called a photo-electric resistor,

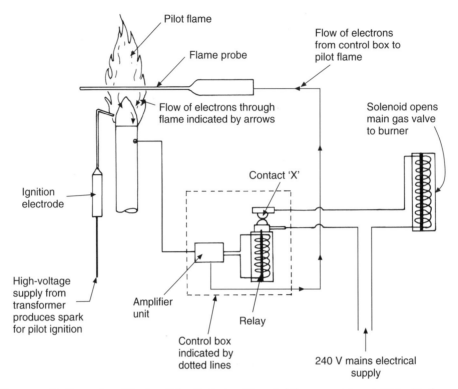

Pilot flame

Flame probe

Flow of electrons
from control box to
pilot flame

Flow of electrons through
flame indicated by arrows

Solenoid opens
main gas valve
to burner

Contact 'X'

Ignition
electrode

High-voltage
supply from
transformer
produces spark
for pilot ignition

Amplifier
unit

Relay

Control box
indicated by
dotted lines

240 V mains electrical
supply

Fig. 2.44 Flame conduction and rectification fail-safe device. The pilot flame must be established before a flow of electrons can pass through the flame to complete the electrical circuit which energises the relay. This causes contacts 'X' to close simultaneously completing the circuit to the solenoid which opens the main valve admitting gas to the combustion chamber. Unless a pilot flame is established no gas can be passed through the main gas valve.

looks very similar to an old-fashioned radio valve. It is housed in a suitable casing fitted into the blower tube of the burner so that its light-sensitive face can detect the ultraviolet rays in the flame. Electricity will only flow through the cell when it is exposed to these rays. The sequence of operations is as follows: on initial light-up the gas is ignited by a spark from the electrode which only operates for a limited period. If all is well and a flame is established, the electronic eye will 'see' the ultraviolet rays, permitting the flow of electricity to open the main gas solenoid valve. Should the gas not ignite; no ultraviolet rays will be produced and the electronic eye will be unable to permit the flow of current to the gas valve which will close. This is known as 'lock out' and is indicated visually by the appearance of a red light in the control box of the burner. The burner must then be reset manually,

usually by depressing a button on the control box. If a fault exists with the burner or one if its components, it will again lock out and steps must be taken to rectify the fault. Typical examples may be:

(a) Gas turned off.
(b) Electrode gap incorrect.
(c) Breakdown of electrode insulation.
(d) Electronic eye requires cleaning or is not correctly fitted.

Figure 2.45(a) shows the cell and its casing while Fig. 2.45(b) illustrates its position in the blower tube of the gas burner — see also Fig. 6.21.

Oxygen depletion valve
This valve is an additional safety device incorporated into many modern gas appliances,

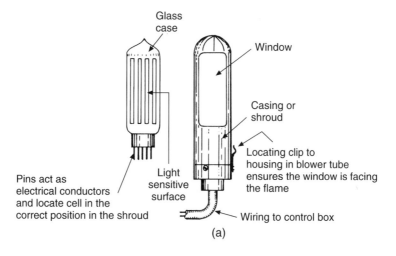

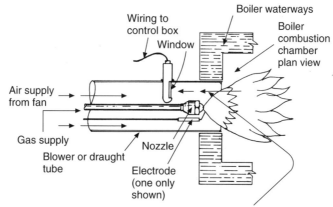

Fig. 2.45 Photo-electric fail-safe device.

especially fires. They are mandatory on open-flued appliances fitted in sleeping areas. It is designed to cut off the gas supply to the appliance if vitiation or lack of air for combustion causes oxygen depletion. The pilot flame will be extinguished, deactivating the thermalcouple.

Thermistors
These are very accurate non-metallic heat-sensing devices that will alter their ability to conduct a flow of electricity when subjected to changes in temperature. The higher the temperature to which they are subjected the greater will be the flow of electricity they will pass. Thermistors are used in many gas modulating controls where the supply of gas to a burner is variable. They are also found in programmable room thermostats.

Gas fires and back boiler units

General
For many years now gas boilers, usually combined with a gas fire, have been used to replace the solid

fuel appliances installed prior to the introduction of central heating. The production of gas appliances for this purpose has rapidly developed due to their efficiency, and possibly more so with the current trend to produce appliances using gas as a fuel to simulate solid fuel fires. The reader should note that not all these fires are suitable for use with gas back boilers and fall into three basic groups covered by BS 5871 entitled and listed as follows: *The Installation of Gas Fires, Convector Heaters, Fire Back Boilers and Decorative Fuel Effect Gas Appliances.*

Part 1 relates to gas fires, convector heaters and fire/back boilers 1st, 2nd and 3rd family gases. Part 2 concerns inset live fuel effect gas fires of heat input not exceeding 15 kW 2nd and 3rd family gases, and Part 3 decorative fuel effect gas appliances of heat input not exceeding 15 kW 2nd and 3rd family gases.

All this seems rather complicated, but is necessary because of the wide range of differences between the working principles of these fires and the methods of flueing, flue lining and hearth requirements relating to them.

Because the rapid development of this type of appliance often outstrips relevant British Standards, it is important to note that appliance manufacturers' instructions take precedence over any British Standard and such instructions must be adhered to. To be safe with gas there are no short cuts! While this is true of all gas appliances it is especially true of gas fires, as it is very easy to go to sleep in front of the fire, and possibly never wake up if the fire is improperly fitted.

Although the installation requirements of modern gas fires vary widely, depending on their type, there are some features which are common to all and careful note should be made of the following. Before any such appliance is fitted the instructions given previously relating to the use of existing flues should be applied. Briefly they must be clean, have no obstruction and sufficient updraught. Any ventilation requirements and the suitability of existing hearths must be investigated. This is important as the recommendations vary widely depending on the type of fire used. Generally speaking, live fuel and decorative fuel fires are

subject to the same hearth and flueing requirements as those for solid fuel appliances. This information can be found in the Building Regulations Part J.

Gas back boilers with combined fires
Unless the flue is already lined with an acid-resistant lining (most buildings constructed after 1965 will meet this requirement) the flue must be lined with a suitable material which must be sealed at the top and bottom. Failure to do this will render the lining ineffective, resulting in excessive condensation. The top is sealed as shown in Fig. 2.26 and Fig. 2.46 illustrates various methods of sealing the bottom of the flue pipe.

The opening into which the boiler is to be situated must have a level base, the top of which must be level with the top of the hearth if applicable. Although it is not mandatory, it is good practice to render the boiler opening with cement and sand mortar, unless the opening has been specially prepared and has good clean surfaces. Assuming the holes for the flow and return, gas and electric supplies have been prepared, the boiler unit can be placed in the opening. It will be seen from Figs 2.47 and 2.48, which illustrate a typical gas boiler and its installation, that the boiler base is extended to carry the gas fire and in most cases provision is made to secure the top of the fire to the boiler housing. It is essential the boiler unit is in the correct position in relation to the surface of the wall or surround to which the fire is to be fixed before screwing it down in the prepared opening. Failure to do this will make effective fixing of the fire impossible. Next the flue lining should be fixed to the flue outlet socket with self-tapping screws prior to sealing with fire cement. Manufacturer's instructions will indicate whether or not a flue restrictor ring is required, which usually depends on the height of the flue.

At this stage the water, gas and electrical connections can be made and the boiler unit can then be commissioned following the manufacturer's instructions.

The fires supplied with back boilers are those which comply to BS 5781:Part 1, sometimes called conventional gas fires. These are either

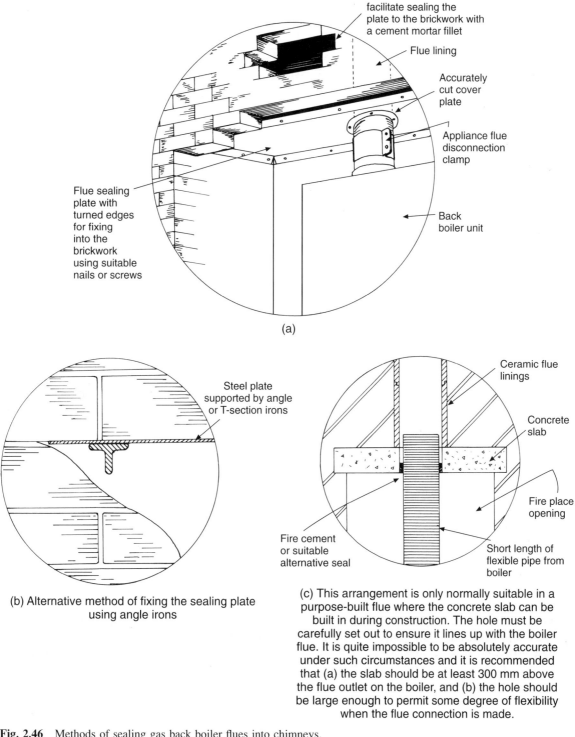

Hole cut into flue to facilitate sealing the plate to the brickwork with a cement mortar fillet

Flue lining

Accurately cut cover plate

Appliance flue disconnection clamp

Back boiler unit

Flue sealing plate with turned edges for fixing into the brickwork using suitable nails or screws

(a)

Steel plate supported by angle or T-section irons

Ceramic flue linings

Concrete slab

Fire place opening

Fire cement or suitable alternative seal

Short length of flexible pipe from boiler

(b) Alternative method of fixing the sealing plate using angle irons

(c) This arrangement is only normally suitable in a purpose-built flue where the concrete slab can be built in during construction. The hole must be carefully set out to ensure it lines up with the boiler flue. It is quite impossible to be absolutely accurate under such circumstances and it is recommended that (a) the slab should be at least 300 mm above the flue outlet on the boiler, and (b) the hole should be large enough to permit some degree of flexibility when the flue connection is made.

Fig. 2.46 Methods of sealing gas back boiler flues into chimneys.

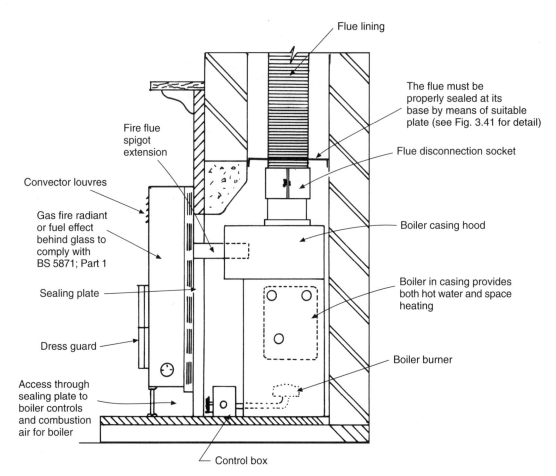

Fig. 2.47   Section through gas back boiler and fire. These are supplied as a complete unit, the fire being made to be removed to allow the boiler to be serviced. It should be noted that adequate ventilation must be provided as the gas input for both the fire and boiler may range from a total of 8 kW to 30 kW.

of the radiant type or those with a glass-fronted sealed live fuel effect and are made in such a way that access to the back boiler controls is underneath the fire, the control for the gas fire being quite separate. To service the back boiler the fire must be removed. Generally the main installation points relating to the fire are similar to those used without a back boiler.

Gas fires — conventional type
Figure 2.49 illustrates a section through an existing fireplace into which the fire is installed. In new buildings where the opening is designed for a gas fire the firebrick back is

unnecessary, but the measurements shown always apply. Before the fire is fixed the closure or sealing plate, supplied with the fire, must be sealed with heat-resistant tape against the opening to which the fire is fitted. Its main purpose is to prevent warm air from the convector entering the flue. This plate has a pre-cut hole through which the flue spigot from the fire passes and a small aperture in its base which serves mainly to ventilate the flue, but also provides an element of flue gas dilution. As with gas back boilers, it may be necessary to fix the flue restrictor plate (supplied by the manufacturer) in cases where the updraught on the flue is excessive.

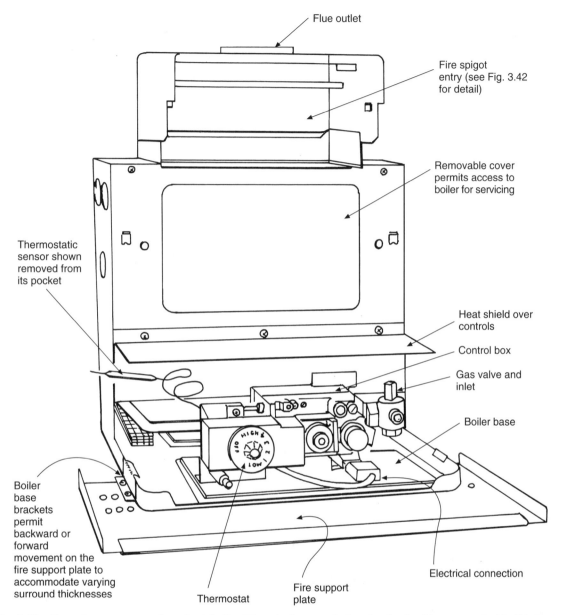

Fig. 2.48 General arrangement of gas back boiler. These boilers are designed to be fitted into a standard builder's opening for a fireplace and are normally used in conjunction with a gas fire. Both the boiler and fire are marketed as a complete unit.

The fire must be firmly fixed, as any movement may result in slackening the joints on the gas supply, thus causing a gas leak. Connection to the adjacent gas point is made using 8 or 10 mm copper pipe which may be bent using springs or one of the small bending machines marketed for this purpose.

A suitable valve should also be provided to isolate the fire for servicing purposes.

Live fuel effect fires (LFE)
Basically these are an open fire incorporating a form of convector, a section of which is illustrated

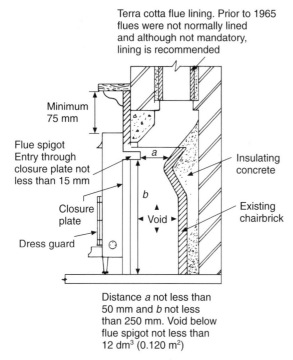

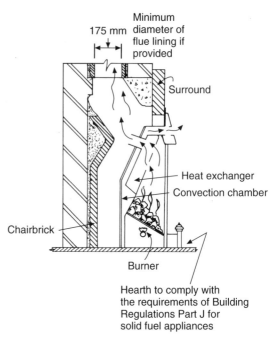

Fig. 2.49 Arrangements for radiant or glass enclosed fuel effect fires complying to BS 5871:Part 1. Note that any damper or flue restrictor in an existing flue must be removed or, if this is not possible, fixed in the open position.

Fig. 2.50 Section through a typical LFE gas fire constructed to the requirements of BS 5258:Part 16. Fitting and installation must comply with BS 5871:Part 2. Supplementary ventilation is not normally required for fires of up to 7 kW input. Any flue lining for this type of fire must be of the same specification as those required for solid fuel. In buildings constructed prior to 1965 no linings were normally provided and while they are not considered essential they are preferred. The existing firebrick back (chairbrick) may sometimes have to be removed for some types of fire.

in Fig. 2.50. Efficiency varies depending on the type of fire, but generally they are less efficient than the fires previously described. As the products of combustion are of a higher temperature than conventional gas fires, flues must be the same in diameter or similar to those for solid fuel appliances. The same also applies to hearths, as fires of this type are not supplied with guards. Hearths must be at least 50 mm high to discourage the laying of carpets immediately adjacent to the fire, but the surface area may vary from fire to fire. With fires of this type having an input of up to 7 kW, no supplementary ventilation is normally necessary.

Decorative fuel effect fires (DFE)   These are similar to LFE fires without a convector (see Fig. 2.51). They are made to simulate open fires and their effect is very realistic, but like the

solid fuel appliances they are designed to replace, their efficiency is very low. Most DFE fires are designed to be installed in a normal fireplace having a chairbrick (firebrick) back, but basket types are also available for fitting into a large inglenook type fireplace. These often require modification to gather in the flue (i.e. to make it smaller) so that the products of combustion are effectively removed. It must be stressed that the manufacturer's instructions relating to these fires must be strictly adhered to. It must also be borne in mind when any gas appliance is commissioned that a further spillage test will be necessary. Any fire not covered by BS 5258 must be subject to the Building Regulations Part J, relating to flues and hearths.

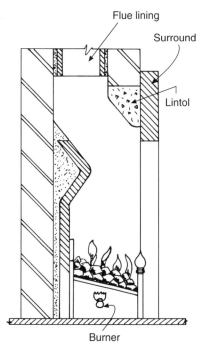

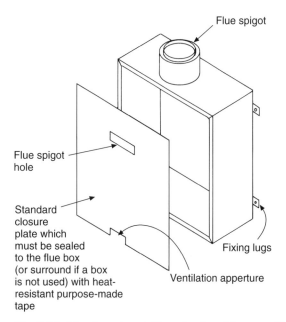

Fig. 2.52 Flue box. A metal box developed for use with certain types of gas fires to in effect simulate a builder's opening. They are primarily designed for situations where a false chimney is constructed in existing properties using timber and plasterboard. They are also useful in conjunction with fire surrounds constructed of brick or stone to which it may be difficult to make an effective seal to the closure plate.

Fig. 2.51 Section through a DFE fire constructed to the requirements of BS 5258:Part 12. Fitting and installation must comply with BS 5871:Part 3. The throat of the flue affects the efficiency of these fires and may require some modification in accordance with the manufacturer's instructions. These fires require a ventilator of at least 100 cm² for fires up to 15 kW input. Where this type of fire is fitted into a large open fireplace, e.g. (inglenook) manufacturers may specify certain requirements to ensure a positive updraught in the flue. Flue requirements as for LFE fires.

Flue boxes

Figure 2.52 shows a typical insulated metal flue box which is primarily designed for installations where no flue is available. They can also be used in most existing fireplaces, providing a clean environment at the back of the fire, and although originally produced for conventional gas fires some companies are producing them for open flame effect fires. A check should be made on the suitability of any such fire for fitting into a fire-box of this type.

Hearths for gas fires

Hearths will vary as to the type of fire installed. Figure 2.53(a) and (b) illustrates the general requirements of BS 5871. These dimensions may vary, however, depending upon the manufacturer's requirements. Hearths for decorative and LFE fires will generally have to comply with the Building Regulations Part J. A hearth should be constructed having a fire-resistant surface of at least 12 mm and must be 50 mm above floor level unless a raised kerb or fender of this height is provided. This ensures that non-fire-resisting floor coverings are not laid over the hearth.

Servicing gas appliances

Gas appliances should be serviced on an annual basis to ensure they are safe and working to maximum efficiency. However, because of the wide variety available it is very difficult to compile a complete list of servicing requirements for each appliance. This is one of the reasons why installation and maintenance manuals *must* be left with a responsible person on completion of a new

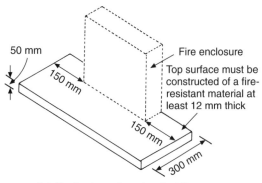

(a) Radiant or glass-fronted fires

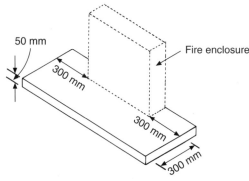

(b) Live and decorative fuel effect fires. Note the increase of hearth extension on both the front and sides of the fire. The dimensions shown are the minimum between any part of the incandescent fire bed

Fig. 2.53 Hearths for gas fires. It should be noted the hearth dimensions shown are based on the relevant British Standards, but in some instances the manufacturers instructions which must be complied with require an increase in these dimensions.

installation. There are, however, some servicing points which are common to all appliances listed as follows and can be used as a general guide.

1. Isolate any electrical connection if applicable.
2. Turn off any gas valve controlling the supply of gas to the burner.
3. Floor-standing boilers and fires tend to become very dusty and the interior of the appliance may need cleaning. An industrial vacuum cleaner will be very useful here.
4. Remove the burner and control assembly.
5. Inspect any flueing arrangement for damage, defective joints, and cleanliness. With fires

ensure the void behind the sealing plate is clean.
6. Clean the flueways or heat exchanger with a bristle brush and clean the bottom of the combustion chamber, preferably using the vacuum cleaner.
7. Inspect and clean the burner and the gas injectors taking care not to damage them.
8. Clean the pilot light assembly.
9. Reassemble any dismantled parts, taking care to renew any seals that have been broken with proper replacements.
10. Reconnect the electrical and gas connections and check for soundness. Light the appliance and after approximately 10 minutes, with the thermostat set at its highest position. Check the gas pressure at the test point on the burner.
11. If applicable check the electrical system for damaged cables and security of electrical connections.
12. It is wise to ensure visually that the appliance ventilation requirements have been complied with, especially if the appliance was originally fitted by another installer.
13. Finally, conduct a spillage test on the flue if applicable.

Tracing gas leaks
This is sometimes a difficult job, especially in existing properties. The best procedure in such circumstances is as follows: ascertain from the customer

(a) Where the smell is more noticeable.
(b) Is it persistent or does it occur only when certain appliances are in use?
(c) Has it got worse over a period of time?
(d) Was the smell noticed before or after a new appliance was installed?

If the cause of the leak is not obvious check any valves, especially on gas cookers; the heat often dries out the grease on the plug. Valves on gas fires are another possible cause of leakage. Look for any mechanical damage on exposed soft copper pipes. It may be necessary to isolate the appliances one by one, testing the remainder of the installation each time. It may be possible to isolate individual branches using the same procedures. In the event of failure to find the leak the National Gas Provider or gas supplier should be contacted.

Decommissioning

In the event of premises being unoccupied, depending on the circumstances, the gas transporter may cap off the main valve and remove the meter. Any pipework that is unlikely to be used again should be removed as far as is possible. All open ends must be securely plugged or capped where the pipework remains and an air test should be conducted to ensure this procedure is effective. It is an offence to leave any open ends on a gas supply pipe, even when it is no longer in use.

Further reading

Much useful information can be obtained from the following sources:

The Gas Safety in the Installation and Use Regulations.
British Standards:
BS 6891 Gas installation
BS 5871 Installation of gas fires, convectors, and fire back boilers, Parts 1, 2, 3.
BS 5440:Part 1 Flues.
BS 5440:Part 2 Air supply.
BS 6798 Installation of gas fired hot water boilers.
Confederation of Registered Gas Installers, 4 Elmswood, Chineham Business Park, Crockford Lane, Basingstoke, Hants RG24 8WG.
Publications Tel. 01256 707060.
Registration enquiries 01256 708133.

Flues
Security Chimneys UK Ltd, Dalilea House, St Mary's Road, Portishead, Bristol, BS20 9QP Tel. 01275 847609.
Chimflue, Tel. 01707 266244, 01264 332878.

Testing equipment
Testo Ltd, 3 Oriel Court, Omega Park, Alton, Hants, GU34 2QE Tel. 01420 544434.

Fires
Baxi Ltd, Brownedge Road, Preston, Lancashire Tel. 0800 085 2397.

Valor Heating, Wood Lane, Erdington, Birmingham B24 9QP Tel. 0121 373 811811.

Self-testing questions

1. State the physical properties of natural gas.
2. (a) Explain the term cross-bonding and state why it is necessary.
   (b) Describe the main cause of flame lift-off and explain why it must not be tolerated in gas burners.
3. (a) State the recommended pressure for testing a new gas installation and describe the procedures employed.
   (b) Why is it necessary to test the main gas cock before testing an existing installation?
4. (a) Describe how to purge gas installations.
   (b) Explain why it is necessary to remove all traces of air from a gas installation.
5. (a) Explain why a supply of air is essential for combustion and what is meant by vitiated air.
   (b) Sketch a downdraught diverter and describe its function.
6. (a) List the combustion products of gas.
   (b) Name the toxic gas which is present in the products of incomplete combustion.
7. (a) Make a simple sketch illustrating the principle of room-sealed, natural draught appliances.
   (b) Sketch and describe three methods of terminating a gas flue in a suitable position.
8. (a) Explain the working principle of an appliance governor and state its purpose.
   (b) Describe the procedure for regulating the working pressure of gas to an appliance.
9. (a) Describe the methods of igniting gas appliances.
   (b) List the essential controls for the correct functioning of a gas boiler.
10. (a) Explain the term *fail safe* in relation to gas appliances.
    (b) Make a simple sketch illustrating the working principles of a thermoelectric valve.

11. State the reason for sealing a flue lining at both the top and bottom.

12. A sealing plate must be used with gas fires complying with BS 5871:Part 1. State its purpose and why it is necessary to seal the edges to the fireplace opening.

13. From Table 2.2 determine the pipe diameter necessary to supply an appliance with an input rating of 1.5 m/hr (pipe length is 9 m).

14. List the procedures that must be carried out when commissioning a gas appliance.

15. Determine the area in a ventilating brick of free air necessary for combustion where an open-flued appliance rated at 16 kW is fitted in a room.

16. State the maximum permissible pressure drop between a meter and any appliance.

17. State the recommended procedure where, during a service call, an appliance is found to be 'at risk'.

18. State the minimum distance that the terminal on a room-sealed natural draught appliance must be from a corner of the building.

3 Cold Water Supply

After completing this chapter the reader should be able to:

1. Explain the main purpose of regulations relating to water supply.
2. Identify water classifications and types of backflow prevention devices.
3. Identify the cause of noise in cold water systems and state the methods of prevention.
4. State the causes and means of prevention of contamination of water in domestic and industrial supplies.
5. Describe the methods of preventing frost damage to water pipes and fittings.
6. Describe the basic principles and operation of water treatment appliances.
7. Understand the basic principle of pump boosted cold water supplies.
8. Describe the various water systems used for fire protection in buildings.

The Water Regulations (1999)

These regulations superseded the 1986 Water By-laws in England and Wales. The water authorities in Scotland began to enforce new by-laws in April 2000 which mainly mirror the regulations for England and Wales. They will also be adopted by the Department of the Environment for Northern Ireland. The new regulations embody most of the requirements of the 1986 Water By-laws; the main differences relate to backflow protection, differences in the categories of water, and WC flushing. More emphasis is also placed on conservation.

As with previous by-laws, the main reasons for the need for regulations may be summarized as avoidance of: waste, contamination, misuse and undue consumption. Undue consumption may be interpreted as unnecessarily using more water than is actually required. It is also an offence to fit a supply of water to a metered premises in such a way that it does not pass through the meter.

Every practising plumber or fitter whose work entails the installation or repair of hot and cold water supply systems must be conversant with the regulations, as the penalty for their contravention is very severe.

Restrictions on the use of water fittings

The following is an abbreviated list of the principle requirements for any appliance or component used with water supplied by water authorities. For more detailed information reference should be made to the actual regulations, the Water Fittings and Materials Directory, and the Water Regulations Guide.

Requirements of water fittings

Every water fitting or appliance must comply with any relevant standard relating to quality and suitability for the purpose for which it is required. These include BS and EN harmonised standards or those having European technical approval which meet these requirements. Every water fitting must be installed, altered or repaired in a workmanlike manner and must comply with the relevant standards mentioned.

Notification of work

Except for minor extensions or alterations to a water system, permission must be applied for and granted in connection with the following:

1. The erection of a structure, not being a pond or swimming pool, requiring a supply of mains water.
2. Any alteration of, or extension to, any building except a dwelling.
3. A change of use of any premises.
4. Installation of:
   (a) Baths having a capacity of more than 230 litres.
   (b) Bidets with inlets below the spillover level or fitted with a flexible hose.
   (c) A single shower unit, not being a drench shower, which may consist of one or more shower heads (under review).
   (d) A pump or booster connected to a supply pipe drawing more than 12 litres per minute.
   (e) Any equipment which incorporates reverse osmosis processes.
   (f) A water treatment process which produces a waste water discharge or which requires the use of water for regeneration or cleaning.
   (g) Any mechanical device used for protection against a fluid category 4 or 5, for example an RPZ valve.
   (h) A garden watering system unless designed to be operated by hand or
   (i) Any water system or pipe laid externally of the building less than 750 mm or more than 1,350 mm below ground level.
5. Ponds or swimming pools with a capacity in excess of 10,000 litres which is designed to be replenished automatically and filled with water supplied by a water undertaker.

All notices to the water authority must include the following:

(a) The name and address of the person giving the notice.
(b) A description of the proposed work.
(c) Particulars of the location and premises to which the proposals relate and the intended use of the premises.

(d) Except in the case of work falling into category C.

There are exceptions to the foregoing: a plumber who is an 'approved contractor' will not require permission to install such appliances as those listed in 4(b) or 4(g).

General requirements of Water Regulations

The use of lead pipes and solders containing lead have been prohibited for use with potable water for many years. The only exception to this is in situations where lead services are still in use. Any repairs to such services, short of renewal, must be made using copper or a suitable plastic pipe and approved fittings.

Most water authorities also have reservations about the use of galvanised steel tubes, especially in areas where the action of the water on zinc causes dezincification. In most cases, where this material is permitted, its use is limited to distribution and hot water services only where the use of copper or plastic materials may be subject to damage.

The use of storage cisterns with purpose-made covers and overflow screens as illustrated in Book 1 is mandatory in all new buildings where any water is stored for drinking and domestic purposes, and any replacements to existing water-storage vessels must comply with the 1986 water by-laws. Any pipe-jointing compound used for making joints on pipework or storage vessels must be non-toxic and resistant to bacteriological growth.

Coal tar substances such as bitumen can no longer be used to protect the internal surfaces of pipes or cisterns against corrosion, but suitable anti-corrosion paints are available and a list of these may be obtained from the local water authority. It is not unknown for water to leak from the primary part of a central heating system into the secondary water, i.e. water drawn off via a hot tap. This can occur due to a leak in the connection or coil in the hot storage vessel. Any inhibitor used to protect the heating system from corrosion must therefore be of a non-toxic nature. It is not permissible to run underground service pipes in soil which may be contaminated by sewage or refuse of any description. While pipes made of plastic are very resistant to corrosion, they can be degraded and softened when subject to contact with petroleum products, oils and phenols, i.e. materials derived from coal tar such as bitumen and creosote. If there is any suspicion of such contamination the service may have to be rerouted or passed through a watertight duct. The local water authority should also be advised where any such doubts arise in relation to underground services.

Types of cold water supply

Services may be supplied direct from the supply pipe and water main, and this is common in small domestic properties. The indirect type of supply is preferred however, mainly because some storage is available and there is less risk of contamination. These two basic systems are described in Book One of this series. In the case of large commercial and industrial buildings, for example hotels, schools etc., an indirect supply is essential. In such cases the water supplier should be consulted as to the capacity of any large storage cistern, as they will be aware of any difficulties in the volume and pressure of the main supply. Cisterns having a storage capacity in excess of 1,000 litres should be provided with a valved wash-out pipe at the lowest level. The wash-out pipe must not be connected directly to a drain except through a tun dish providing an AA type air gap. When not in use the valve must be securely plugged. Table 3.1

shows the recommendations of BS 6700 for water storage in various types of building. Circumstances may vary but it provides a good general guide.

Pumped systems cold water supply

Many high-rise buildings are unable to be supplied with water direct from the mains. The fact that main pressure varies (i.e. depending on whether the building is at the top or bottom of a hill) also has some influence on the pressure available. Daytime pressures are also lower than those at night due to the larger volume of water consumed. If for example a pressure of 350 kPa is available at the main (this is the equivalent of 35 m head) it will in theory serve a building 35 m higher than the main. After making an allowance for pressure loss, due to the resistance of the pipe and fittings, the effective pressure may only be approximately 32–33 m. For this reason water boosting is employed to supply cold water to upper drinking water draw-offs, storage cisterns and if necessary for fire-fighting in high-rise buildings. With the exception of installations where an interruption of the supply would not be serious, dual pumps are essential to allow for the possibility of mechanical failure or periodic maintenance. There are two main types of installation:

(a) Direct boosting, where the water is pumped directly from the mains.

Table 3.1   Water storage requirements in various buildings. REF BS 6700.

Type of building		Storage (litres)
Dwelling-houses and flats	(per resident)	90
Hostels	(per resident)	90
Hotels	(per resident)	200
Offices without canteens	(per head)	40
Offices with canteens	(per head)	45
Restaurants	(per head, per meal)	7
Nursing/convalescent homes	(per bed space)	135
Day schools — nursery-primary	(per pupil)	15
Day schools — secondary-technical	(per pupil)	20
Boarding schools	(per pupil)	90
Childrens home or residential nursery	per bed space	135
Nurses' home	per bed space	120
Nursing or convalescent home	per bed space	135

(b) Indirect systems, where the water is pumped from a break tank which is fed from the main via a float-operated valve.

Direct boosting is rarely permitted due to:

(a) The volume of water drawn from the main leading to the loss of supply, or at least lowering of pressure to other consumers.
(b) The possibility of backflow or cross-connection being much greater.

Under no circumstances is it permissible to connect a pump directly to a pipe connected to the main without the written consent of the supplier, except if it draws less than 12 litres/minute.

Indirect systems

Indirect systems are so called because water is pumped from a break tank which is supplied from the main through a float-operated valve. They fall into two main types: those which pump to a drinking water header shown in Fig. 3.1, and those which employ a low-level pressurised storage vessel shown in Fig. 3.2.

In all cases of pumped systems, draw-offs within reach of mains pressure are connected directly to the main. This reduces the volume of water that has to be pumped, enabling the use of smaller pipes, pumps, and in some cases, storage vessels.

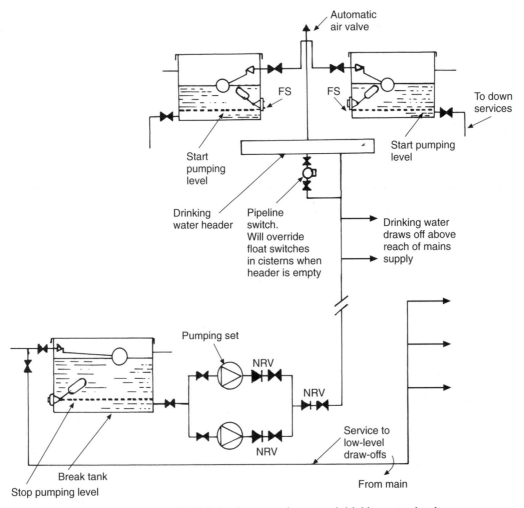

Fig. 3.1 Indirect pump booster system with high-level storage cisterns and drinking water header.

Indirect boosting with a header

The capacity of the break tank shown in Fig. 3.3 requires careful consideration. To avoid stagnation they are normally sized to provide 1 hour's supply, but conversely they should hold sufficient water to enable the pumps to function for 15 minutes before the low-level cut-out operates. In all cases of pumped supplies, the system must be designed to reduce the number of pump stop/starts to pump small quantities of water; failure to do this will shorten its working life. All pumping systems are therefore controlled in such a way that when the pump starts it will continue pumping for a preset period of time. To supply drinking water points above the reach of mains pressure using this system, a header is employed which is sized to provide 5–10 litres per day per dwelling served. When the pump is not running and drinking water is drawn off, the header begins to empty until the

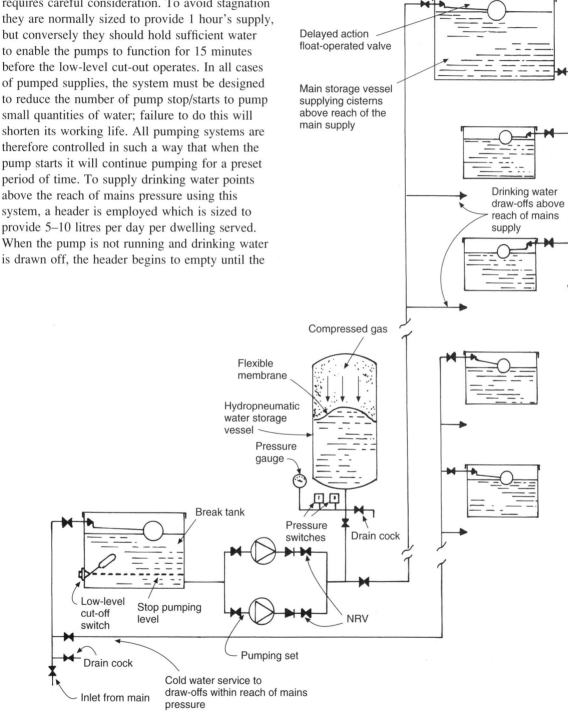

Fig. 3.2 Hydropneumatic boosted cold water system.

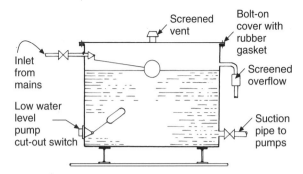

Fig. 3.3   Detail of break tank.

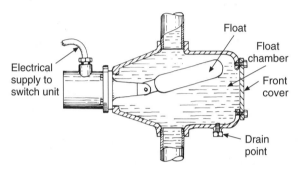

Fig. 3.4   Pipeline switch. When the chamber empties the float will fall and operate the switch to activate the pump to refill the header.

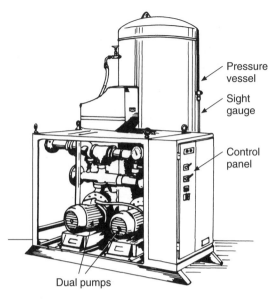

Fig. 3.5   Typical hydropneumatic packaged pumping set.

pipeline switch, illustrated in Fig. 3.4, activates the pump, and by means of a timing switch causes it to operate for a set period of time. If the water is replaced in the header before the pump timing cycle is incomplete, the excess water will be pumped into the storage cistern through the float-operated valve. Should this be closed while the pump is operating, no serious damage to the pump will take place, as being of the centrifugal type it can operate against a closed outlet for a limited period of time. If the pipeline switch is not activated, but the water level in the cistern falls to such a level that topping up is necessary, the float-operated switch will fall to the 'start pumping' level shown, restarting the pump. These switches are arranged to override the timing cycle, as the pump will run for a longer period to replenish the water in the storage cistern. From the foregoing it will be seen that the pump is activated in two ways: by a fall in the drinking water level, and by a fall in the water level in the storage

cistern. An automatic air valve must always be provided to allow air into the drinking water header when water is drawn off, closing when it is full.

Indirect hydropneumatic systems
This is the most common type of installation now in use, its main advantage being that all the equipment and component parts are usually supplied as a complete package unit (see Fig. 3.5) for both large and small installations. It is also more convenient in situations where a number of storage vessels at different levels are to be served, which would be impracticable using several water-level-operated switches. The indirect hydropneumatic system operates on the principle of pumping water into a pressure vessel, causing the air it contains to be compressed. When a tap on the riser is opened, water is forced upward due to the pressure exerted by the air. Continued draw-off will lower the pressure in the vessel until it falls to a predetermined level, when the pump will restart on a signal from the pressure switch. Modern pumping units, use a pressure vessel having a flexible membrane similar to those used with unvented hot water and sealed heating systems, its capacity depending on the size of the

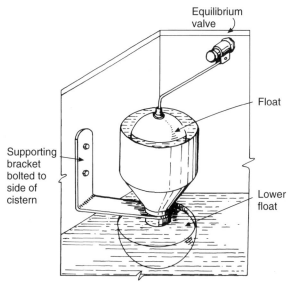

Equilibrium valve

Float

Supporting bracket bolted to side of cistern

Lower float

Fig. 3.6 Pictorial illustration of a delayed action float-operated valve.

installation. When the system is serviced, the only maintenance requirement with this type of vessel is to check the pressure of the gas, which may be air or nitrogen.

Delayed action float-operated valves These are designed to ensure the valve is either fully open or closed, and while they may be used with advantage in many other situations, they are essential with hydropneumatic pumped systems, as the float does not fall until a large volume of water is required to replenish the contents of the cistern. This ensures the number of pump/stop starts are reduced to a minimum. Figure 3.6 illustrates a typical valve of this type and its operating cycle is illustrated in Fig. 3.7.

Non-return valves These must be fitted to prevent the possibility of back pressure caused by a moving column of water being suddenly halted. This happens when a pump stops and in some cases has been known to burst the pump casings. Pump manufacturers usually specify the use of a non-return valve on the riser and each individual pump. The normal disc or flap type of non-return valves are suitable where back pressure is unlikely to be

excessive, but a spring-assisted recoil valve, shown in Fig. 3.8 is recommended where high pressures are anticipated. No provision is shown against backflow in Figs 3.1 and 3.2. A double check valve would be necessary on each of the branches serving drinking water points and any other branches taken from the main riser. As a further precaution against contamination, a type BA or CA device may be necessary on the main inlet at the discretion of the water supplier; see Tables 3.2 and 3.3.

Pumps, pump components and siting
Pumps and the associated controls and components should be housed in a room as close as possible to the point where the main enters the building. The room should be dry, ventilated, protected against frost and flooding and of sufficient size to allow for maintenance and the replacement of component parts. Access should be restricted to authorised personnel only. In buildings which rely solely on the pumps for a supply of water, pumps must be duplicated. The previous comments relating to pump noise should be noted. Most packaged units are supplied with water hammer arrestors, but where a pumping set is constructed on site, provision must be made to include protection against water hammer, as the surge developed when a pump stops or starts can give rise to very high pressures. Reliable control systems are essential — pressure and float switches must be suitable and adequate for their purpose. Most pumping sets are electrically operated and should be controlled by a pump selector switch so the pumps can be operated alternatively. On most modern equipment switching arrangements are fully automatic. Where the failure of the main power supply could be serious, back-up plant such as a generator set must be provided.

Maintenance and inspection
The user should make arrangements for servicing the plant at regular intervals. Any work carried out and the dates of inspection should be recorded in a log book, which is usually retained in the plant room. As the equipment is almost entirely automatic, maintenance consists mainly of keeping the components clean and checking that the controls are functioning correctly.

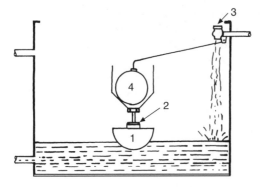

(a) The cistern is filling with both valve 2, which is carried by the semi-hemispherical float, and valve 3 open

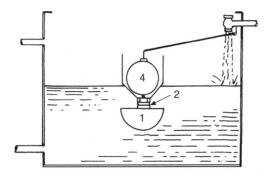

(b) As the water level rises float 1 is lifted closing valve 2. Valve 3 is still fully open

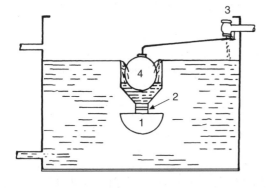

(c) The cistern continues to fill until water flows over the edge of the tun-dish. This lifts the ball float 4 closing valve 3. As the tun-dish holds little water the closure is very sudden and for this reason an equilibrium float-operated valve and the use of water hammer arrestors is essential to reduce noise

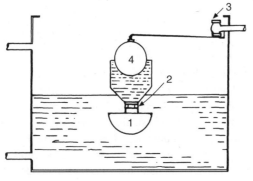

(d) As the cistern empties valve 3 remains closed until the water level falls to a level where it no longer supports float 1. This causes it to fall and open valve 2 allowing the tun-dish to empty. Float 4 will now fall opening valve 2, restarting the sequence as shown in (a)

Fig. 3.7   Delayed action float-operated valve sequence of operation.

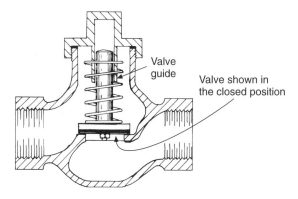

Fig. 3.8 Spring-loaded non-return valve. Used as an alternative to the normal flap or disc NRV on boosted cold water systems where a quick-acting valve is necessary to avoid back pressure on pumping equipment.

Noise from pumps or boosters

All pumps make some degree of noise due both to the motor which 'hums' and to the rotation of the impeller which causes vibration. This vibration is transmitted to the floor supporting the pump and its pipework connections. By introducing flexible pipe joints and pump fixings as shown in Fig. 3.9, this type of noise is almost eliminated. Manufacturers produce these pumps having regard for the noise factor, and motor speeds are kept to a minimum in relation to the output of the pump. To avoid the 'hum' generated by electric motors becoming a nuisance, pumps should be situated as far as possible from living apartments — in some cases they are contained in a soundproof room.

Large water storage cisterns and associated pipework

In buildings, where large quantities of water are consumed it may be necessary to use two or more cisterns to ensure a more equal distribution of the mass of water over a greater load-bearing surface or because a cistern large enough to meet the requirements is not available. In some instances additional storage may be required in an existing building; Fig. 3.10 shows two arrangements for coupling two or more storage cisterns together. To avoid stagnation, it is essential to ensure that the water flows through both cisterns (as though they were a single cistern) and that the outlet connection is on the opposite side to that of the inlet.

Where it is practicable at least one outlet from the cistern should be connected in the bottom to avoid the build-up of sediment, which can allow bacteria and other harmful organisms to multiply. Where large storage accommodation is necessary it should be broken down into two or more cisterns in such a way that each can be isolated for maintenance to be carried out without an interruption of the supply. Cisterns having a capacity of more than 1,000 litres must be provided with a wash-out pipe with a full way valve permanently capped when not in use.

Storage cistern overflows

Book One of this series deals with the size and siting of overflow pipes for cisterns in domestic properties and those having capacities of up to

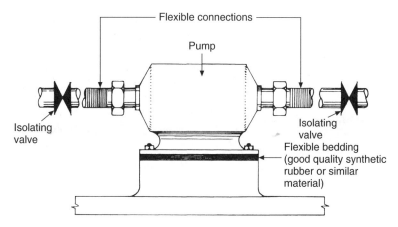

Fig. 3.9 Noise reduction in pumping equipment.

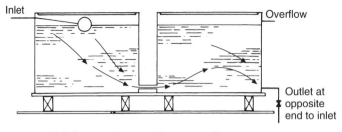

(a) Connecting two cisterns end on

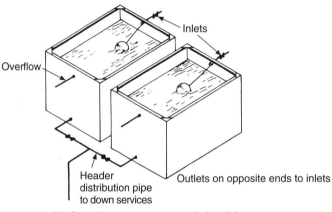

(b) Coupling two cisterns side by side

Fig. 3.10 Coupling large water storage cisterns to avoid stagnation.

1,000 litres. It is recommended in the Water Regulations that cisterns having an actual capacity greater than this are fitted with both an overflow and a warning pipe; see Fig. 3.11(a). This is mandatory where the capacity is in excess of 5,000 litres, the invert of the overflow pipe being 50 mm above the normal shut-off level. The warning pipe may be omitted if a level indicator is fitted which provides either visual or audible warning if the water exceeds the normal cut-off level. Fig. 3.11(b) illustrates a typical method of achieving this. It is important that the relationship between the inlet valve and overflow pipe in a cistern complies with the Regulations against backflow, see Tables 3.2 and 3.3.

Ventilation pipe terminations
Any ventilating pipe terminating in a cistern must comply with the requirements of Fig. 3.11(c).

Noise in cold water systems

It is unavoidable that a certain amount of noise is created by plumbing systems, either by the flow of water through the pipes, or by actuating the fittings, a typical example of the latter being the use of a flushing cistern and its subsequent refilling. For some people a certain amount of noise is acceptable, but to others the same noise level would be intolerable, and every plumbing system should be as noise-free as possible. The following text will indicate the more common causes of noise in cold water supplies and its prevention.

Water hammer
This is not only undesirable from the point of view of noise, but in some forms it can cause damage to plumbing systems. It is always associated with high-pressure supplies, whether direct from the main or

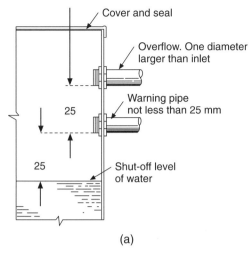

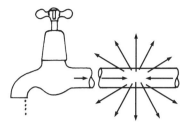

Fig. 3.12 Water hammer. Sudden closure of tap causes opposing pressure to that of the incoming water resulting in a loud bang in the pipework.

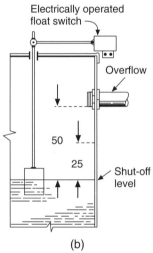

(a)

(b)

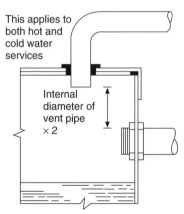

(c) Termination of vent pipes in cisterns

Fig. 3.11 Positioning overflow and warning pipes in large storage cisterns (distances in mm).

boosted by pumps, and in some cases pressure-reducing valves may have to be employed to lower the pressure in certain sections of a building. Water hammer usually occurs when a high-pressure flow of water is suddenly arrested, as in the case of a tap that is turned off quickly (see Fig. 3.12). This has the effect of causing a loud bang or series of bangs throughout the pipework, and momentarily subjecting the whole system to a pressure almost double that of the incoming water. If this is allowed to persist the excessive pressure can cause a pipe, perhaps already weakened by frost, to start leaking.

The plug cock, classified as a quarter-turn tap, was the first type of tap used in plumbing systems, its history going back many hundreds of years. However, due to the increase in piped water supplies in the early nineteenth century and subsequently higher water pressures, it was found, because of their quick closure, they were a major cause of water hammer. This prompted water authorities to insist, prior to the 1986 by-laws, on screw-down taps (which are designed to effect gradual closure) to be fitted on mains water supplies. Quarter turn taps are now permissible for use as servicing valves on such appliances as storage cisterns, flushing cisterns and clothes- or dish-washing machines, and as they are used infrequently they are unlikely to cause persistant water hammer. Comparatively recently ceramic-disc taps have become more common and like the plug cock it is a quarter-turn tap which in high-water-pressure areas can give rise to water hammer. Even screw-down taps, when the spindle and gland become worn, can be responsible for this problem, especially when the washer is replaced with a soft rubber washer of the domed type sold by some

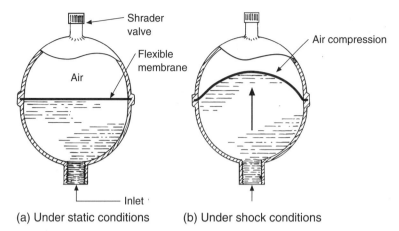

Shrader valve

Flexible membrane

Air

Air compression

Inlet

(a) Under static conditions   (b) Under shock conditions

Fig. 3.13   Water hammer arrester.

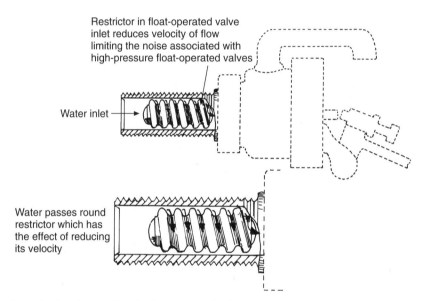

Restrictor in float-operated valve inlet reduces velocity of flow limiting the noise associated with high-pressure float-operated valves

Water inlet

Water passes round restrictor which has the effect of reducing its velocity

Fig. 3.14   Reducing velocity of water flow to float-operated valves.

ironmongers. These will almost certainly cause water hammer and in such cases the gland should be repacked and the washer replaced by one of the traditional flat type made of harder rubber. In some cases the thread on the spindle is so worn that the flow of water through the tap causes it to move rapidly up and down making a humming or a singing sound. In these circumstances, although the gland may be repacked as a temporary measure, it is better to replace the tap.

Some appliances, such as washing machines and dishwashers, employ electrically operated valves which are also noisy in operation. Only those having the approval of the local water authority should be used, as apart from the noise they cause they may not conform to the back siphonage regulations.

Special measures are required to limit the effects of water hammer in larger buildings or a series of dwellings supplied by the same service pipe.

In such instances where there are long pipe runs of high pressure water, it may be necessary to fit an air vessel similar to that shown in Fig. 3.13.

They operate on the principle that gases are easily compressible. This enables any momentarily high pressure caused by a sudden closure such as a draw-off to be absorbed.

Float-operated valves are often responsible for noise in plumbing systems due to high-velocity supplies. This is caused by the following:

(a) The water passing through the valve orifice.
(b) Splashing as it falls into the cistern.
(c) Creating ripples or waves on the surface of the water, causing the float to bounce.

When the valve is nearly closed, these waves cause it to open partly and close very quickly, giving rise to a persistent banging, often terminating in a shrill whine just prior to fully closing. While it is not possible entirely to eliminate noise caused by float-operated valves, it can be minimised by the following methods:

(a) The use of velocity restrictors shown in Fig. 3.14.
(b) The use of spray-type outlets in the valve shown in Fig. 3.15.
(c) This is a development of the silencing pipe principle and is shown in Fig. 3.16.
(d) An equilibrium float operated valve is a very effective method of preventing noise provided suitable provision is made for an appropriate air gap, as most of these valves are of the Portsmouth pattern with a bottom outlet.

The arrangement shown in Fig. 3.11 is acceptable to most authorities, provided that a suitable air gap is arranged.

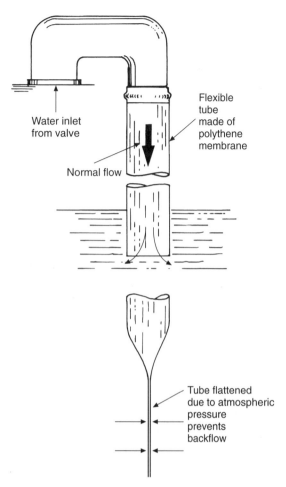

Fig. 3.16 This arrangement permits the flow of water to discharge below the surface of the water in a manner similar to the original rigid silencing pipes. In the event of backflow causing a negative pressure, atmospheric pressure will flatten the tube as shown inset, thus preventing possibly polluted water entering the main. It should be noted that some authorities may not permit this arrangement on water supplied from the main and it should not be used in industrial premises.

It has already been stated that water hammer is always due to high pressure and velocity. In some cases a reduction of noise can be achieved by partly turning down the outside stop valve so that it passes just sufficient water for the householders' requirements. This does not have, as many think, the effect of reducing the pressure, but it does reduce the velocity, and in some instances can cure the problem.

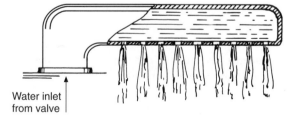

Fig. 3.15 This attachment has the effect of breaking up the water into a number of finely divided jets which reduce the velocity of the incoming water and the associated noise.

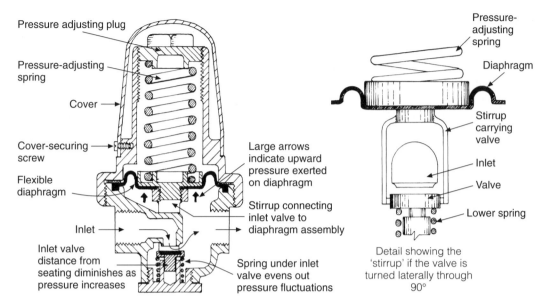

Fig. 3.17   Pressure-reducing valve.

Pressure-reducing valves

These valves may be fitted in areas having very
high water pressures. Those produced for domestic
purposes are not expensive and their use solves
many of the noise problems caused by modern taps
of the quarter-turn pattern. They are also effective
in preventing splashing due to the reduction in
pressure when the water is discharged into a sink
or washbasin. A simple pressure-reducing valve is
shown in Fig. 3.17 where it will be seen that its
working principles are similar to those of other
pressure-regulating devices with which a plumber
is familiar, such as gas governors and regulators
for high-pressure gases. As water flows through
the inlet, pressure is exerted on the diaphragm
which lifts the valve, via the stirrup, closer to its
seating. The outlet pressure is varied by adjusting
the tension of the large spring. The more the spring
is compressed the greater will be the outlet pressure.
The purpose of the spring under the valve assists in
evening out any pressure fluctuations. It is important
to note that the correct pressure adjusting spring is
supplied with this valve and the supplier should be
informed of the pressure range over which it is
required to work. A pressure gauge fitted at a
convenient point on the outlet side of the

installation will enable fine adjustments to be
made when commissioning.

Noise due to high-velocity flow

The flow of high-velocity water through water
pipes can also be responsible for unnecessary noise,
especially pipes of copper or stainless steel. These
two materials, being thin walled and rigid, tend to
vibrate with the passage of water, and if the pipes
are fixed on a hollow surface, such as a partition
wall constructed of plasterboard and timber, the
wall reacts as a sound box, magnifying the
vibration. If fixing pipes to such surfaces cannot be
avoided, enough clips or brackets must be used to
avoid too much pipe movement, and in some cases
a rubber sponge backing placed behind the clips
will improve the situation. In extreme cases the use
of plastic pipes will almost certainly prevent any
noise due to their flexible nature and the fact that
the pipe walls are relatively thicker than copper and
stainless steel.

Pollution of water supplies

When water has been discharged from the main
water pipe into a cistern or sanitary appliance of

any sort, it can be assumed that it is polluted (i.e. has been in contact with possibly harmful bacteria) and must in no circumstances come into contact with water in the main or service pipe. The only exceptions to this are pumped schemes for the supply of drinking water in high buildings, and in such cases there are special requirements to prevent the water becoming contaminated. There are many examples on record of water contamination, all of which were a possible danger to health, not only to the occupants of the premises concerned, but to the community as a whole.

One of the most common causes of water contamination occurs where water for drinking and culinary purposes is drawn from uncovered cisterns. It is well known that birds, rodents and all sorts of insects have been found in such cisterns, quite apart from impurities such as dust which settles on the bottom forming an unpleasant sludge. A fine wire mesh screen must be fitted in the overflow and sealed covers made of the same materials as the cistern must be fitted. It is not necessary or advisable that they should be airtight, simply that the ingress of debris is prevented. Covers made of wood or hardboard must not be used as these materials, being organic, produce a mould in damp conditions which can fall into the water and in itself cause contamination.

The 1999 Water Regulations list five water categories in ascending order of risk to the consumer. They also specify the minimum acceptable methods of preventing contamination of water supplies, which usually occurs due to 'backflow'. The following text is based on the recommendations of the Department of the Environment, Transport and Regions (DETR).

Water Categories 1–5
Previous by-laws categorised water in three groups according to risk. The Water Regulations now contain five groups as listed:

1. Water supplies direct from a main supply, generally without being stored before use. One exception to this is the water in a break tank fitted with pumped systems.
2. Water which is pure and wholesome like that of category 1, except it has undergone a change in taste, smell, temperature or appearance, none of which are considered to be a hazard to health. Typical examples are domestic hot water, a mixture of category 1 and 2 water discharged from mixer taps or showers, and water softened in appliances using salt during the regeneration process. In connection with water softeners a drinking water tap should be supplied directly from the main, especially if used by people with certain medical conditions.
3. This category may pose a slight health hazard and is not suitable for drinking and culinary purposes. Primary water in indirect heating systems, and water used for ablutionary purposes, clothes or dishwashing machines are typical examples.
4. Water falling into this category is not suitable for drinking or culinary purposes and may contain substances causing cancer, micro-organisms, bacteria and viruses, all of which constitute a positive danger to human health. Typical waters classified as category 4 are listed as follows:
    (a) Primary water in commercial or industrial heating systems whether or not any treatment using additives has taken place.
    (b) Water treatment processes using materials other than salt.
    (c) Domestic irrigation systems using perforated hoses or sprinklers fitted less than 150 mm above ground level.
    (d) Commercial clothes and dishwashing machines.
5. This is the most dangerous of all hazard categories and human exposure to water in this category is very serious indeed. Pathogenic organisms are a general term given to the many types of bacteria, viruses and parasites capable of causing serious illness; salmonella and cholera are two of the most commonly known but there are many others. The types of water falling into this category are in many cases similar to those listed in category 4, but they are classified as category 5 if the period of exposure to the risk is longer or if the concentration of the toxic substances in the water is higher. Discharged water from sinks, bidets and WCs, both domestic and commercial, is always classified as category 5,

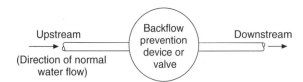

Fig. 3.18 Illustration of the terms 'upstream' and 'downstream'.

together with and what is termed 'greywater'. This is the discharged water from ablutionary appliances which instead of passing into the foul drain is stored, and after treatment may be used for flushing WCs.

Backflow

Water supplied through the 'undertaker's' mains is pure and wholesome and it is important that it is not contaminated in any way. Schedule 2 section 6–2 of the Water Regulations deals specifically with backflow and its prevention. Backflow may be defined as the reversal of the normal flow direction in supply pipes and water mains. Two terms are commonly used in connection with the flow of water in pipes: (a) upstream and (b) downstream, both of which are illustrated in Fig. 3.18. Backflow may occur due to either back pressure or back siphonage. If the pressure of the water upstream of a valve or fitting falls below that of the water downstream, backflow may take place. A typical example of back pressure is where a conventional hot storage vessel is fitted, and due to the expansion of the water it is forced back into the supply pipe. It is uncommon in domestic premises and is far more likely to happen in an industrial environment. It has been known to happen in unvented hot water systems and combination boilers provided with storage vessels, but the volume of water involved in these cases is small and would not exceed a category 2 risk. Provided the temperature of the water is not in excess of 25 °C this risk can be ignored. Back siphonage is far more common and will take place if a negative pressure occurs in the water service or mains pipework. This is by no means unusual and can happen for example if a washout on the main is opened for cleaning, or the main is shut down and drained for repair work. In the event of a major fire, the connection of fire

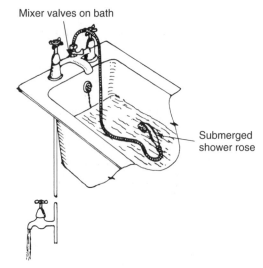

Should a tap be opened when there is a negative pressure in the service pipe, water in the bath could be drawn off through a tap fitted at a lower level.

Fig. 3.19 Back siphonage due to a submerged inlets.

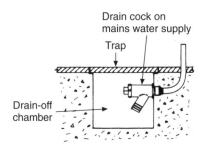

Floor surface arrangements of this nature could lead to back siphonage if the drain-off chamber becomes filled with water when the cock is in the open position, i.e. during a draining-down operation.

Fig. 3.20 Back siphonage due to submerged drain-off.

brigade pumps to a hydrant can also cause an appreciable drop in mains pressure. A classic instance of backflow due to back siphonage is shown in Fig. 3.19 and can still happen if the suitable safeguards are not applied.

Yet another cause of contamination can arise where a drain-off cock is fitted in a floor well (see Fig. 3.20). When the valve is opened and becomes submerged, should a negative pressure occur in the main at that moment, back siphonage can cause the

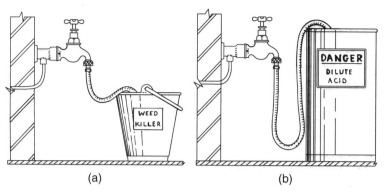

(a)   (b)

Fig. 3.21 Illustrating the danger of using hoses connected to taps with inadequate or incorrect backflow protection.

contaminated water to enter the main supply. The seriousness of this problem is further illustrated by the fact that situations have arisen whereby it was possible for water from gutters, and in one instance a urinal, to enter the water supply system.

Hose-union taps have always been a possible source of contamination in both domestic and industrial situations. The end of the hose can be left in vessels that contain or have contained unpleasant and dangerous substances. They can be left in garden ponds during refilling or put into inspection chambers to wash out a system of drains, these being only a few of the possible risks of contamination due to hose pipes. To illustrate this point further consider the use of hosepipes which are often used with attachments containing solutions for washing down cars or chemicals used in the garden (see Fig. 3.21). These are perfectly safe while water is discharged through the hose, but if a negative pressure occurs on the main, all sorts of dubious substances could find their way both into the service pipe and water main with possibly very serious results. It must also be borne in mind that the number of hot water systems supplied directly from the main will increase in the future. This means care must be taken to prevent any possible pollution, not only via cold water supplies but also those of hot water too. It cannot be stressed enough that compliance with the Water Regulations must be strictly observed to ensure that any danger to public health by water supplies is minimized.

Backflow prevention

There are many devices listed to prevent backflow. Some are mechanical, a typical example being a check valve. Others are classified as non-mechanical, the traditional air gap being one of the most commonly known. They are all classified as being suitable for certain types of fluid category. Tables 3.2 and 3.3 show the main types of backflow prevention methods used against the relevant water category. They are not exhaustive but carry most of the common risk levels encountered for domestic and commercial applications in the UK. Reference to these tables will be necessary when reading the following text to determine the degree of protection offered by each method of backflow device shown against back pressure and back siphonage. The fluid catergories for which they can be safely used should also be noted. Note that many of the devices listed relate to situations that are unusual in the normal course of work.

It is the responsibility of the plumber or fitter to ensure the correct type of device is fitted to satisfy the requirements of the Water Regulations. If any doubt exists the water supplier must be consulted.

Non-mechanical backflow devices

Air gaps
An air gap must be clearly visible and constitutes a positive break between the level of water in a cistern or appliance and the lowest level of

Table 3.2 Schedule of non-mechanical backflow prevention arrangements and the maximum permissible fluid category for which they are acceptable. (Crown Copyright 1999 Reprinted with permission of HMSO.)

Type	Description of backflow prevention arrangements and devices	Suitable for protection against fluid category	
		Back pressure	Back siphonage
AA	Air gap with unrestricted discharge above spillover level	5	5
AB	Air gap with weir overflow	5	5
AC	Air gap with vented submerged inlet	3	3
AD	Air gap with injector	5	5
AF	Air gap with circular overflow	4	4
AG	Air gap with minimum size circular overflow determined by measure or vacuum test	3	3
AUK1	Air gap with interposed cistern (For example, a WC suite)	3	5
AUK2	Air gaps for taps and combination fittings (tap gaps) discharging over domestic sanitary appliances, such as a washbasin, bidet, bath or shower tray shall not be less than the following:	X	3
	Size of tap or combination fitting Vertical distance of bottom of tap outlet above spill-over level of receiving appliance		
	Not exceeding $G^1/2''$ 20 mm Exceeding $G^1/2''$ but not exceeding $G^3/4''$ 25 mm Exceeding $G^3/4''$ 70 mm		
AUK3	Air gaps for taps or combination fittings (tap gaps) discharging over any higher risk domestic sanitary appliances where a fluid category 4 or 5 is present, such as: a. any domestic or non-domestic sink or other appliance; or b. any appliances in premises where a higher level of protection is required, such as some appliances in hospitals or other health care premises, shall be not less than 20 mm or twice the diameter of the inlet pipe to the fitting, whichever is the greater.	X	5
DC	Pipe interrupter with permanent atmospheric vent	X	5

Notes: X Indicates that the backflow prevention arrangement or device is not applicable or not acceptable for protection against backpressure for any fluid category within water installations in the UK.
Equivalent to Table S6.1 in DETR Guidance document

discharge from the inlet. It must not be less than 20 mm or twice the internal diameter of the inlet, whichever is the greater. The angle of flow must not be more than 15° from the vertical centre line of the water inlet. Figure 3.22 and Fig. 3.23 illustrate diagrammatically various configurations of the air gap principle. Reference must be made between the illustrations and Tables 3.2 and 3.3 so that a clear picture of the requirements of the various categories of water can be seen. The

degree of protection for both back pressure and back siphonage is also shown in the tables.

It will be seen that types AA and AB give good protection against both types of backflow. Such protection is necessary when installing cattle drinking troughs, supplies to a sewage treatment plant, or vats containing toxic chemicals. Types AC, AD and AF are again mainly required in industrial or commercial premises. It will be seen that some types of protection require the overflow to discharge

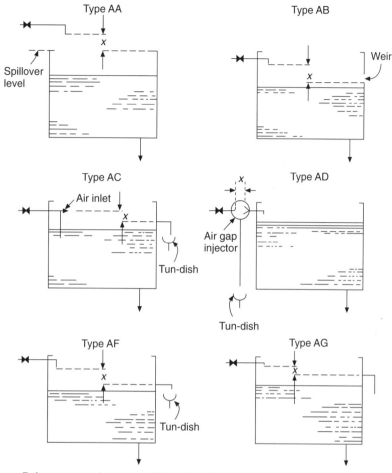

Reference must be made to Table 3.2, which shows the types of risk for which the air gaps are suitable. Some are used only for certain industrial processes. The air gaps for systems with tun-dishes must comply with Table 3.2.

Fig. 3.22   Types of air gap. In all illustrations, the air gap is indicated by x.

into a tun-dish. The reason for this is because the overflow may discharge into, e.g. a contaminated pond or river. If backflow via the overflow took place from such a source, it could contaminate the contents of the cistern. The AG air gap is suitable for lower risks and it will be seen from the table it offers protection up to category 3 for both types of backflow. It complies with the requirements of BS:6281 PT2 and was formerly known as the class B air gap, which satisfied the requirements of previous by-laws in relation to flushing cisterns fitted with BS:1212 PT2 or 3 float-operated valves.

In most cases air gaps complying to type AUK 1,2,3 are built into an appliance, e.g. wash basins and baths, and a plumber will usually be familiar with these. Care must be taken, however, in relation to type AUK 1 and close-coupled WC suites with flushing cisterns fitted with valves. If the internal overflow pipe is shortened in any way, a check must be made to ensure the minimum distance of 300 mm is maintained between the spill-over level of the WC and the overflow weir or invert. If a siphon is used as shown in Fig. 3.23(a) this measurement is usually built in, but it is recommended that a measurement

Table 3.3 Schedule of mechanical backflow prevention arrangements and the maximum permissible fluid category for which they are acceptable (Crown Copyright 1999 Reprinted with permission of HMSO.)

Type	Description of backflow prevention arrangements and devices	Suitable for protection against fluid category	
		Back pressure	*Back siphonage*
BA	Verifiable backflow preventer with reduced pressure zone	4	4
CA	Non-verifiable disconnector with difference between pressure zones not greater than 10%	3	3
DA	Anti-vacuum valve (or vacuum breaker)	X	3
DB	Pipe interrupter with atmospheric vent and moving element	X	4
DUK1	Anti-vacuum valve combined with a single check valve	2	3
EA	Verifiable single check valve	2	2
EB	Non-verifiable single check valve	2	2
EC	Verifiable double check valve	3	3
ED	Non-verifiable double check valve	3	3
HA	Hose union backflow preventer. Only permitted for use on existing hose union taps in house installations	2	3
HC	Diverter with automatic return (normally integral with some domestic appliance applications only)	X	3
HUK1	Hose union tap which incorporates a double check valve. Only permitted for replacement of existing hose union taps in house installations	3	3
LA	Pressurised air inlet valve	X	2
LB	Pressurised air inlet valve combined with a check valve downstream	2	3

Notes:
1 X Indicates that the backflow prevention device is not acceptable for protection against Backpressure for any fluid category within water installations in the UK.
2 Arrangements incorporating a Type DB device shall have no control valves on the outlet of the device. The device shall be fitted not less than 300 mm above the spillover level of an appliance and discharge vertically downwards.
3 Types DA and DUK1 shall have no control valves on the outlet of the device and be fitted on a 300 mm minimum Type A upstand.
4 Relief outlet ports from Types BA and CA backflow prevention devices shall terminate with an air gap, the dimension of which should satisfy a Type AA air gap.
Equivalent of Table S6.2 in DETR Guidance document

check is made to ensure compliance with the regulations.

Pipe interrupter
This is a device with no moving parts and is shown in Fig. 3.24. Its working principle is very simple: should a negative pressure take place on the inlet (upstream) of the valve, air is drawn in via the air ports. This has exactly the same effect as when air is admitted to a siphon bend: it stops siphonic action. This type of arrangement can be made as an integral component of such equipment as washing machines or dishwashers. It can also be used when controlled by an upstream valve for delivering water to or cleaning vehicles, ships or large cisterns where the supply is operated by a solenoid valve. Where flushing valves are permitted, they must be provided with a pipe interrupter, often built into the outlet of the valve itself. They are suitable for category 5 risks against backflow.

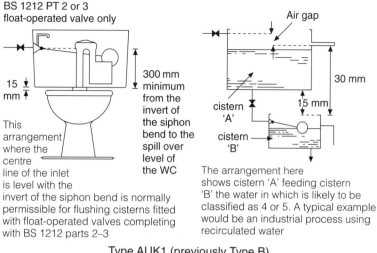

BS 1212 PT 2 or 3
float-operated valve only

15 mm

300 mm minimum from the invert of the siphon bend to the spill over level of the WC

This arrangement where the centre line of the inlet is level with the invert of the siphon bend is normally permissible for flushing cisterns fitted with float-operated valves completing with BS 1212 parts 2–3

Air gap

30 mm

15 mm

cistern 'A'

cistern 'B'

The arrangement here shows cistern 'A' feeding cistern 'B' the water in which is likely to be classified as 4 or 5. A typical example would be an industrial process using recirculated water

Type AUK1 (previously Type B)

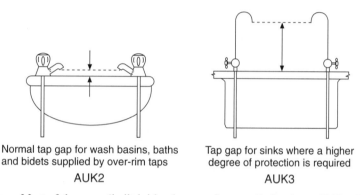

Normal tap gap for wash basins, baths and bidets supplied by over-rim taps
AUK2

Tap gap for sinks where a higher degree of protection is required
AUK3

Fig. 3.23   AUK air gaps. Most of these are 'built-in' by the manufacturer. Refer also to Tables 3.2 and 3.3.

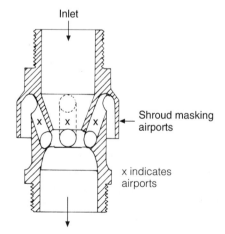

Inlet

Shroud masking airports

x indicates airports

Fig. 3.24   Pipe interrupter type DC. This device is non-mechanical and has no moving parts. It is fitted in line with the pipe run where there are no control valves downstream of the outlet.

Mechanical backflow protection

In most cases this type of backflow prevention has some moving parts, usually spring-loaded valves. Some are classed as 'verifiable' which means if necessary they can be checked for effectiveness. To test a verifiable double check valve for example, turn off the main supply and open the valve or test screw between the two internal valves. If only a small quantity of water emerges it can be assumed that all is well. If a larger quantity is discharged it can be assumed that one of the valves is letting by and needs either servicing or replacing. It should be noted that it is not possible to verify the first valve embodied in a double check valve. One other point that must always be taken into account with mechanical backflow valves is the fact that it may

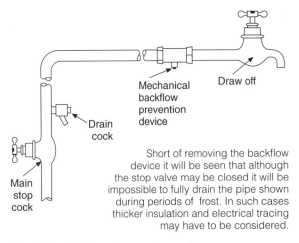

Short of removing the backflow device it will be seen that although the stop valve may be closed it will be impossible to fully drain the pipe shown during periods of frost. In such cases thicker insulation and electrical tracing may have to be considered.

Fig. 3.25 Draining down water services.

be impossible, in periods of frost, to completely drain a system (see Fig. 3.25). Exposure to very cold weather in some industrial and commercial installations may require some form of tracing, and frost protection in such situations must be seriously considered. All mechanical backflow devices must be accessible for inspection, maintenance and testing. Except for double check valves fitted to hose union bib taps, all such devices must be fitted within the premises. Strainers must be fitted immediately upstream of any backflow prevention device used to protect an installation against category 4 or 5 risks. This is necessary to avoid any debris becoming deposited on the valve seats preventing their closure. Service valves must also be provided both upstream and downstream of the device. Any relief outlets must be provided with a type AA air gap not less than 300 mm above floor level. Table 3.3 lists most of the mechanical backflow devices currently available and the following text and illustrations describe those most commonly used in the UK. **The water supplier must be informed if it is proposed to install any mechanical device to prevent backflow risks from category 4 or 5.**

The reduced pressure zone backflow prevention valve (RPZ)

These valves are currently the only mechanical device suitable for both types of backflow against category 4 risks and can be used for both whole site

protection or point of use application. They are rarely used or indeed necessary in domestic dwellings as the contamination risks do not warrant it. Figure 3.26 illustrates diagrammatically an RPZ valve, and it will be seen that it has similar characteristics to those of a double check valve. Both check valves are designed to progressively reduce the water pressure in two stages so that under normal flow conditions pressure is always less on the downstream side than it is upstream, hence the name 'reduced pressure zone valve'. This pressure imbalance and the check valves themselves act to prevent backflow. The other main characteristic of the RPZ valve is that if a negative pressure occurs upstream, even if one of the check valves A or B are not operating correctly, it will open to the atmosphere via the relief valve C creating an air gap. This will prevent water which may have become contaminated entering the upstream part of the system. It will be seen that valve C is normally closed, as the upstream pressure of water acting on the diaphragm overcomes the spring and keeps the valve closed. Only when the upstream pressure falls and backflow occurs will the relief valve open, creating an air gap due to the action of the spring and a lowering of the pressure on the diaphragm.

Installation Manufacturers of RPZ valves specify the installation requirements necessary for servicing and testing. Generally they should be fitted no less than 300 mm from floor level with a minimum clearance behind the assembly of 100 mm. Suitable methods of discharging waste water must be provided for the air gap to always be maintained. The usual method of achieving this is to provide a tun-dish as shown in Fig. 3.26(c).

Maintenance These valves must be inspected annually to ensure they are functioning correctly. A differential pressure gauge is connected to each of the three test points in the valve body (not shown in the illustration). The pressure shown on the gauge will indicate whether all the valves are functioning correctly. Most manufacturers of these valves recommend that maintenance is carried out by plumbers or fitters who have undertaken an approved course and are suitably qualified.

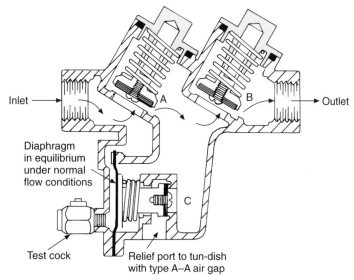

(a) Normal flow condition: valves A–B open, valve C closed

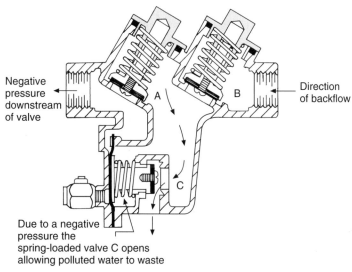

(b) Backflow condition: valves A–B closed, valve C open

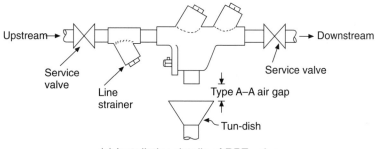

(c) Installation details of RPZ valve

Fig. 3.26 RPZ valve (verifiable backflow preventer with reduced pressure zone).

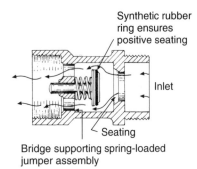

Fig. 3.27 Non-verifiable single check valve type EB.

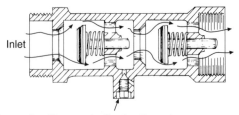

Test cock, with no water flowing through the valve both jumpers should be seated. If a persistent flow of water can be seen when the test cock is opened, one or both valves are defective. The cause of the trouble should be investigated or the complete unit replaced

Fig 3.28 Double check valve type EC.

Non-verifiable disconnector with different pressure zones (type CA)

This is similar to the RPZ valve in that it provides a positive disconnection area between the upstream and the downstream flow of water. In a similar way to the RPZ valve, the area between the two main valves will open to the atmosphere in the event of an upstream pressure drop. Any water discharging through the outlet is allowed to run to waste through a tun-dish. It is only suitable for use with water up to a category 3 risk.

Check valves

These are basically a spring-loaded valve which will close due to pressure of the spring if the upstream pressure falls below that of the downstream.

Single check valves   Type EA is a single verifiable check valve, the term 'verifiable' meaning that all valves of this type can be checked for effective functioning. Type EB valves are non-verifiable. However, both types are suitable for category 2 risks. They will permit water to flow upstream but will close against downstream pressure. They are suitable for use against both back siphonage and back pressure. They are mainly used with mixing valves supplied with hot and cold water at differing pressures, and are often built in by the manufacturer. Figure 3.27 shows a typical non-verifiable valve of this type.

Double check valves   These are similar in construction to single check valves, but because

they incorporate two valves, in the event of one failing to close under backflow conditions, the other will still afford the necessary protection. The classification EC indicates the valve is verifiable and is shown in Fig. 3.28 Non verifiable types without a test cock are classified as the ED type. Both are suitable for protection against category 3 contamination risks.

Prior to commissioning a system it should always be flushed out, but it is especially important if any mechanical devices such as check valves are fitted. Flushing out should be undertaken with the valves out of position, as any detritus such as filings or swarf could lodge on the valve seatings and prevent them closing. Verifiable valves must always be checked during commissioning to ensure they are working correctly. Taking as an example the verifiable double check valve, it will be seen that it has a test cock between the two valves. By shutting off the water supply and opening the test cock, little or no water should appear. This indicates the downstream valve is closed and in working order. If it is not then the complete unit should be removed for repair or replacement, as although the valve upstream may be holding, there is no way of checking it unless the valve incorporates two test cocks.

Hose union taps – backflow protection

Because the category of risk will depend upon the circumstances for which they are used, the type of backflow protection will vary. For domestic

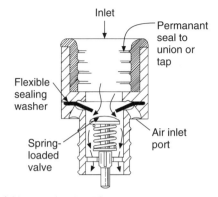

(a) Hose union backflow preventer type HA

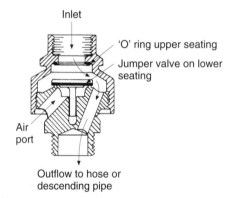

(b) Check and vacuum breaker type HA

Both valves shown work on the same principle except that one is spring-loaded. In the event of a negative inlet pressure the valves close and admit air to the outlet allowing any water in the hose to drain.

Fig. 3.29 Combination check and vacuum breakers.

properties using hand-held hoses, category 3 backflow protection devices are normally acceptable. For new build work a double check valve situated inside the building must be fitted to the supply. Book 1 of this series illustrates hose union taps type HUK1 which embody two check valves and can be used as a replacement only in existing properties. Another acceptable method of protection, again only for existing properties, is the use of a combined check and anti-vacuum valve (type HA in Table 3.3). Two very similar types are shown in Fig. 3.29(a) and (b) and are designed to screw onto the tap outlet. The reason that HUK1

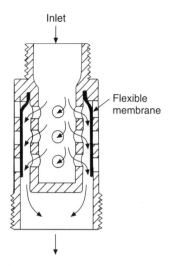

Fig. 3.30 Pipe interrupter with vents and moving element, type DB.

and HA types should not be used on new work is because of the difficulty of draining down in frosty weather, making them very prone to damage.

Pipe interrupter with vent and moving element
Classified as DB this is currently an unfamiliar device in the UK. Figure 3.30 illustrates a typical valve of this type. The flexible membrane closes off the air ports when in normal use, but in the event of a negative pressure in the supply the membrane will be drawn towards the inlet ports, closing them and thus preventing any backflow and the possibility of contaminated water entering the supply. It must be installed in a vertical position as shown with no valves or restriction on its outlet and fitted no less than 300 mm above the spillover level of the appliance it supplies. It will be seen that it has similar characteristics to a pipe interrupter and in some circumstances it is used as an alternative.

Diverter with automatic return
This device is usually built into combined bath–shower mixers and is classified in Table 3.3 as the HC type. When the taps are opened the valve is lifted manually to allow a mixed supply of water to the shower head, the flow of water keeping it open while the shower is operating. It is designed to drop under its own weight if the supply is interrupted.

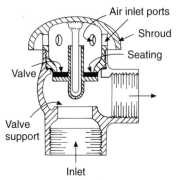

A negative pressure upstream allows the valve to drop maintaining equilibrium in the pipe work system.

Fig. 3.31 DA type anti-vacuum valve.

Anti-vacuum valves (vacuum breaker)
A typical valve of this type is shown in Fig. 3.31. Under normal conditions the valve is held in the closed position by the water pressure. If the pressure upstream falls to or below that downstream of the valve, it will fall open under its own weight to prevent backflow. When the valve is used on its own it is listed in Table 3.3 as type DA. It is often used in this form in conjunction with unvented hot water cylinders to prevent their collapse. When it is used as a backflow device it is more likely to be

used in conjuction with a single check valve. The various configurations shown diagrammatically in Fig. 3.32 illustrate this. These are quite often built into domestic washing machines and dishwashers, and into many types of commercial appliances with similar applications. They are also used in connection with large-scale irrigation systems. As it is not unknown for leakage to occur due to failure of the valve to re-seat properly after operating, it is not advisable that they are used in situations where this is likely to cause serious damage.

Application of backflow-prevention devices in domestic premises

Figure 3.33 illustrates a typical arrangement of both hot and cold water services in a dwelling. It will be seen that it does not vary a great deal with traditional practice. All draw-off taps are protected by a type AA air gap apart from the external hose-union tap which must be provided with either a check and anti-vacuum valve at the hose connection, or a double check valve on the supply pipe. Float-operated valves should be of the diaphragm type complying to BS 1212 Part 2 or 3. The storage cisterns must be provided with a type AG air gap and flushing cisterns with type AUK1. The gap between the float-operated valve and

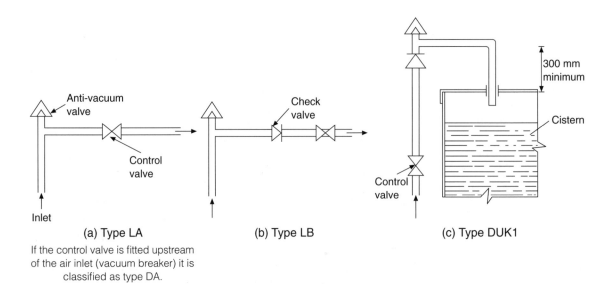

If the control valve is fitted upstream of the air inlet (vacuum breaker) it is classified as type DA.

Fig. 3.32 Backflow configurations with air admittance and check valves.

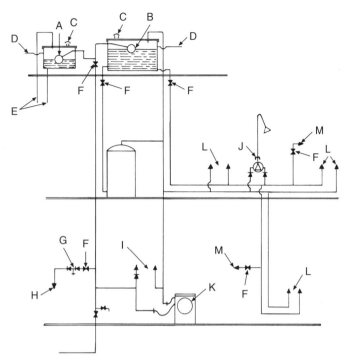

Fig. 3.33 Water supply for domestic premises complying with the Water Regulations. Both storage cisterns (A) and (B) have type AG air gaps; (C) cistern vents; (D) screened overflow; (E) feed and vent pipe to heating system; (F) service valves; (G) double check valve; (H) outside tap; (I) sink mixer of the biflow type with a single check valve on the cold supply; (J) bath shower mixer tap with single check valves on inlets and flexible hose; (K) washing machine with integral AG air gap; (L) all draw-offs including the over-rim bidet have type AUK air gaps. (M) Flushing cisterns incorporate type AUK1 air gaps.

overflow is the same with both types, but the AUK1 type also incorporates a vertical gap not less than 300 mm from any contaminated water, in this case the WC. Service valves must be fitted so that float-operated valves can be repaired or replaced without shutting down supplies to other appliances. These service valves may be of the quarter-turn ball type, stop- or gate valves, bearing in mind the latter are not suitable for mains water supply. Where a sink mixing tap is subject to unequal pressures, unless it is of the byflow type it should be fitted with a single check valve to prevent possible backflow from the hot water supply into the cold service pipe. If, for example, a hose was connected to the outlet of the mixer, backflow could take place if the main supply, or the water main, were to be subjected to a negative pressure. Such hoses are available for users of washing machines not permanently connected to

the hot and cold supplies. Permanently plumbed-in clothes- or dish-washing machines should comply with the Water Research Centre's requirements and BS 6614 which specifies the requirements for connection of these appliances to water supplies. If this is the case they will be provided with a type A6 air gap or pipe interrupter made integrally in the machine. Although it is not mandatory at present, some manufacturers of these machines may provide single check valves as an integral part of the hoses.

Showers having flexible hoses do present a serious risk of contamination, which can be addressed as follows:

(a) The flexible hose may be attached in such a way that the shower rose is prevented by a restraining ring from becoming submerged in the appliance it serves, thus maintaining an air gap above the flood level of the appliance.

(b) The mixer is provided with single check valves to both the hot and cold supplies. A further check valve is fitted in the flexible hose outlet. All these valves are normally built into the mixer unit and give the equivalent protection of double check valves on each supply. A further form of protection is the use of a hose restraining ring which prevents the shower rose falling below the flood level of the appliance.

It may be asked why, in the system shown, is it necessary to go to these lengths as many are in current use without any backflow protection. The reason for this is that at some future date the premises might be fitted with an unvented hot water system with both hot and cold supplies taken from the main. It will be seen that in such circumstances a very real danger of contamination could exist. Reputable manufacturers are aware of this and produce taps and valves with built-in backflow protection where possible.

Appliances with submerged inlets
The only type of domestic appliance normally falling into this category is a bidet, which due to the very nature of its use poses a serious risk of contamination. In previous by-laws these appliances have never been permitted to be directly connected to a main supply pipe and the same prohibition is maintained in the Water Regulations. When the cold supply to a bidet is provided by a storage vessel, a separate distribution pipe from that supply to other draw-offs is necessary, except that where applicable the same pipe can be used to supply a WC or urinal. The methods of supplying hot water may vary. A separate water heater supplying the bidet only is the ideal solution. The heater would have to be a low pressure type as it would not be permissible to connect it to the main supply. Figure 3.34(a) illustrates an acceptable solution. It is not mentioned in the Water Regulations but was accepted by the 1996 Water By-laws. Yet another alternative is shown in Fig. 3.34(b) where both the hot and cold supply are connected to a mixer valve. Although several types of backflow devices are

listed as being suitable for this arrangement, the pipe interrupter is probably the most suitable, but remember there must be no restriction to flow downstream of the water. From the foregoing text it will be seen that to meet the requirements of the Water Regulations, any fitting having an inlet below flood level is both difficult and expensive to install, and the use of bidets having over-rim supplies is likely to become normal practice in domestic properties. As these comply with the requirements of a type AUK2 air gap they are the only type of bidet suitable for use with unvented hot water systems.

Flexible hoses
The use of flexible hoses attached to sanitary appliances has always constituted a serious backflow risk. A typical example is shown in Fig. 3.19, which would be considered a category 3 risk. Figure 3.34(c) illustrates an even greater danger where the hose may fall into water classified as category 5. A hose restraining ring would definitely not be suitable here as it can be detached; a fixed shower head would be more practical, indeed the only option if it is connected to a mains water supply. Even if the hose is connected to a distribution pipe serving other appliances the danger is still very real, and a type AA air gap would be the only solution complying with the Water Regulations.

Backflow protection in multi-storey public and commercial buildings

This is a very wide area and to list every instance of possible contamination cannot be done within the limitation of this volume. The following text, however, illustrates some examples of good practice so that the reader will be able to recognise installations that do not comply with the Water Regulation. One of the most common instances of water contamination that may occur in industrial and agricultural premises, apart from back siphonage, is a cross-connection between water from the main and water from some other source. Surprising as it may seem, mains water pipework

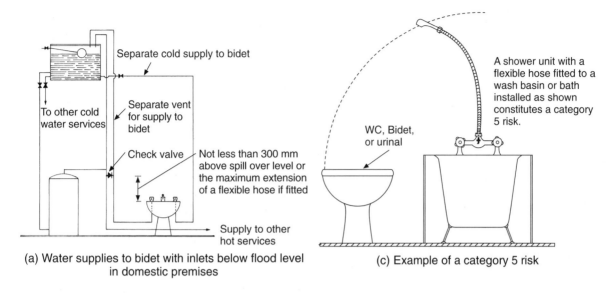

(a) Water supplies to bidet with inlets below flood level
in domestic premises

(c) Example of a category 5 risk

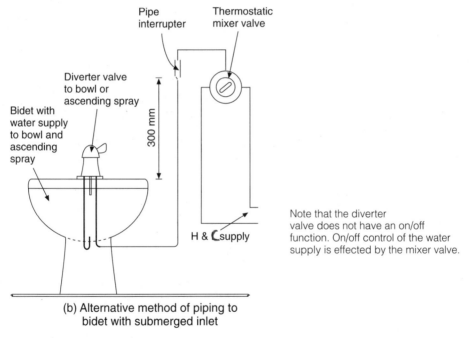

(b) Alternative method of piping to
bidet with submerged inlet

Fig. 3.34   Appliances with submerged inlets, and an example of a category 5 risk.

has been found to be connected to supplies, the original source of which was subsequently discovered to be from wells, and in one case, a pond. Possibly even greater dangers exist in industrial premises dealing with dangerous chemicals and a good plumber must always investigate an existing system very carefully before making any connection to a mains water

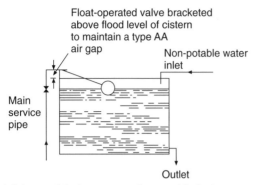

(a) A type AA air gap must be provided where any supply of non-potable water is used in conjunction with water supplied by a water authority

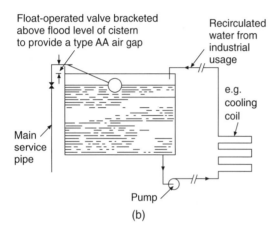

(b)

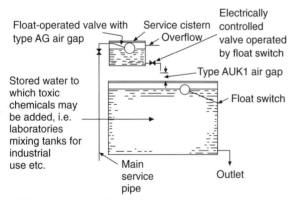

(c) As an alternative to the arrangement shown the outlet pipe could discharge into a tun-dish fitted in such a way that the air gap is maintained. Water level control in this case would be via a solenoid valve operated by a float switch

Fig. 3.35 Backflow protection in public and commercial buildings.

supply. The illustrations in Fig. 3.35 are mainly concerned with the prevention of back siphonage. Figure 3.35(a) shows how water from another source is prevented from contaminating mains supplies, and Fig. 3.35(b) shows an arrangement where water is used for cooling in industrial premises. Typical examples include cooling for plastic moulding machines, milk-cooling and air-conditioning plant. Note that in both cases the type AA air gap *must* be used. Figure 3.35(c) shows how water is supplied to cisterns and tanks to which toxic chemicals may be added, i.e. laboratory mixing tanks, etc. The risks are so great here that a tun-dish, or a service cistern as shown is fitted.

Standpipes in industrial premises are subject to strict control and should be supplied with water from a cistern. In certain circumstances where there is little risk of contamination, it may be possible, with the prior agreement of the water authority, to supply such taps direct from the main, providing a double check valve is fitted in the pipeline. Where bidets of the submerged inlet type or those having flexible hoses are fitted in hotels, hostels or nursing homes having several floors, the arrangement shown in Fig. 3.36 is an acceptable method of installation. Both the hot and cold water services must be completely separate from any other water supplies. The pipework shown is designed to prevent siphonage from the bidets on upper floors from contaminating the supplies from those fitted below. The single check valves and 300 mm upstands provide primary backflow protectection here, the vents on the distribution pipes secondary protection.

Secondary backflow protection

This term relates to any device fitted to supplement primary or point of use protection such as an air gap on each tap or drawoff. Figure 3.37(a–c) shows some typical methods of providing secondary backflow protection in multistorey domestic dwellings where risks in excess of category 3 are unlikely to be encountered. In Fig. 3.37(a) and (b) a verifiable double check valve is fitted on each branch downstream of the stop valves. An alternative to this is shown in Fig. 3.37(c). 300 mm upstands and a vented distribution

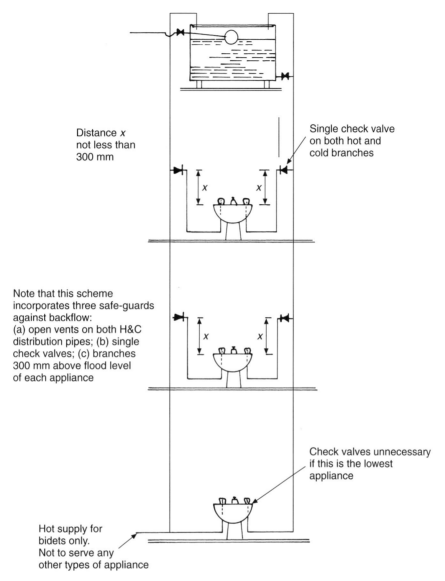

Distance *x*
not less than
300 mm

Single check valve
on both hot and
cold branches

Note that this scheme
incorporates three safe-guards
against backflow:
(a) open vents on both H&C
distribution pipes; (b) single
check valves; (c) branches
300 mm above flood level
of each appliance

Check valves unnecessary
if this is the lowest
appliance

Hot supply for
bidets only.
Not to serve any
other types of appliance

Fig. 3.36 Backflow protection for bidets with inlets below flood level in a public or commercial building.

pipe provide the secondary backflow protection here. The Water Regulations Guide suggests that ventilated distribution pipes are only suitable for two-storey buildings. Care must be exercised with pipe sizing to ensure adequate protection of the lower floor appliances. Two other provisions are also necessary: (a) the branch pipe must not run above the level of its junction to the distribution pipe at any point, and (b) a minimum of 300 mm must be maintained between the branch and the flood level of the highest appliance. One of the reasons why the water supplier must be informed prior to undertaking certain types of work is to enable them to assess the possible backflow risks. In some situations, especially in commercial and industrial buildings, category 4 or 5 risks may

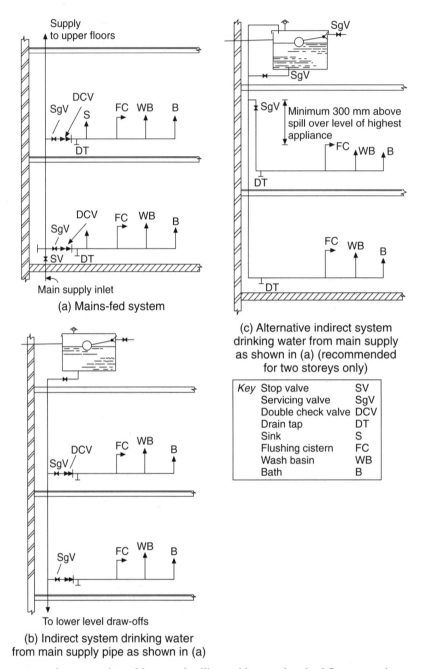

Fig. 3.37   Cold water supply systems in multi-storey dwellings with secondary backflow protection.

exist and the supplier will make the necessary recommendations. Generally an RPZ valve will satisfy category 4 risks. No direct connection to a supply pipe is permissible in the case of a category 5 risk. Where such risks exist water must be supplied via a storage cistern, the inlet having an AA, AB or AD type air gap. Suitable protection must be provided with sprinkler systems and is dealt with in the section on fire-fighting.

Sterilisation of cold water systems

Where large water installations are carried out, often over a period of many months or even years, it is possible for storage vessels and pipework to become polluted. In such cases the system must be sterilised to ensure that all harmful bacteria are destroyed before it is commissioned.

This procedure is usually carried out by water authority employees or it can be done by the plumbing contractor. The procedure is as follows. The cistern is cleared of all debris, after which the system is thoroughly flushed out. It must then be refilled and at the same time a sterilising chemical containing chlorine is added, care being taken to ensure that it is thoroughly mixed with the water. The recommended dosage is 50 parts chlorine to 1 million parts water. If other chemicals are used the manufacturer's instructions must be rigidly adhered to regarding the quantity to be used with a given volume of water. When the system has been completely filled the supply of water must be shut off and the taps on the distributing pipe opened, those nearest the cistern first. When the water from each tap begins to smell of chlorine they should be closed. The cistern should then be topped up with water, the topping-up water also containing the same proportion of chemical. The whole system is then allowed to stand for a period of 3 hours, after which a test is made by smell for residual chlorine. If none is found the whole procedure must be repeated. When the treatment is concluded the system should be emptied and flushed out with clean water to remove any taste or smell of the chlorine.

Provision for fire-fighting in buildings

The reader should note that water supply for fire-fighting and the associated equipment is very specialised, and it is not possible to deal with this subject fully within the confines of this book. The aim here is to deal with the basic principles only.

In the event of a fire in low-rise dwellings and small commercial premises, the fire service uses the fire hydrants connected directly to the water mains for a supply of water for fire-fighting. In larger and multistorey buildings, special arrangements are necessary for fire protection, and are mandatory in many cases. Methods of fire protection fall into two main categories:

(a) Equipment that can be used by the layman, such as portable fire extinguishers or hose reels.
(b) Systems of pipework which can only be used by the fire service.

Hose reels

Figure 3.38 illustrates a typical hose reel and its associated components. They are designed as a first-aid measure, and in the event of a fire, can be operated by the occupants of the building. They should be sited in such a way that the user can escape if the fire becomes out for control.

Suitable siting positions are near fire exits, adjacent to landings or along recognised escape routes. The following lists the essential requirements of hose reel installations:

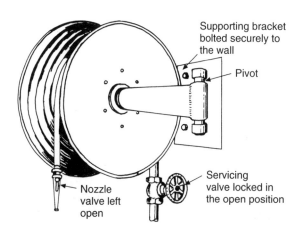

Fig. 3.38 Hose reel. The type shown is designed for housing into a wall or cupboard. When in use it can be pulled through 90° on the pivot. Most hose reels are designed to open an integral valve when the hose is pulled out. Lengths of hose available: 18 m, 24 m, 30 m, 36 m. Diameters 20 or 25 mm.

(a) When the hose is fully unrolled, the nozzle must not be more than 6 metres from any part of the floor area it covers.

(b) When the two highest or most remote hose reels are in simultaneous use, the minimum flow rate through each reel must be 0.4 litre/second. A water pressure of 2.5 bar at the nozzle is necessary to give the water jet a range of 8 m when used horizontally. This distance will be less when used in the vertical position.

(c) The diameter for a pipe serving a single hose reel is normally 25 mm, depending on its length and the number of bends on the pipe run, and in some cases a larger pipe diameter may be necessary. In buildings having several storeys, the fire main servicing the hose reels should not be less than 50 mm in buildings up to 15 m in height. If this height is exceeded, the main must be increased to 64 mm. Branch pipes serving a hose reel must not be smaller in diameter than the hose on the reel.

Water supply to hose reels   In small low-rise installations, hose reels may be connected directly

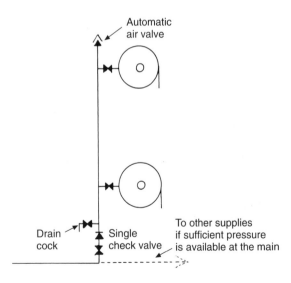

Drain cock

Single check valve

Automatic air valve

To other supplies if sufficient pressure is available at the main

Fig. 3.39   Hose reel connected directly to the main. Subject to the permission of the local water authority where the mains pressure and the service pipe size is adequate, hose reels may be connected directly to the main or possibly an existing service pipe. This arrangement may be possible for small low-rise buildings.

to the main, subject to the pressure being sufficient to meet the foregoing requirements. Water in hose reels supplied direct from the main is unlikely to constitute more than a category 2 risk in the event of backflow. This is due to the possibility of the stagnant water in the hose reel re-entering the supply pipe or main.

A typical installation of this type is shown in Fig. 3.39. In multistorey buildings, fire mains to hose reels must be boosted in a similar way to that used for cold water supplies. Figure 3.40 shows a typical pumped hose-reel installation with a break tank. In some circumstances, water authorities will permit direct boosting from the main for fire-fighting purposes. In such cases the requirements of the system will be the same, except that no break tank will be necessary. If the capacity of the high level storage vessel meets the requirements of the hose-reel installation, water can be pumped downward as the alternative shows. Pumped hose-reel installations must be fully automatic, so that when a hose reel is operated the pressure on the pipeline is lowered, causing the pressure switch to activate the pump. The purpose of the small pressure vessel shown is to prevent any pressure loss, due for example, to small leaks starting the pump when a hose reel is not in use. An alternative to a pressure switch is a flow switch which will detect a flow of water when a hose reel is operated.

Fire protection systems for use by fire service personnel

In buildings of up to 61 m in height, dry risers are normally provided — except in very special circumstances. They are usually not charged with water and are intended for fire service use only. Their purpose is to avoid running out long lengths of fire service hose up stairways or passageways, thereby obstructing what may be a means of escape for the occupants of the building. The lower end of the riser terminates external to the building in a purpose-made box containing the inlet breeching. The box is usually provided with a wire reinforced glass door, which can be opened from the inside when the glass is smashed in the event of an emergency. A breeching piece serving a 100 mm riser should have two connections for fire service

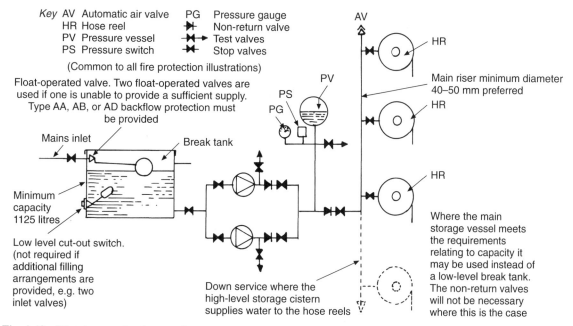

Key AV Automatic air valve   PG Pressure gauge
     HR Hose reel   ⊁ Non-return valve
     PV Pressure vessel   ⊁▸ Test valves
     PS Pressure switch   ⊁ Stop valves
     (Common to all fire protection illustrations)

Float-operated valve. Two float-operated valves are used if one is unable to provide a sufficient supply. Type AA, AB, or AD backflow protection must be provided

Mains inlet

Break tank

Minimum capacity 1125 litres

Low level cut-out switch. (not required if additional filling arrangements are provided, e.g. two inlet valves)

Down service where the high-level storage cistern supplies water to the hose reels

Main riser minimum diameter 40–50 mm preferred

Where the main storage vessel meets the requirements relating to capacity it may be used instead of a low-level break tank. The non-return valves will not be necessary where this is the case

Fig. 3.40 Wet riser serving hose reels.

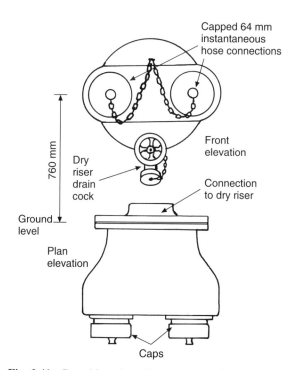

Capped 64 mm instantaneous hose connections

Front elevation

760 mm

Dry riser drain cock

Connection to dry riser

Ground level

Plan elevation

Caps

Fig. 3.41 Breeching piece. To ensure an adequate supply of water to the dry riser provision is made to connect two fire service hoses from the fire engine pumps.

pumps (see Fig. 3.41), a 150 mm riser should have four. They are fixed with their centre line being 760 mm above ground level. As with any fire-fighting installation, where any part of the pipework extends above roof level, it must be provided with a lightning conductor. Figure 3.42 illustrates a typical dry riser installation.

Wet risers A typical diagrammatic installation is shown in Fig. 3.43. These are similar to dry riser systems, but are permanently charged with water. They are normally only provided in buildings higher than 61 m. As with dry risers, there are some standard regulations relating to wet risers, but any installation will be subject to the approval of the local authority fire control officer, who should be consulted regarding the suitability of any system. The main requirements are listed as follows:

(a) The installation should be capable of maintaining a pressure at the top outlet of 4 bar with a flow rate of 22.7 litres per second.

(b) The maximum working pressure when only one outlet is in use is 5 bar.

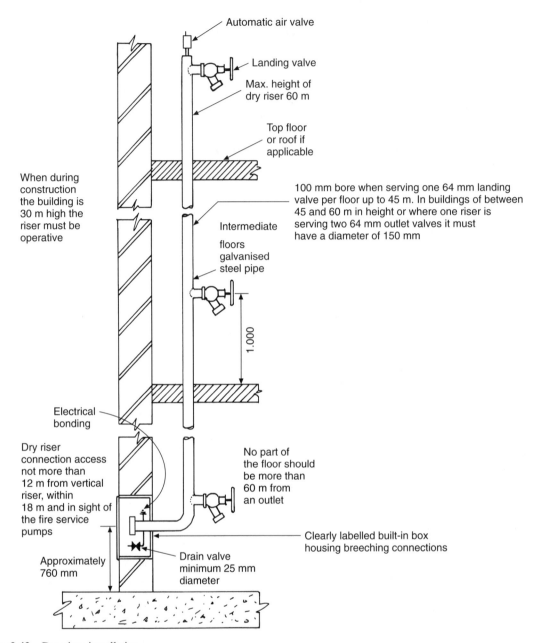

Automatic air valve

Landing valve

Max. height of dry riser 60 m

Top floor or roof if applicable

When during construction the building is 30 m high the riser must be operative

100 mm bore when serving one 64 mm landing valve per floor up to 45 m. In buildings of between 45 and 60 m in height or where one riser is serving two 64 mm outlet valves it must have a diameter of 150 mm

Intermediate floors galvanised steel pipe

1.000

Electrical bonding

Dry riser connection access not more than 12 m from vertical riser, within 18 m and in sight of the fire service pumps

No part of the floor should be more than 60 m from an outlet

Clearly labelled built-in box housing breeching connections

Approximately 760 mm

Drain valve minimum 25 mm diameter

Fig. 3.42   Dry riser installation.

(c) Because of the very high pressure involved, and to prevent damage to the water mains, pumping directly from the town supplies is not permissible. Therefore water for wet risers must be pumped from a break tank having a minimum capacity of 45.5 m³ (45,500 litres),

and the valve feeding the tank must be capable of delivering water at a minimum of 7.6 litres per second.

(d) The requirements for landing valves relating to their fixing heights, and the areas protected, are the same as those for dry risers. They are, however, provided with an adjustable regulator which limits the pressure to 4.5 bar. This avoids the possibility of bursting the fire service canvas hoses. Additional protection is provided by means of a pressure relief valve in the outlet of the landing valve, which will open at 6.5 bar. A piped supply from these valves is returned to the break tank as shown.

Sprinkler installations

There are two basic systems of sprinkler installation:

(a) The wet system, which is permanently charged with water and most frequently used.
(b) The alternative wet and dry system which is used in unheated buildings, or where the heating system only operates when the premises are occupied.

Sprinkler systems are fully automatic and usually fitted in large public and commercial premises such as cinemas, department stores and warehouses, where a fire, and sometimes the method used to extinguish it, could cause a great deal of damage.

The purpose of the sprinkler system is to limit the extent of the fire and possibly to extinguish it before it gets out of control. Because this method of fire protection is so effective, its installation in commercial buildings reduces the cost of fire insurance premiums. A typical wet system is shown

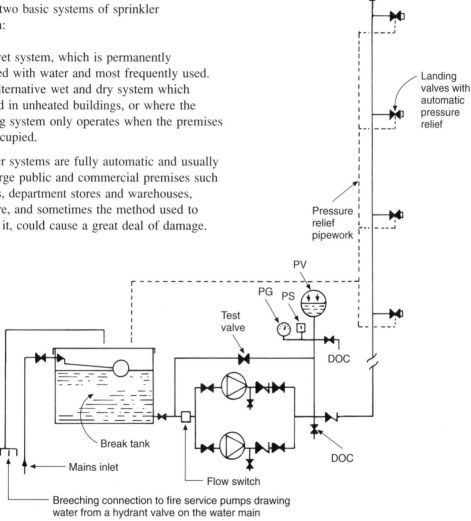

Fig. 3.43 Wet riser pumping system. In schemes of this type where total reliance for fire-fighting depends on internal pumping equipment provision must be made for an alternative supply of electricity such as a standby generator due to the possibility of power failure. Due to the very high pressures developed in this type of system water can only be pumped from the break tank. To ensure sufficient water is available for fire-fighting, two float-operated valves are often provided in addition to the fire service connection. (See Fig. 3.40 for key to abbreviations.)

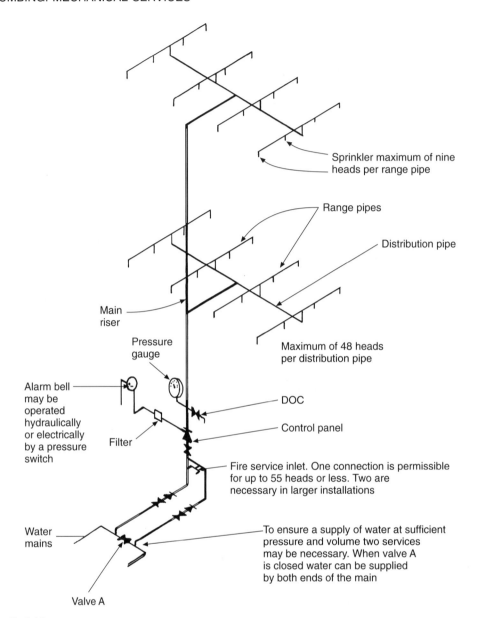

Fig. 3.44   Sprinkler system.

in Fig. 3.44. It consists of a system of pipework fitted with sprinkler heads, shown in Fig. 3.45, which react to a rise in temperature, causing a sprinkler valve to open and spray the area it covers with water. The type of head illustrated employs a glass bulb, which bursts due to the expansion of the fluid it contains when it detects an increase in temperature. These bulbs are manufactured covering a wide temperature range, which allows

for variations in the ambient temperature in both public and industrial buildings. Seven variations are listed covering temperatures from approximately 57 to 290 °C, each variation is identified by its colour.

Another type of sprinkler head employs a series of metal plates joined together with low-melting-point solders. These plates hold the valve in the closed position, and in the event of a fire the solder

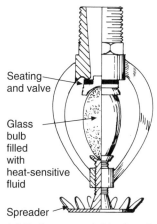

(a) Bulb-type sprinkler head

When the fluid in the bulb senses a temperature rise it
expands and breaks the glass, releasing the valve.
The ensuing flow of water is diverted outward covering
a specified floor area.

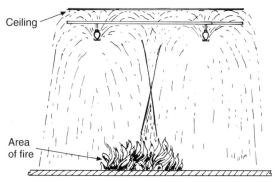

(b) The area covered by sprinklers and the overlap

Sprinkler head should preferably be within 75 and 150 mm
of ceiling, but not more than 300 mm below combustible
ceilings or roofs.

Fig. 3.45 Bulb-head sprinklers.

Table 3.4 Maximum area covered by a sprinkler.

Hazard class	General (m^2)	Special risk areas or storage tanks (m^2)
Very light hazard	21	9
Ordinary hazard	12	9
Extra high hazard	9	7.5–10

melts at a specified temperature. Variations in the
temperature at which the sprinkler head operates
is achieved in this case by using solders having
differing melting points. The number of sprinkler
heads, their diameters and the area each covers
depends on the fire hazard classification shown in
Table 3.4.

Such buildings as offices and libraries would be
classified as light hazards. Ordinary hazards would
cover industrial and commercial premises such as
shops and warehouses containing or handling
combustible materials unlikely to burn intensely
during the early stages of a fire. Industrial or
commercial buildings having a very high fire
hazard, would include premises having high piles
of combustible stocks or those handling very
flammable materials.

When a sprinkler system is installed it normally
covers the whole building. There are exceptions,
due to low fire risk, where this rule may be waived.
For example rooms having fire resistance walls and
doors which are separated from the area covered by
sprinklers.

Wet and dry sprinkler installations This type
of system is filled with water during the summer
months only. When periods of frost are expected,
the system is emptied and charged with compressed
air at a greater pressure than that of the water
supply. In the event of a fire, the sprinkler head will
operate releasing the air in the system, allowing it
to become charged with water. An alternative to the
wet and dry system is to charge it with a solution of
anti-freeze fluid. If this is the case it is considered
to be a category 4 backflow risk which would
necessitate the use of a type BA device such as a
RPZ valve. In some instances a sprinkler system
may be supplied by a storage cistern in a similar
way to that shown in Fig. 3.40. The inlet of all
cisterns supplying water to fire-fighting equipment
must be provided with a type AA AB or AD air
gap. A single check valve must also be fitted in the
supply pipe.

All sprinkler systems should comply with the
rules of the Fire Officers Committee (FOC). The
purpose of this body is to control the standards
of fire protection equipment, installation and
practice. Insurance companies responsible for
insuring against fire risks, will insist that the FOC
recommendations and rules are complied with.
It is therefore prudent to contact this body prior
to the installation of any fire protection systems.

Water treatment

The causes of hardness of water and the treatment of hard water by public authorities have been dealt with at some length in Book 1. While it is true that water authorities make considerable efforts to minimise the total hardness of the water they supply, it would be too expensive in those parts of the country where the water is initially very hard to reduce the hardness to the levels required by the consumer.

Conversely, in areas where the water is very soft, water authorities increase the hardness content of the water they supply. This is due to the fact that soft water, possibly slightly acidic in nature, can, when conveyed in lead pipes, take lead into solution which could result in lead poisoning in extreme cases. Although lead pipe for the conveyance of water has been discontinued for many years and its use for this purpose is now forbidden by the water by-laws, many old properties are supplied with water through pipes made of lead. As the renewal of all existing lead pipes would be very expensive, most water authorities in areas where the water is naturally soft, increase its total hardness content, substantially reducing any risk of lead poisoning.

There are many advantages in the use of soft water, greater economy is obtained from the use of soap and detergents for ablutionary and laundering processes. There is a reduction of scale in kettles and culinary utensils and an absence of unpleasant scum which forms in ablutionary appliances due to the reaction between soap and the hardening salts in the water. Largely due to the foregoing, water softeners and scale reducers are being fitted in increasing numbers and the following text deals with the equipment in common use for domestic and small industrial premises.

Base exchange water softening

The principal method of softening water for both industrial and domestic use in most cases is that known as the *base exchange process*. It employs the use of zeolites and sodium chloride (common salt) in the water. The principles of base exchange, to be understood fully, require some knowledge

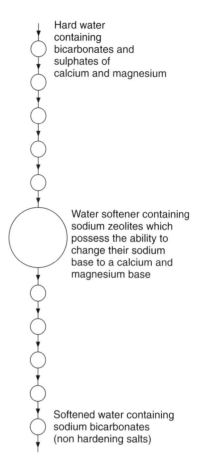

Hard water containing bicarbonates and sulphates of calcium and magnesium

Water softener containing sodium zeolites which possess the ability to change their sodium base to a calcium and magnesium base

Softened water containing sodium bicarbonates (non hardening salts)

Fig. 3.46 Working principles of a water softener.

of chemistry, but it is sufficient to say at this stage that the sodium zeolite base is exchanged for one of calcium as shown in Fig. 3.46.

A domestic simple water softener is shown in Fig. 3.47. Hard water enters at the top, passes through the zeolite bed and emerges through the outlet with the hardening salts completely removed. After a period of time, depending on the capacity of the softener, the zeolite becomes exhausted or so saturated with calcium that it is no longer capable of softening the water. The softener must then be regenerated, first by back washing, i.e. reversal of the flow of water through the softener prior to the addition of salt. This is accomplished by a control system of automatic valves which permit the flow of water through the softener to be reversed, washing out the hardness salts and simultaneously drawing a measured quantity of brine from the brine

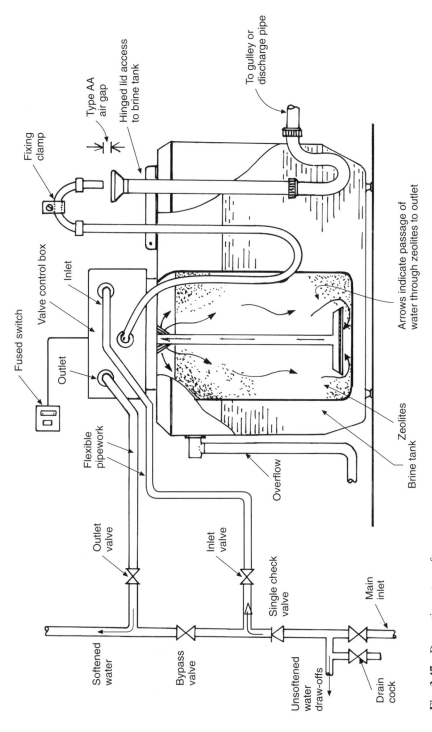

Fig. 3.47   Domestic water softener.

reservoir. This in effect regenerates the zeolites enabling another softening cycle to be commenced. The softener shown here is designed to be plumbed into the cold water supply and is automatic in operation. The control system incorporating an electrical timer, operates the valves which activate the regeneration process including the addition of brine. The time switch should be set so that the regeneration process takes place when no water is being used, usually at night. Time control only is the simplest form of automation, the problem being that it does not take into account the volume of water that has passed through the softener. This means that regeneration takes place whether or not it is necessary and leads to waste of salt. More sophisticated systems of control are based on measuring the volume of water used or monitoring the hardness of the water. The initial cost of softeners incorporating these systems of control is, however, more expensive than for those with simple time control.

Risks of water contamination due to the use of water softeners is not considered to be high and as such a single check valve, usually supplied on the inlet to the softener, is all that is normally necessary. It is important, however, that a tun-dish providing a type AA air gap is provided at the waste connection to prevent any possible contamination from the drain or sewer. The removal of the hardening salts may leave the water soft and acidic. The usual practice is to mix some hard water with the softened water to give a blended supply. At least one drinking water point should be provided directly from the main, as in most cases it is more pleasant to the taste than soft water. To reduce running costs, supplies to standpipes and WCs should bypass the softener.

Scale reducers

These are not water softeners in the true sense of the word, and although the technology relating to their working principles is not new, it is only recently they have been produced for domestic use. It was discovered that passing water containing calcium crystals (hardening salts) through a magnetic field causes a change in the shape of the crystals, an increase in their size and a decrease in their ability to dissolve in water. An increase in the size of the crystals has two beneficial effects. First, they lose their ability to 'coagulate' or join together as easily as those of a smaller size thus preventing the build-up of rock-like scale in the plumbing installation. Secondly, the presence of these larger crystals disrupts the equilibrium between the water, and tends, in very simple terms, to dissolve any existing scale. Unlike the base exchange process of water softening, the calcium remains in the water, but instead of forming scale or fur it is drawn off via the taps or deposited in the form of sludge in the base of boilers or hot storage vessels where it can easily be washed out. Figure 3.48 shows in

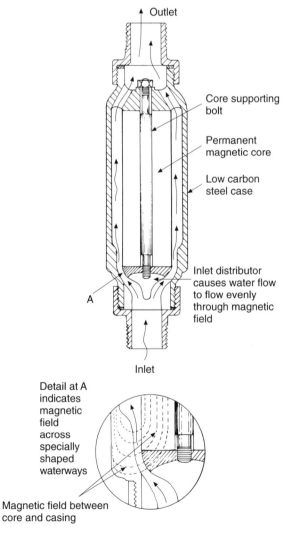

Fig. 3.48 Scale reducer.

diagrammatic form the basic principles of scale reducers, the heart of which is a permanent magnet made mainly of iron with additions of cobalt, nickel, aluminium and copper. The resulting alloy is coated with PTFE which protects it from corrosion. The magnet is housed in a metal case which is designed to increase the turbulence of the water as it flows round the magnet.

Electronic water conditioners
These produce similar physical changes to calcium crystals to those produced by magnetic type scale reducers. In simple terms the calcium loses its solubility and ability to coagulate to form the hard, rock-like fur on the inside of hot water pipes and storage vessels. The principle of electronic conditioners differs from the magnetic type in that they employ magnetic hydrodynamics (MHD), which is effected by passing a form of radio signal through a coil wrapped around the pipe to which it is fitted. The signal is inaudible and sets up a dynamic field around and through the coil, pipe and water. Because the signal field changes at high audio frequences the scale-forming crystals are changed but remain in the water as a fine sediment which is washed away with the flow of water. An illustration of a typical unit of this type is shown in Fig. 3.49, and it will be seen that an electrical supply via a fused switch or socket outlet is

necessary. The control panel contains a step down transformer giving an output, in most cases, of 6 volts. Unlike water softeners these units do not produce immediate results; usually a period of between 3 and 6 months must elapse before any significant change can be seen. This also applies to scale reducers of the static magnetic type.

Chemical methods of scale prevention
As with scale reducers this method of treating water to protect plumbing installations against the build-up of scale is not new, and similar equipment to that shown in Fig. 3.50 may be found on the inlet of gas water heaters fitted in areas of water supply known to have a high hardness content. Because of their relatively low cost, these scale inhibitors as they are called, are sometimes used as an alternative to water softeners and are ideal for use in individual appliances such as drink-vending equipment, washing machines and electric instantaneous water heaters. It must be emphasised that like scale reduction chemical treatment of water is not 'softening' in the true sense. Water to be treated passes through a nonferrous metal dispenser

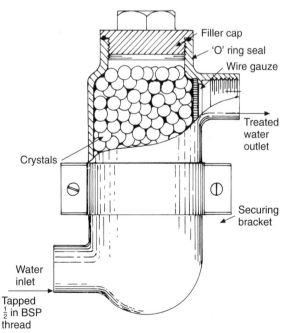

Fig. 3.50 Scale inhibitor.

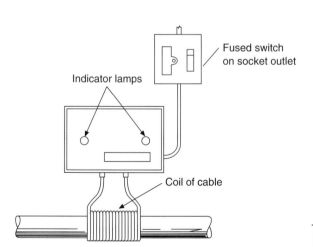

Fig. 3.49 Electronic water conditioner, normally fitted to the incoming main water supply.

containing the chemical crystals which have the effect of 'inhibiting' or suspending in the water the hardness salts responsible for scale formation. This prevents the hard scale building up in pipes and fittings. The working life of the chemicals varies with the volume of water used. In a domestic system it can be anything from 9 to 15 months. Replacement of the crystals is a simple operation and can be carried out by the householder by unscrewing the top of the container and topping up as necessary. It must of course be fitted in a reasonably accessible position. The suppliers of the chemical crystals state there are no harmful effects from their use and as the natural minerals are not removed from the water, its beneficial qualities are not impaired.

Hydraulic pressure testing

It is often necessary on large installations to test sections of the work while the job is progressing,

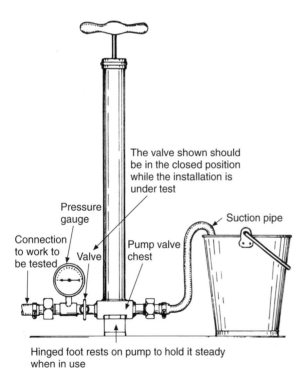

The valve shown should be in the closed position while the installation is under test

Pressure gauge

Connection to work to be tested

Valve

Pump valve chest

Suction pipe

Hinged foot rests on pump to hold it steady when in use

Fig. 3.51 Typical force pump used to pressure test water services.

especially if it is likely to be difficult to gain access to the work on completion. Typical examples occur when pipes are fitted under floors or in wall ducts where they will be covered when the job is finally commissioned. Pressure testing is carried out with the equipment illustrated in Fig. 3.51 on both hot and cold water services. Installations are normally tested to at least one and a half times the normal working pressure, and in cases where access is extremely difficult, twice the working pressure. In all cases the pressure should be maintained for at least 30 minutes.

Testing plastic pipes
The method employed differs from that used when testing rigid pipe systems. Due to its elasticity the diameter of the plastic pipe will expand slightly under pressure. The Water Advisory Service recommend the following two methods of testing plastic pipe:

(a) The test pressure is maintained by periodic pumping over a period of 30 minutes. The pressure is then reduced by a third and no drop in pressure should occur over a period of 90 minutes.
(b) As with (a) the test pressure is maintained for 30 minutes after which pumping stops and the pressure is noted. A pressure drop of less than 0.6 bar is permissible for a period of the next 30 minutes, then a further 0.2 bar during the next 2 hours.

Provided the test meets the specifications of (a) or (b) and there is no visible leakage, it is satisfactory.

Inspection, commissioning and testing
Reference should be made to BS 6700 which details the procedures that must be followed in relation to the water supplies in large buildings. The following text lists the main points to which attention must be given:

(a) The materials, equipment and workmanship must comply to relevant standards and job specifications.
(b) The installation should comply with the current Water Regulations and any other appropriate regulations. This includes the

Health and Safety at Work Act, which applies not only to those installing the system but also to those using it.

(c) The installation must be visually inspected on completion for damage to pipework and protective coatings, and for security of fixings.

(d) Storage vessels must be inspected to ensure they are clean and securely supported. Cistern covers must be correctly fitted.

(e) The system should be slowly filled with water, with the highest draw-off open allowing any air to escape. After inspecting for leaks, the installation should be tested as previously described (see Hydraulic pressure testing p. 144).

(f) Each draw-off, shower unit and float-operated valve, should be checked for delivery of the specified flow rate when the system is permanently connected to the main supply, see Fig. 3.52.

(g) Float-operated valves must be correctly adjusted to achieve the correct water level in cisterns.

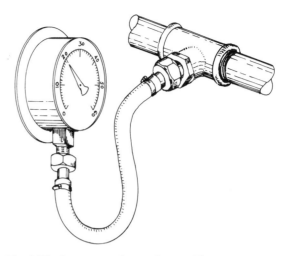

Fig. 3.53 Pressure testing equipment. The water pressure in any pipeline may be tested by using a Bourdon-type gauge as shown or by using a special adaptor which can be connected directly to a draw-off tap. Most pressure gauges give alternative readings in both p.s.i. and bars.

(h) Any backflow devices fitted in the installation must be checked for efficient working.

(i) Any insulation specified should be fitted at this stage, and where necessary colour coded to BS 1710. All pipeline control valves should be labelled showing their purpose.

(j) On completion all relevant information, such as drawings, test results where applicable and manufacturers' installation and maintenance documents, should be given to the building owner for safe keeping.

Pressure testing

It is sometimes necessary to test the pressure on existing supplies, for example, where it is proposed to install an unvented hot water system or to make checks on the pressures of fire services.

Figure 3.53 shows a portable pressure gauge which may be connected to a test point as shown, or connected directly to a draw-off by means of a rubber adaptor.

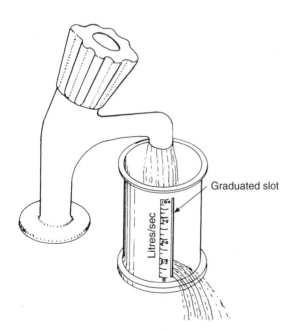

Fig. 3.52 Testing the flow rate at draw-offs. A reading is taken at the highest point of water flow in the slot. In the example a discharge rate of 2.3 litres per second is shown. Gauges are also available with graduations from 2.5 to 20 litres per second.

Decommissioning

When properties are left unoccupied for long periods, all water systems must be valved off at the mains and drained down when not in use. Failure to observe this procedure may result in frost damage in winter, with the possibility of an undetected water leakage over a long period of time. All systems, including hot water and heating, must be clearly labelled 'EMPTY'. Any electrical or gas services related to such systems must also be effectively isolated or capped off.

Further reading

BS 6700:1997 Design, installation, testing and maintenance of services supplying water for domestic use with buildings and their curtilages.
BS 5306:1988:Part 1. Hydrant systems, hose reels and foam inlets.
BS 5306:1990:Part 2. Specifications for sprinkler systems.

The Building Regulations
EN 806 Parts 1–5 Plumbing installations in building (currently in preparation).
Water Regulations Guide, Water Regulations Advisory Service, Fern Close, Pen-y-fan Industrial Estate, Oakdale, Newport, NP11 3EH, Tel. 01495 248454.

Water treatment and conditioning
Hydropath UK Ltd, Unit F, Acorn Park Industrial Estate, Redfield Rd, Nottingham, NG7 2TR Tel. 0115 986 9966.
Scalemaster, Unit 6, Emerald Way, Stone Business Park, Staffs, ST15 OSR, Tel. 01785 811636.
Fast Systems Ltd, Dalton House, Newtown Road, Henley-on-Thames, Oxfordshire, R69 1MG, Tel. 01491 491200.
Tap Works, Mill Rd, Stokenchurch, Bucks, HP14 3TP, Tel. 01494 484000.

Hose Reels
Norsen Ltd, Unit 17, Airport Industrial Estate, Kenton, Newcastle-on-Tyne, NE3 2ET, Tel. 0191 2866167.

Water and Fire Pumps
Pullen Pumps Ltd, 58 Beddington Lane, Croydon, Surrey, CR9 4PT.

Delayed Action Float Valves
H Warner and Son Ltd, Arclion House, Hadleigh Rd, Ipswich, 1P2 OEQ, Tel. 01473 253702.

Water Controls
Reliance Ltd, Worcester Rd, Evesham, Worcestershire, WR11 4RQ, Tel. 01386 47148.

Self-testing questions

1. Define the difference between indirect and direct cold water systems.
2. Describe how stagnation is avoided in cold water supplies.
3. List the causes of noise that can occur in high pressure cold water supply.
4. (a) Explain the term backflow in relation to cold water services and mains.
   (b) List three possible causes of backflow.
5. Describe the difference between a type AUK2 and AUK3 air gap.
6. (a) State the water category for which the RPZ valve is suitable.
   (b) Explain why a tun-dish is necessary on the discharge pipe.
7. (a) Explain why it is essential that potable water does not come into contact with water of category 5.
   (b) State the requirements regarding a supply of water to a bottle washing plant which is a category 5 risk.
8. Describe the procedures and equipment used to pressure test a cold water system where large sections of pipework are to be concealed in floor ducts and suspended ceilings.
9. Describe the special arrangement that must be made regarding overflow and warning pipes in cisterns containing more than 5,000 litres of water.
10. (a) Define the term secondary backflow protection and the type of building in which it would be required.
    (b) Describe the methods and components necessary for secondary backflow protection.
11. Explain the action of zeolites used in the water softening process.

12. State the maximum height of a dry riser and its minimum diameter.
13. Sketch and describe a suitable system of water supply for a hose reel installation where the main supply is of insufficient pressure.
14. State the minimum pressure necessary for the effective operation of hose reels.
15. Evaluate the advantages and disadvantages of the various methods of treating hard water.

16. State the vertical distance from the spillover level of an appliance and the outlet of a $\frac{3}{4}$ inch tap in a domestic property.
17. State the type of backflow device that must be fitted to the outlet of pressure flushing valves.
18. List the procedures which must be carried out when commissioning cold water systems.

4 Hot Water Supply

After completing this chapter the reader should be able to:

1. Identify the main cause of lime scale formation and corrosion in hot water systems
2. Understand the working principles and advantages of unvented hot water systems and the associated operational and safety controls.
3. Recognise and state the working principles of pumped hot and cold water supplies.
4. Calculate the boiler power required to heat a given quantity of water.
5. Understand the principles and limitations of circulating pressure in connection with gravity hot water systems.
6. Describe and sketch the methods of supporting and making connections to cylindrical vessels fitted in a horizontal position.
7. Select the methods of connecting towel rails and space-heating equipment in various circumstances and to different types of systems.
8. Identify the systems, applications, advantages and principles used in relation to gas and electric water heaters.

Corrosion and scale formation in hot water systems

It is assumed that the reader is conversant with the basic principles and design factors relating to small hot water systems which are fully dealt with in Book 1 of this series. These also apply to more complex systems, and if they are fully understood no difficulty will be experienced in the study of more advanced work.

One of the main factors which the plumber has to bear in mind in relation to all water supply work is corrosion, to which is added in the case of hot water supply the problems associated with temporary hardness. Both temporary and permanent hardness are undesirable in water supplies, and water having a high temporary hardness content will cause scale or fur to form on the internal surfaces of the boiler and circulation pipes. The effect will be to lower the efficiency of the boiler and will sometimes result in noise in the circulating pipes.

Temporary hardness in water is produced when water having a high carbon dioxide gas content comes into contact with carbonate rocks. These carbonates are only soluble in water due to the presence of carbon dioxide which enables the water to take them into solution. It is a physical fact that when water is heated to temperatures of approximately 65–70 °C, all traces of gas including the carbon dioxide are given off, and as the water is then no longer able to contain the carbonates in solution, they are deposited in the boiler and circulating pipes as fur or scale. The fur builds up in the pipes and can cause a serious obstruction.

In areas where soft water is encountered, corrosion problems are more common than furring and scaling, due mainly to oxidation and electrolysis.

Oxidation
This is caused by the oxygen contained in the water attacking unprotected ferrous metal surface. In most

hot water systems this will apply to the boiler, which is generally made of cast iron or low-carbon steel. Oxidation brings about the formation of red rust or black magnetic oxide of iron, both of which can be the cause of discoloured water becoming discharged from the hot draw-offs. In combined hot water and heating schemes employing sheet steel radiators for space heating it can also, in conjunction with other forms of corrosion, result in their complete destruction.

Electrolysis
Due to the fact that soft water is capable to varying degrees of dissolving all metals, when such water is conveyed, for example, through copper pipes, it can take into solution a small percentage of this metal. When water containing dissolved copper comes into contact with zinc, which is sometimes present in hot water schemes in the form of a galvanised coating on cisterns and cylinders, the zinc will be destroyed by electrolysis leaving the steel surfaces unprotected against further attack by the water, resulting in rapid corrosion of the steel. Some degree of immunity can be achieved by 'cathodic protection' and it is quite common to fix a sacrificial anode in any hot storage vessel made of galvanised steel. It consists of a block of magnesium or aluminium hung or bolted inside the vessel, and these metals, being lower on the electrochemical scale than zinc, are attacked in preference to the galvanised coating.

The best protection against electrolysis is to use only one metal in hot water schemes, but this is usually economically impossible. For instance, in the case of cylindrical hot storage vessels, while it is common to use copper for the smaller sizes, to make larger ones of this material would be too expensive. This is due to both the quantity and thickness of the copper sheet necessary to cope with the higher internal pressure to which these vessels would be subjected. Cylinders or hot water storage having capacities in excess of 240 litres, are almost invariably made of galvanised steel plate, or in the case of gas heated vessels, vitreous enamelled steel.

In combined systems of hot water supply and heating it is almost impossible to avoid mixtures of metals. For example, in most cases copper pipes are used in conjunction with thin sheet steel radiators in heating installations, and unless some preventive

measures are taken serious corrosion will take place in the radiators.

It has been found that corrosion and scaling problems are more common in direct schemes of hot water than in the indirect type. This is due to the fact that all the water in a direct system, including that in the boiler and circulating pipes, is constantly changed, thus introducing more corrosive or hard water (depending on the locality) into the system. This gives rise to continual corrosion attack on any ferrous metal components it may contain. Further reference is made to this subject in Chapter 5 (page 189).

Supplementary storage systems

When a system of hot water supply is considered in tall buildings with the main storage vessel at low level, it is sometimes necessary to install a supplementary storage vessel at high level, as shown in Fig. 4.1. This is really an enlargement of the pipe and ensures an adequate supply of hot water to draw-offs on the upper floors in the building which might otherwise be starved of water when those at lower levels are in use. One could, of course, increase the pipe size to ensure that an adequate supply of water is available at all draw-off points, but this would increase the heat losses. A little thought will show that the larger the pipe diameter, the greater will be its surface area capable of dissipating heat, which in most cases would be in excess of that presented by the surfaces of the supplementary storage vessel. It is important that this vessel is not too big, causing unnecessary heat loss. The maximum capacity should not be more than one-fifth of the total storage content, i.e. if the total storage is 1,000 litres, not more than 200 litres should be stored in the supplementary storage vessel. It is necessary to ensure that both hot storage vessels and the secondary circulation are well insulated to avoid excessive operating costs.

High-level flow systems

The following relates to the use of solid fuel boilers where a gravity circulation, usually the primary flow and return to the hot water storage vessel, must be provided to avoid overheating. This does not apply

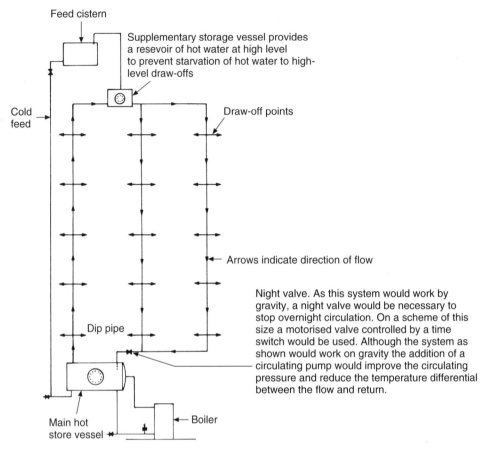

Feed cistern

Supplementary storage vessel provides
a resevoir of hot water at high level
to prevent starvation of hot water to high-
level draw-offs

Cold
feed

Draw-off points

Arrows indicate direction of flow

Night valve. As this system would work by
gravity, a night valve would be necessary to
stop overnight circulation. On a scheme of this
size a motorised valve controlled by a time
switch would be used. Although the system as
shown would work on gravity the addition of a
circulating pump would improve the circulating
pressure and reduce the temperature differential
between the flow and return.

Dip pipe

Main hot
store vessel

Boiler

Fig. 4.1   Supplementary storage system.

to fully automatic gas- or oil-fired boilers where
fully pumped systems can be employed. There
are situations where a door or window interrupts
the most suitable way of running the flow and
return.

A simple but effective method of overcoming
this problem is by using what is called the *high-
level flow system* illustrated in Fig. 4.2. The hot
storage vessel must not be fitted too low or reverse
circulation may take place for the following reason.
If the water temperature in the boiler falls below
that of the hot storage vessel (this could happen if
the fire dies down overnight) a circulation could
take place from the hot storage vessel to the boiler.
This would cause not only unacceptable overnight
heat losses, but due to reversed circulation the
system will be unacceptably noisy when the fire is
relit and normal direction of circulation is resumed.

To ensure an acceptable circulation with this
system, the horizontal runs should be as short as
possible and the $\frac{4}{5}:\frac{1}{5}$ ratio from the centre of the
boiler (shown in Fig. 4.2) should be adhered to.
Failure to do so may result in very sluggish
circulation.

It is always necessary to provide what is called
a *heat leak* with solid fuel boilers. This means that
any heat produced by the mass of fuel in the boiler,
even when the thermostat is closed, must be
accommodated by a gravity circulation. Failure
to do this will result in overheating and possible
boiling in the system.

Dual boiler systems

Some systems employ two boilers, a typical
example being where a traditional range is

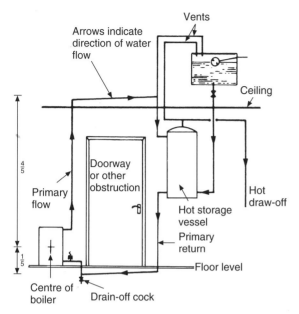

Fig. 4.2 High-level flow system used to overcome obstacles which would preclude a normal flow and return pipe run. Providing the radio of $\frac{4}{5}$ above, $\frac{1}{5}$ below, the centre line of the boiler is observed and the horizontal runs are not excessive, an acceptable circulation can be expected.

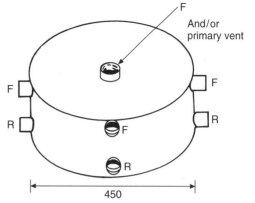

(a) Designed to fit under the HWSV

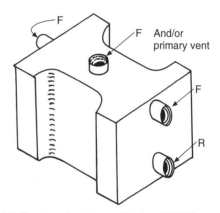

(b) Can be sited alongside the HWSV

F — flow connections
R — return connections

Fig. 4.3 Neutralisers for dual boiler installations.

employed for cooking and possibly domestic hot water supply, and is interconnected with a boiler for space heating. Unless the system is very carefully installed, water may circulate through the boiler not in use, and the updraught through the flue will rapidly dissipate any heat into the open air — not a very economic proposition. Although the output from range boilers is usually very low and their recovery rates are slow, they can be used to augment a space heating appliance. This may be a wood or solid fuel appliance if a plentiful supply of such fuel is available, but many customers like them for their aesthetic appeal. The type of fuel used with dual boiler schemes is not important, except in the case of solid fuel where a suitable heat leak must be provided. The main problem encountered with this type of installation occurs when only one boiler is operating. This system employs a component called a circuit neutraliser (see Fig. 4.3) which, in effect, provides an area in which there are no positive or negative pressures. It acts in a similar way to a boiler (except those with a low water content) in a combined hot water and heating system, where the

boiler is the neutral point. The neutraliser serves the same purpose and is fitted where the circulating pipes from the two boilers interconnect. It may be fitted in the cupboard housing the hot store vessel if it is of sufficient size, but it can be installed at any convenient point, providing the base is at least 300 mm above the highest boiler. Figure 4.4 illustrates a typical layout using in this example two solid fuel boilers.

In order to preserve the 'heat leak' when using solid fuel boilers, it is generally an accepted practice that no electrical controls are fitted in such a way that the heat leak is closed. The one possible exception is hopper-fed boilers, which usually incorporate forced draught. These are normally thermostatically controlled, and it is possible using

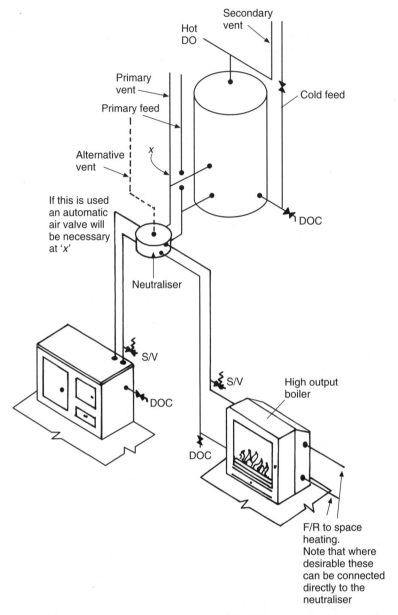

Fig. 4.4   Interconnecting two boilers. This system can be adapted for fully pumped schemes with gas or oil boilers.

the system shown in Fig. 4.5 to exercise a degree of control over the stored water. Although the original concept of the neutraliser was to intercouple solid fuel appliances, it has been developed for fully automatic gas and oil boilers and can be usefully employed in any situation where two or more boilers are interconnected. It is recommended that the manufacturer of the components used should be

contacted for more detailed information if a system of this type is contemplated.

Unvented hot water systems

Traditionally, hot water storage systems in the United Kingdom have been of the low-pressure type, water being supplied to the storage vessel

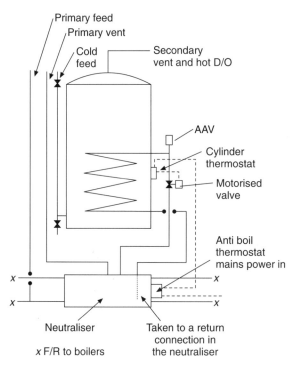

Primary feed
Primary vent
Cold feed
Secondary vent and hot D/O
AAV
Cylinder thermostat
Motorised valve
Anti boil thermostat
mains power in
x
x
Neutraliser
x F/R to boilers
Taken to a return connection in the neutraliser

Fig. 4.5 Electrical control for dual boiler schemes. This form of control depends on the effectiveness of the boiler thermostats, but it is designed to provide a heat leak in the event of a pump or power failure.

from a feed cistern usually situated in the roof space. Until the revision of the Model Water By-laws 1986 all previous legislation by the water authorities precluded mains feed storage systems having storage capacities of more than 15 litres; their main objection being:

(a) Greater possibility of contamination.
(b) The fact that in some areas the distribution system might not be able to meet the demand of all draw-offs, both hot and cold being taken from the main.
(c) The possible dangers due to explosion in such systems.

Most of the foregoing problems have been resolved, although the installer of unvented systems must be sure that the mains supply is of sufficient pressure and volume to meet the draw-off demands, as an unvented hot water system does not possess any magical qualities for improving a poor water supply. Most readers will be aware that for a long

period of time mains pressure hot water equipment has been available mainly in the form of single- or multi-point gas water heaters. While in the right circumstances such heaters are perfectly satisfactory, they suffer from the disadvantage of having a low flow rate in comparison with that of storage systems.

Unvented hot water systems have many advantages and some disadvantages and these should be considered very carefully before a decision is made on which type of system is best for a specific installation. It is a fact that the elimination of a traditional feed cistern and the pipework necessary for its installation saves both material and on-site labour costs, as most unvented systems and their necessary controls are supplied as a packaged unit by the manufacturers. Such systems eliminate the need for water storage and its associated pipework in the roof space. This is a very important advantage. Due to the requirements for ventilation in roof spaces, they have become very cold areas indeed. It should not be forgotten, however, that if the premises are heated by a traditional system having a feed and expansion cistern, this will normally remain in the roof space. Unvented hot water systems give greater flexibility to the design of taps, mixers and shower-heads, some of the latter being dependent for their satisfactory operation on higher pressure than those normally associated with traditional systems. Many modern tap designs incorporate mixing devices and due to the requirements of the water by-laws prior to 1986 it was not permissible to connect such mixers to the main supply, bi-flow types being the only exception. Due to the more stringent requirements to prevent backflow and water wastage, non-bi-flow mixers are now acceptable on mains-fed supplies and this should result in lowering costs for the installation of mixers for all applications. Savings on pipework and fittings may be made as due to the higher pressures involved smaller pipes can be used.

The disadvantages for unvented storage systems are few but somewhat formidable. The storage vessel, which is generally a packaged unit (this is to say all the necessary controls are already fitted by the manufacturer), tends to be expensive in comparison with the components of a traditional vented system. Such a system if properly installed, normally lasts for a long period of time, often the lifetime of the installation, requiring little or no maintenance. Unvented systems rely mainly on automatic controls which require periodic maintenance and if they do become defective they must be replaced. **No attempt should be made to repair them** as they must be regulated and checked under factory conditions. Generally speaking they are most likely to cost more to maintain. Any savings made by the installation of unvented systems is offset by the cost of stronger storage vessels and ancillary controls that must be provided. Another important consideration is the lack of water storage and should the mains be shut down for any reason, no water will be available until the supply is restored. Such circumstances are fortunately rare, but to be in a situation where no water is available at all is, to say the least, very inconvenient. Some authorities have suggested that the possibility of

water contamination is greater with the use of unvented systems as there are more draw-off points liable to backflow. It must be pointed out, however, that in new premises with backflow protection devices conforming to the 1996 Water Regulations this danger should be minimal. The greatest danger exists where unvented systems are installed in existing properties in which backflow prevention devices are not fitted. In such cases the installer must satisfy himself that no danger from contamination exists before the installation is completed. Most existing systems would, for example, have to be fitted with check valves on such fittings as bath and sink mixers in order to meet the requirements of the Water Regulations.

System layout

Figure 4.6 shows the main components required for an unvented hot water system. In practice most of them are fitted to the storage vessel by the

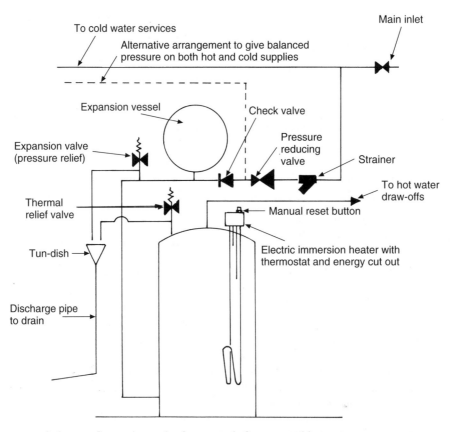

Fig. 4.6 Diagrammatic layout of operating and safety controls for unvented hot water storage systems.

manufacturer, all the plumber has to do on site is to make the necessary connections. Do make sure when installing the storage vessel that sufficient space is available to carry out maintenance work. These systems should be serviced annually and it is a messy and complicated job if the unit has to be completely removed for this purpose.

The water may be heated by an electric immersion water heater or the storage vessel may be of the indirect type, the secondary water being heated by a boiler in the traditional way. Whatever method of heating is used the equipment must be provided with a safety cut-out device which isolates the heat source in the event of failure of the thermostat.

Operating components

Strainers   These are essential to prevent any debris such as silt or shrimps passing into the system possibly causing problems with the valves downstream. A suitable strainer is shown in Fig. 4.7. When the system is serviced the strainer should be inspected and washed in clean water.

Pressure reducing valves   The working principles of these valves has been described in the previous chapter. They are fitted to reduce the inlet pressure to the working pressure of the equipment used and to maintain a constant flow rate to the draw-off points. It is currently recommended that the pressure on these systems is limited to 2 or 4 bar, depending on the type of material, i.e. steel or copper from which the storage vessel is made. Only steel vessels

should be used for the higher pressures. The working pressure of the system is usually two-thirds of the test pressure of the storage vessel, and the pressure relief, or expansion valve (described later), is usually set at this pressure. The outlet pressure of the reducing valve must therefore be closely related to the pressure at which the expansion valve will open. All the valves affected by pressure, when obtained with a packaged unit are factory set, thus ensuring a close relationship with each other.

Check valve   These are single check only and are provided to prevent backflow of hot water into the cold water services, but they also prevent 'implosion', a term used to describe what the plumber understands as cylinder collapse. A study of the system will show that it is possible for the water in the storage vessel to be siphoned out by opening a cold water tap if the main stop valve is shut down.

Anti-vacuum valve   As a further safeguard against implosion an anti-vacuum valve is provided to admit air to the storage vessel should the pressure inside fall below that of the atmosphere. Figure 4.8 illustrates a typical valve of this type, although in some cases they are made as an integral part of the thermal relief valve.

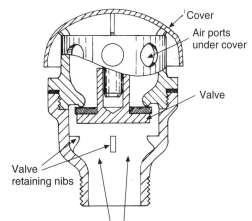

Pressure of water under normal working conditions holds valve in the closed position as shown. If a sub-atmospheric pressure occurs in the storage vessel the valve drops allowing air to enter through the ports to maintain normal atmospheric pressure

Fig. 4.8   Anti-vacuum valve.

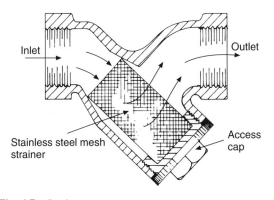

Fig. 4.7   Strainer.

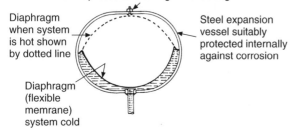

Gas charging point. The gas charge static pressure must be equal to the system working pressure which will be the same as the pressure-reducing valve setting

Diaphragm when system is hot shown by dotted line

Steel expansion vessel suitably protected internally against corrosion

Diaphragm (flexible memrane) system cold

Fig. 4.9   Section through expansion vessel. See also Fig. 5.22.

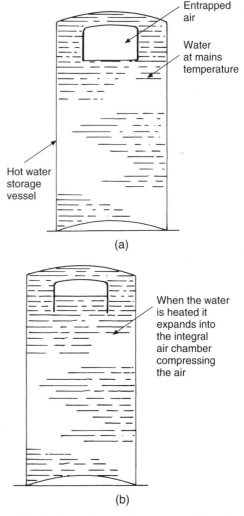

Entrapped air

Water at mains temperature

Hot water storage vessel

(a)

When the water is heated it expands into the integral air chamber compressing the air

(b)

Fig. 4.10   Illustrates the principle of unvented hot water storage vessels having integral expansion chambers (water connections and working components are not shown).

Expansion vessel   This is designed to accommodate the expansion of the water in the system when it is heated and to prevent any operation of the expansion valve which could lead to wastage of water. It is important that it is correctly sized so that it can absorb the required volume of expansion. Figure 4.9 shows a section through a typical expansion vessel and it will be seen that as the water in the system expands it causes the diaphragm to distend, compressing the gas, usually nitrogen or air. If due to a fracture in the rubber diaphragm the gas escapes, the vessel will become waterlogged and because liquids are for practical purposes incompressible, expansion of the water when it is heated will cause the expansion valve to open.

An alternative to the expansion vessel, usually fitted to unvented cylinders, is the internal arrangement shown in Fig. 4.10. Air is trapped in the expansion vessel when the system is filled, and being a gas, is easily compressed by the expansion of the heated water. There are two ways in which the air can be lost: turbulence of the water when the draw-offs are opened, and the fact that water under pressure absorbs air. If this occurs it will cause persistent operation of the expansion valve and is thus easily detected. The remedy is to drain the system so that the air can be replenished.

Expansion valve   As previously explained this valve is used to relieve the pressure in the system due to the expansion of the water if for any reason it is not accommodated in the expansion vessel.

Figure 4.11 illustrates its working features. As it is fitted to the cold inlet side of unvented hot water systems, it is not considered as a safety valve and should not be referred to as such. Any discharge through these relief valves must be passed through a tun-dish to provide a type AUK 3 air gap, thus avoiding contamination via the drain. A typical tun-dish is shown in Fig. 4.12.

Safety controls   Possibly the greatest disadvantage of unvented hot water systems is the possibility of

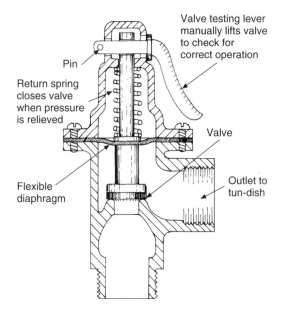

Fig. 4.11 Expansion valve (pressure relief valve). This valve is designed to protect the storage vessel from bursting. It is designed to open only when provision for expansion has failed or the pressure-reducing valve is malfunctioning and the system is subjected to excessive pressure.

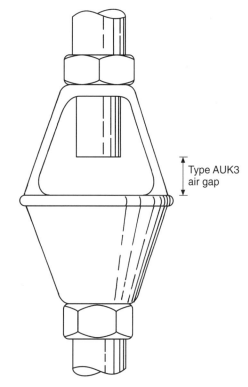

Fig. 4.12 Detail of tun-dish.

serious explosion. It must be pointed out, however, that such a catastrophe is unlikely due to the safety systems built into the installation. The real danger lies not with the system but with the possibility of untrained people interfering with the factory-set controls or incorrectly installing a system. Unfortunately the legislation governing installation of these systems is very difficult to enforce except in new properties which are normally inspected by a building control officer and a water board official. The reader should be aware that at atmospheric pressure water boils at a temperature of 100 °C and is converted to steam. If, however, the water is in a sealed container its temperature can be increased without its conversion to steam. Should the storage vessel burst, the water, if at a temperature in excess of 100 °C when escaping into the atmosphere, will immediately be converted to steam, which because it occupies a much greater space than water will cause an explosion. It is perhaps worth noting that water in a sealed container at 100 °C, if released to the atmosphere would occupy 1,600 times its

volume when it changes to steam. The safety devices built into unvented hot water systems includes three lines of defence against such a disaster, all of which are designed to prevent the water achieving temperatures of 100 °C. These safety controls are designed to act in sequence as the temperature rises and are listed as follows:

Thermostats These are common to all types of heating equipment, their object being to permit varying operating temperatures of appliances. Domestic hot water temperatures are normally between 60 and 65 °C, as higher temperatures may cause scalding, increase of heat loss of the stored water and the formation of lime scale. The use of both invar rod and fluid-expansion type thermostats are suitable with these systems, the working principles of which are described in Chapter 2.

Temperature-operated cut-outs These usually take the form of a second thermostat, factory set at approximately 85 °C, and should the appliance

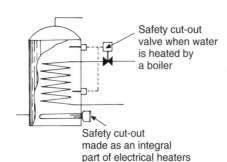

Safety cut-out valve when water is heated by a boiler

Safety cut-out made as an integral part of electrical heaters used with these systems

Fig. 4.13 Safety controls for unvented systems.

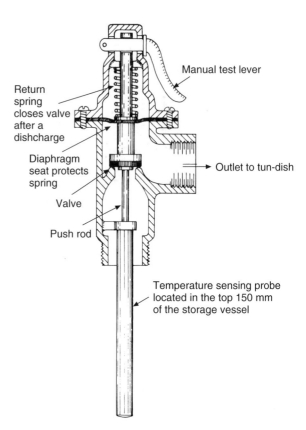

Manual test lever

Return spring closes valve after a dishcharge

Diaphragm seat protects spring

Valve

Push rod

Outlet to tun-dish

Temperature sensing probe located in the top 150 mm of the storage vessel

Fig. 4.14 Combined temperature and pressure relief valve. This valve will open when (a) the design pressure in the system is exceeded and (b) when the temperature exceeds 90–95 °C due to the failure of the thermostat and safety cut-out. The type shown doubles as an expansion relief valve, which saves the cost of using two valves.

thermostat fail, will limit the water to this temperature and automatically cut off the source of heat. All cut-outs of this type are fitted with a reset button which must be operated manually to restore the source of heat. If manual operation of the cut-out is continual the consumer will be made aware that all is not well with the thermostat. Where the water is heated electrically the cut-out is usually located near or in the heater cap. When a boiler is employed a thermostatically operated motorised valve may be used as shown in Fig. 4.13. Both the thermostat and the cut-out sensors must be located at a point in the storage vessel where they are in contact with the hottest water and both should comply with the requirements of BS 3955.

Temperature relief valves These must comply with BS 6283:Parts 2 and 3 and only valves conforming to these specifications must be used. Figure 4.14 shows the operating principles of this component. It has a spring-loaded valve which is opened at a temperature of 90–95 °C due to the expansion of temperature-sensitive fluid acting on a push rod which moves the valve upward causing it to open. The temperature relief valve is designed to operate only after the failure of both the thermostat and the temperature-operated cut-out. The type shown illustrates only the basic principles of these valves. They are obtainable having built-in vacuum relief valves and a pressure relief device that opens if the discharge pipework is obstructed. Some valves of this type also incorporate expansion relief which saves the cost of two valves. As with expansion valves a discharge pipe is necessary and must

conform to current legislation. Both expansion relief and temperature relief valves must be capable of automatic closure and be watertight when closed. This should be checked periodically by operating the levers attached to these valves which will ensure that they function correctly. Figure 4.15 illustrates pictorially a typical unvented hot storage vessel showing the relative portions of all the necessary components.

Discharge pipes (see Fig. 4.16)
In the event of a pressure relief or expansion valve opening, water at possibly 100 °C will be discharged. The safety requirements relating to its disposal are quite specific and are listed as follows:

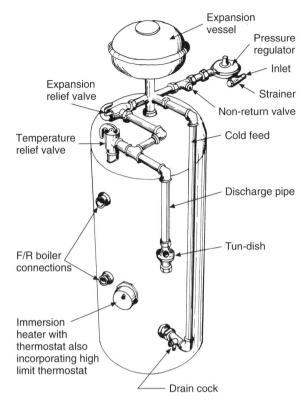

Fig. 4.15 Unvented hot water storage vessel showing arrangements of controls and safety components.

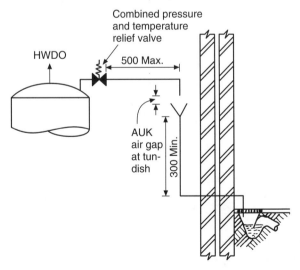

Fig. 4.16 Discharge pipe details (unvented hot water systems).

(a) The pipe must be made of metal with metal fixings.

(b) It should not be longer than 9 m with no more than 3 easy-radius bends, and should have a continuous fall to the point of termination. Sharp bends should be avoided, but if this is not possible the necessary allowance for resistance must be made (see Table x.x, p. xx). Its diameter must be enlarged if it is over 9 m long.

(c) Its diameter must not be less than that of the valve outlet, and one size larger than the tun-dish.

(d) Pipes fitted externally of the building require frost protection (a slight let-by may result in freezing and consequent blockage by ice). The discharge point must be visible, but not so that it could cause scalding. The distance between the point of outlet and the gulley grating should not exceed 100 mm, and if in reach of very young children, it should be guarded with a suitable mesh cover.

(e) Any pipe termination points at high level may discharge onto a roof (providing the covering will not be damaged by high temperature) or into a metal hopper and pipe. The termination point must always be visible.

Legislation relating to unvented hot water systems
The Water Regulations specify the requirements of expansion and temperature relief valves in the context of prevention of water waste. The Building Regulations specify the requirements of these systems in terms of safety and are found in the approved document G3. This states that the components should be supplied in the form of a unit or package conforming to the British Board of Agrément.

A unit is defined as an appliance to which both the operating and safety components are fitted by the manufacturer, a package being supplied only with the safety devices fitted, the operating components being supplied separately and fitted by the installer. In both cases this ensures that the safety devices are 'factory set' and should not be tampered with. The approved document G3 also requires that the installation of these systems is carried out only by an installer who has undertaken an approved course of training. The requirements for the discharge pipe on both expansion and temperature relief valves and also specified in G3 and have been dealt with in the previous text.

Methods of heating for unvented hot water systems
The diagrammatic system shown in Fig. 4.15 is heated by means of an electric immersion heater, but a boiler can be connected in the same way as a traditional vented system providing it is fitted with a suitable thermal cut-off arrangement should the thermostat fail. If the primary system is of the open vented type having a feed and expansion cistern the advantage of avoiding pipes and cisterns in the roof space will be lost, and serious consideration should be given to the installation of an unvented heating system which is described in the following chapter.

Combination boilers

Combination boilers, commonly called combis, incorporate what is, in effect, a central heating low-water-content boiler, and a secondary heat exchanger which indirectly heats a mains water supply for domestic hot water. These boilers combine the basic principles of both gas instantaneous heaters, and generally speaking, sealed heating systems, although some types may be fitted with an open vented heating system. Because they are relatively small and light in weight, most of them are designed to be wall hung. Manufacturers of combination boilers provide flueing systems which may be natural draught open flues, or balanced flues with or without fanned extractors. When fans are provided they are built into the boiler unit in such a way that the burner will not fire until it is operating.

Combination boilers have all the advantages of both sealed heating and unvented hot water systems in that no storage vessels are necessary, and as the hot water supply is at mains pressure they are a positive advantage for showering. The electrical control system is incorporated in the boiler unit itself, thus external wiring is normally limited to the requirements of the room thermostat and

time switch, if externally fitted. The one main disadvantage with most combination boilers is their flow rate in comparison with storage systems of hot water. This is not quite so true as was once the case, as due to improvements of design, manufacturers have been able to increase the flow rates without reducing the water temperature. One of the larger types of these boilers produced by Worcester has a flow rate of 14.5 litres per minute at a temperature rise of approximately 35 °C, which compares reasonably with the requirements of BS 6700 which specifies 18 litres per minute for a $\frac{3}{4}$ in bath tap. Obviously if another tap is open on the system the flow rate will diminish, but in practice it has been found that combination boilers of the larger type normally satisfies the demands of an average domestic property having only one bathroom.

While these boilers differ in some ways, depending on the manufacturer, the working principles are very similar. Figure 4.17 illustrates the basic working principles of combination boilers which are based on the Worcester 350 boiler. Not all combination boilers employ two pumps as shown, some employ one and a three-way diverter valve in a similar way to a fully pumped heating system. It should be noted that no internal wiring or basic gas controls are shown in the diagrammatic illustration. It will be seen that if the reader has some knowledge of the basic principles of both instantaneous gas water heating and sealed systems, the working principles of combination boilers should not be difficult to understand. The components of these systems have been brought together and housed in one unit. The controls, however, are electrically operated and all boilers of this type are fully automatic. The same requirements for filling the primary part of the system, the safety devices employed, e.g. pressure relief, and determining the working pressure, are the same as for sealed systems. In the case of combination units the instantaneous heater forms part of the primary section, water drawn off from the hot tap being passed through a water-to-water heat exchanger, similar to a small indirect cylinder. Unlike an indirect cylinder, however, it is the domestic hot water that passes through the coil, not the primary water as is usual, details of which are shown in

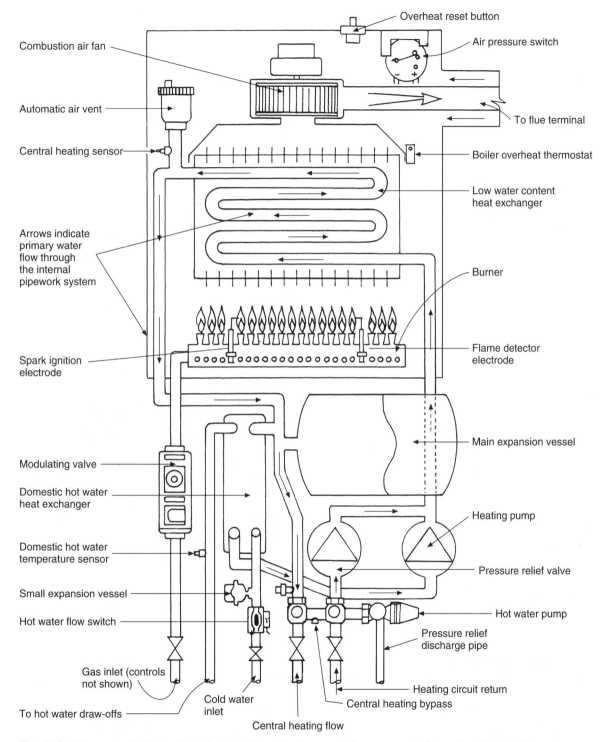

Fig. 4.17   Diagrammatic illustration of a typical combination gas boiler, room-sealed type based on the Worcester 350 model.

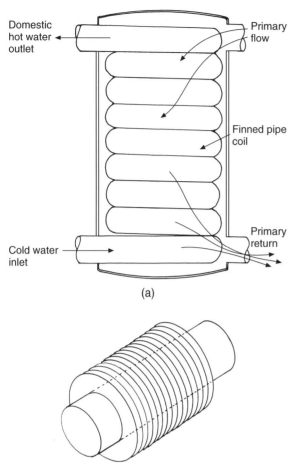

Domestic hot water outlet

Primary flow

Finned pipe coil

Cold water inlet

Primary return

(a)

(b) Section of finned flexible pipe used to conduct the maximum quantity of heat from the primary water to that of the secondary serving the hot draw-off

Fig. 4.18 Water to water heat exchanger as used in combination boilers to heat domestic hot water.

Fig. 4.18. The small air vessel on the cold water inlet accommodates any pressure fluctuation and expansion that may occur in the hot water heat exchanger. It should be noted that the domestic hot water requirements normally takes precedence over the space heating circuit.

When a hot tap is opened, the static pressure on the hot draw-off is lowered and the flow of water (working pressure) causes the flow switch to close, sending a signal to open the domestic hot water circuit, simultaneously starting the hot water pump, and the sensor reacts to the cold water temperature. During this period the heating pump will stop and there will be no flow to the heating circuit, so that all primary hot water produced by the burner is used to heat the domestic water. When the demand for hot water is satisfied the burner will shut down unless a call for heat is made by the space heating controls. It will be seen that the boiler components (e.g. pumps and valves) are activated by sensors which react to either changes of temperature or pressure and operate a system of microswitches. The central heating sensor situated at the top of the main gas to water heat exchanger shown in Fig. 4.17, senses the temperature of the water in the primary circuit and controls the modulating valve to achieve the desired temperature. The low-pressure sensor will detect a loss of pressure in the primary side of the system which will be indicated by a red warning light in the control panel. The modulating gas valve is electrically controlled and automatically matches the burner output to meet the demand. It is best described as a multifunctional gas valve which is adapted in such a way that the governor is actuated not only by the volume of gas it passes but also by means of a solenoid. To give an example, if the heating circuit is (operating) firing and a hot tap is opened, the burner will increase to a larger flame, then modulates down to suit the flow rate at a sensor reading of 55 °C. It also ensures the temperature of the domestic hot water does not exceed 70 °C via the domestic overheat thermostat, which is wired in series with the domestic hot water sensor.

The actual installation of combination boilers should present no difficulties to a competent plumber. Having ensured the unit is correctly fixed on a surface which is suitable for its support and installation of the flue, it is simply a matter of connecting the water, heating and gas services. The manufacturer's instructions, and both gas and electrical regulations must be complied with.

Servicing, maintenance and fault finding on combination boilers is slightly more complicated than most other appliances, and for this reason it is recommended that any repairs and maintenance be carried out by someone experienced in this work. Most manufacturers run special courses relating to their equipment at a very low cost for those who wish to specialise in this area of work.

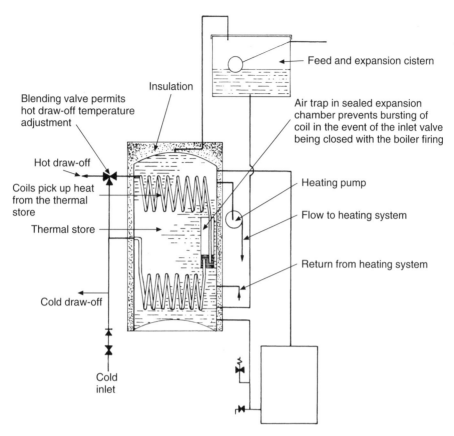

Fig. 4.19   Thermal store system. These systems have the advantage of mains flow hot water without the use of the high pressures that are used in unvented systems. The type shown has an independent boiler and feed cistern. Other models, incorporating both a boiler and feed cistern as a unit, are available for small domestic properties.

Although combination boilers are designed to limit operating temperatures to a maximum of 65–70 °C, most manufacturers recommend fitting a scale reducer where the water has a high temporary hardness content. If a water softener is already installed this, of course, will not be necessary. To avoid damage to the pump a bypass circuit must be provided in a similar way as that shown in Chapter 5, Fig. 5.14. Some combination boilers incorporate this bypass circuit integrally, others do not, so always check the manufacturer's instructions on this point.

Thermal storage systems   Figure 4.19 illustrates a typical system of this type and it will be seen that it is designed like the unvented hot water systems described previously to enable hot water to be delivered direct from the mains. It works in a similar way as an instantaneous heater, but in this case the carefully designed heat exchanger operates on a water-to-water basis, being completely immersed in the 'thermal store'. As can be seen from the illustration a well-insulated body of water is maintained at a temperature of 80–85 °C by the boiler. The volume of domestic hot water delivered at usable temperatures is an improvement on most combination units, and generally for a temperature rise of 45 °C with an inlet pressure of 2 bar from the mains, they will deliver 12–24 litres per minute, depending on the type used. If the mains pressure exceeds 5 bar a pressure-reducing valve is recommended, as like combination units, if the water pressure is excessive, water will pass through the heat exchanger too quickly and will not pick up sufficient heat for domestic purposes. It should also be noted that the water in the top part of the thermal store must be maintained at 80–85 °C to give the quoted temperature rise for domestic hot water. As the stored water is only subject to pressure from the feed and expansion cistern there is no danger of explosion so the control system is very simple. As with combi units they can therefore be installed by a plumber who is not BBA approved. This does not imply, of course, that they can be fitted by any odd-job man who is unlikely to appreciate the implications of Water Regulations and Building Regulations. Many of the smaller units are supplied as a complete unit including the boiler, others where higher outputs are required can be fitted with a boiler in the usual way. As with unvented systems pipework and cisterns in the roof space can be eliminated, as the feed and expansion vessel supplying the primary water to the radiators, thermal store and boiler is in most cases made as an integral part of a unit. Space heating can be controlled by a room thermostat although better and more economic control would be achieved by using thermostatic radiator valves.

Pumped systems

For many years equipment has been available to 'boost' or pump shower fittings in situations where the static head in the cistern supplying both hot and cold supplies is unable to provide sufficient pressure. This arrangement can be extended to enable all the cistern-fed supplies to be delivered at higher pressures. There are many permutations with this arrangement, dependent upon the type of system employed. Figure 4.20 shows two arrangements:

(a) When the cold water supply is of the direct type where all cold water draw-offs are connected to the main service.
(b) The indirect type where all but one drinking-water tap is supplied from a cistern.

Suitable pumps are available to enable the owners of existing premises to boost the supply of both hot and cold services to permit the use of many of the special mixer fittings and shower heads now available. Unlike combi and thermal storage appliances which are limited to the domestic market, boosted systems can be used in all types of

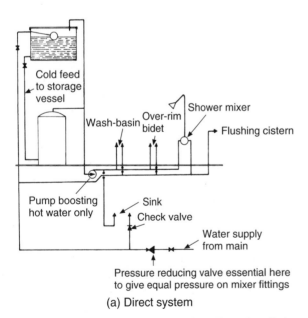

(a) Direct system

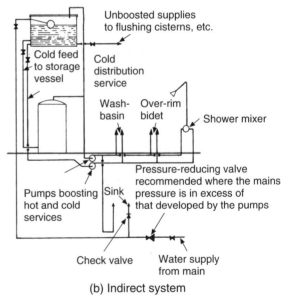

(b) Indirect system

Fig. 4.20 Pumped supplies for multi-appliance installations.

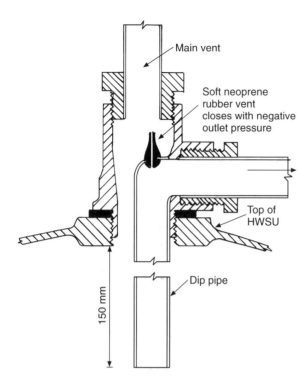

Main vent

Soft neoprene rubber vent closes with negative outlet pressure

Top of HWSU

Dip pipe

150 mm

Fig. 4.21 Surrey flange. Prevents ingress of air via the vent pipe into pumped hot water draw-offs.

buildings. Like other equipment currently developed for modern plumbing systems, the pump, storage vessels and necessary controls may be obtained as a complete unit or, alternatively, suitable pumps are available for fitting into existing systems.

Surrey vented flange
It is sometimes found that even if the pump duty has been selected carefully, air may be drawn down the vent pipe giving an unsatisfactory supply of hot water. This may be because the cold feed is of inadequate size, or more likely the pressure exerted by the feed cistern is insufficient. The problem may be overcome by making an additional connection to the HWSU or by fitting the surrey flange shown in Fig. 4.21.

Storage vessels

Capacities of storage vessels
In industrial or commercial buildings the capacity of hot storage vessels is based on several factors which

include the number of persons to be accommodated, the type of building, the incidence of usage and any peak demand the storage vessel has to meet. In small domestic properties the storage capacity is based on the number of bedrooms the building contains as this is generally an indication of the number of people in occupation. Table x.x in Book 1 lists the capacity of hot storage vessels for small dwellings heated by a boiler. If electricity is used to heat the water and full advantage is to be taken of cheap night rates, larger storage vessels than those listed are recommended. It should be noted that the capacities are for copper direct cylinders only; those for indirect cylinders will be slightly less due to the space occupied by the coil or annulus. The boiler power, unless otherwise specified, should be capable of raising the temperature of the stored water through 50 °C in 2 hours and 30 minutes.

The following illustrates a typical example. To determine the boiler power required (in kilowatts) to heat 120 litres of water through 50 °C in 1 hour the following formula should be used:

$$\text{No. of kilowatts} = \frac{\begin{array}{l}\text{Quantity of water in litres}\\ \times \text{Temperature rise in °C}\\ \times \text{Specific heat of water}\end{array}}{\text{No. of seconds in 1 hour}}$$

$$\therefore \text{kW} = \frac{120 \times 50 \times 4.2}{3,600}$$

$$= \frac{12 \times 5 \times 4.2}{36}$$

$$= \frac{5 \times 4.2}{3}$$

$$= \frac{21}{3} = 7$$

To heat 120 litres in 1 hour requires 7 kW, but a firing period of 2.5 hours is allowed.

$$\frac{7}{2.5} = 2.8 \text{ kW (in round figures 3 kW)}$$

In fact, 2.8 kW would be required to heat the water in the time given.

The recovery chart in Table 4.1 is not comprehensive, but it may be useful to give a rough guide for sizing smaller heaters or boilers.

Table 4.1 Recovery chart: approximate time in minutes to heat water.

Loading (kW)	Litres heated through 50 °C															
	5	8	10	15	30	60	80	100	150	200	250	300	400	600	800	1000
1.0	18	28	35	53	105	210	280									
2.0	9	14	18	26	53	105	140	175	263							
3.0	6	9	12	17	35	70	92	115	173	230	288					
4.0	5	7	9	14	27	54	72	90	135	180	225	270				
6.0	3	5	6	9	18	36	48	60	90	120	150	180	240	360	480	600
8.0	2	4	5	7	14	27	36	45	68	90	113	135	180	270	360	450
9.0	2	3	4	6	12	24	32	40	60	80	100	120	160	240	320	400
12.0	2	3	3	5	9	18	24	30	45	60	75	90	120	180	240	300
15.0	1	2	3	4	8	15	20	25	38	50	63	75	100	150	200	250
18.0	1	2	2	3	6	12	16	20	30	40	50	60	80	120	160	200
24.0	1	2	2	2	5	9	12	15	23	30	38	45	60	90	120	150
36.0	1	1	1	2	3	6	8	10	15	20	25	30	40	60	80	100

Table 4.2 Approximate volume and temperature of water required for domestic appliances.

Appliance	Approximate quantity of water in litres	Suitable temperature (°C)
Bath	114	43
Washbasin	4.5–6.0	43
Sink	4.5–9.0	60
Shower	4.5–9.0	40–43

To use the chart as a check to the foregoing calculation, draw a horizontal line beneath the 3 kW loading. Then look at the numbers of litres of water being considered on the top row. Unfortunately, 120 litres is not shown so look at the two columns 100 and 150; where your horizontal line from 3 kW cuts beneath these columns, observe the numbers given, i.e. 115 and 173. The average of these two numbers is 144 which is the number of minutes taken to heat the water through 50 °C. Our example of 120 litres would fall approximately at 144 minutes, or 2 hours 24 minutes, which is roughly equal to the time calculated.

Hot water requirements for domestic appliances
Table 4.2 will enable the reader to recognise the approximate volume and temperature of water required for domestic appliances.

Stratification
This relates to the temperature differences of the water at the top and bottom of a hot storage vessel. The term *stratification* is used because the water forms layers or strata of different temperatures, the hottest at the top and the coolest at the bottom. When a storage vessel has cooled overnight, the layers are more distinct and can be felt by running a hand down the side of the vessel before any water is drawn off. This is a physical fact and it is significant in that if the temperature of the stored water were the same throughout the storage vessel, no circulation would take place. Quite apart from this, every time a hot draw-off tap is opened the water that is drawn off is replaced by cold water from the feed cistern which has the effect of reducing the water temperature in the lower part of the storage vessel.

The effects of stratification on the temperature of the hot draw-off are not so noticeable with vertically fitted cylinders as those lying in a horizontal position. For this reason, where possible, cylindrical vessels should be fitted in an upright position, although this is not always practicable. A rough estimate of the average temperature of water stored may be made by adding the temperature of water at the top of the vessel to that at the bottom and dividing the result by two. A more accurate method would be to find the average of a series of temperature readings taken over the height of the vessel.

Support and connections to horizontal hot storage vessels

When very large hot storage cylinders are necessary to meet the hot water requirements of a building, it may not be possible to position them vertically due to their length and the need to provide sufficient circulating pressure by gravity (see Fig. 4.22). By fitting the cylinder in a horizontal position the circulating pressure can be increased without the use of a pump. It cannot be ignored, however, that in large commercial buildings greater use is made of pumps, which despite their initial and running costs have many advantages. Their use provides much greater circulating pressures enabling smaller circulation pipes to be used, better control systems and, possibly the most important, greater flexibility

in system design. This means the relative positions of the boiler and hot store vessel are not important, nor the position of a hot store vessel in relation to a secondary circulation. A typical example is shown in Book 1, Fig. x.xx. Such a system would not circulate by natural means.

Figure 4.23 shows the connections to both direct and indirect cylinders fitted in a horizontal position. The reader should assume that the vessels shown are made of galvanised steel, as those made of copper lack the strength to withstand the stresses to which they would be exposed in this position,

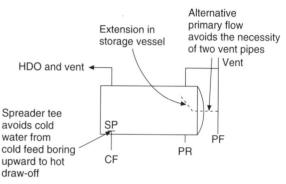

(a) Direct cylinder with no secondary circulation

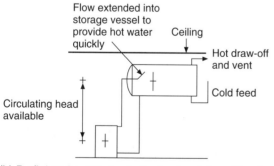

(a) Insufficient circulating head available when cylinder is in vertical position

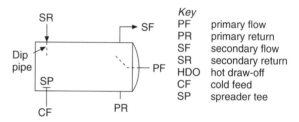

(b) Direct cylinder with secondary circulation

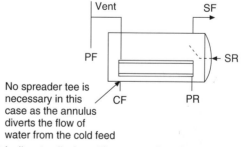

(c) Indirect cylinder with a secondary flow and return

Fig. 4.22 Circulating head can be increased by fitting the hot storage vessel in a horizontal position.

(b) By fitting the cylinder in the horizontal position the circulating head is increased providing a more effective circulation

Fig. 4.23 Positions of connections for cylinders fitted in the horizontal position.

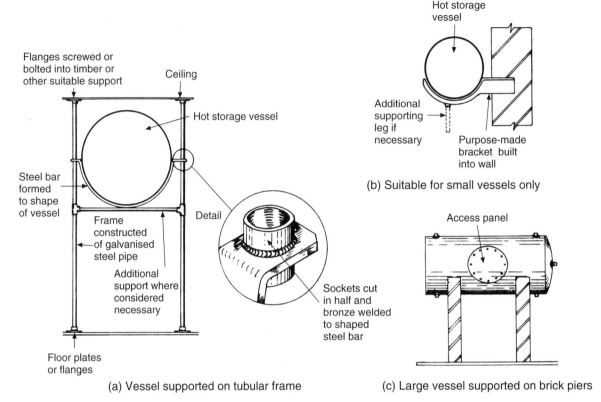

Fig. 4.24 Alternative methods of supporting horizontally fitted hot water storage cylinders.

and indirect copper cylinders employ coils as heat exchanges which would result in the coil becoming airlocked. Small cylinders of limited capacity are sometimes fixed on specially shaped brackets cantilevered into the wall, but the safest method is to provide support from the floor such as brick piers or a suitably constructed cradle. Figure 4.24 shows alternative methods of support for horizontally fitted cylinders. Care must be taken to ensure that maximum advantage is taken of the standard tappings.

It is sometimes necessary to fit a cylindrical vessel in a position, possibly a narrow airing cupboard, where its diameter may be limited, but where there is no restriction on its height. If its required capacity and diameter are known its height can be calculated. First ascertain how much water will be contained in a cylinder of such a diameter having a height of 1 m. The capacity required is

then divided by this result, which will give the height of the cylinder. For example, assuming that the diameter of the vessel is to be 0.6 m and the required capacity 425 litres, the volume can be found in the following way. Using the formula $A = \pi r^2$ to find the area of the base of the cylinder (where radius = 0.6 m ÷ 2 = 0.3 m):

$$\text{Base area} = 3.142 \times 0.3 \times 0.3$$
$$= 0.283 \text{ m}^2 \text{ approx.}$$

The volume of a cylinder with this base area and height of 1 m = 0.283 m^3.

There are 1,000 litres in 1 m^3, therefore volume = 283 litres. The required capacity (425 litres) is now divided by 283 to find the required height.

$$\text{Height} = 425 \div 283 = 1.502$$

Therefore the height of the cylinder will be approximately 1.5 m.

Towel rails

Before central heating became commonplace it was unusual to find any form of heating in the bathrooms of small dwellings. In buildings with a secondary circulation of hot water supply it was realised that a heated rail for towel airing could be fitted into it very easily. As central heating came within the reach of more property owners, both towel rails and the method of heating them changed.

A traditional towel rail is made of brass or copper tube brazed or soldered together, after which it is chromium plated. Some of these rails are fitted with a small inset radiator made of sheet steel or cast iron which, due to the possibility of corrosion, precludes their connection to a secondary circulation. Due also to the increase of dezincification, towel rails made of brass tube rapidly corrode if fitted into direct systems of hot water supply, and for this reason rails should be made of copper with zinc-free joints. Bearing all this in mind, some careful thought must be given to any situation requiring the use of a towel rail.

Figure 4.25 demonstrates what might be called the traditional method of fitting towel rails. The building might be a hotel or boarding house where the bathrooms are heated by radiators connected to the heating system. To enable towels to be aired when the heating is off during the summer months, the towel rails are fitted into the secondary circulation.

Figures 4.26 and 4.27 show piping alternatives for both direct and indirect domestic systems without a secondary circulation. In both cases the towel rails are connected to the primary part of the system.

In buildings where indirect heating systems are installed, a radiator rather than a towel rail is to be preferred, due to its greater heating surfaces. They can be fitted with a non-heated towel rail which clips over the top edge, giving it all the advantages of a towel airer. The alternative flow connection shown in each case may be considered if the primary circulation is not in a favourable position in relation to the towel airer. It also has the advantage of having no venting problems.

It has been assumed in Figs 4.25–27 that the circulation to the hot storage vessel functions by gravity so that both hot water and towel airing are available without the necessity of switching on the pump. Only full-way radiator valves should be used.

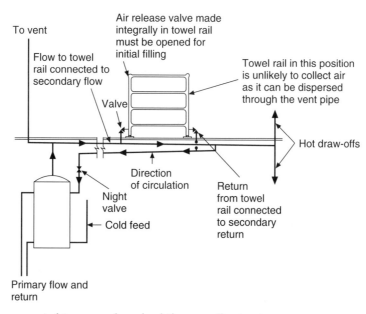

Fig. 4.25 Towel airer connected to a secondary circulation on a direct system.

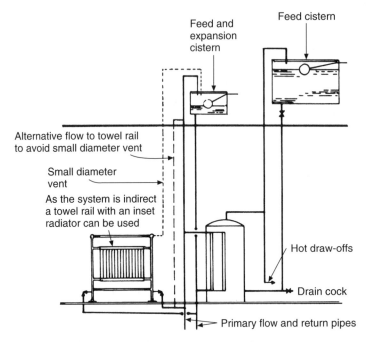

Fig. 4.26 Towel airer connected to an indirect system of hot water supply.

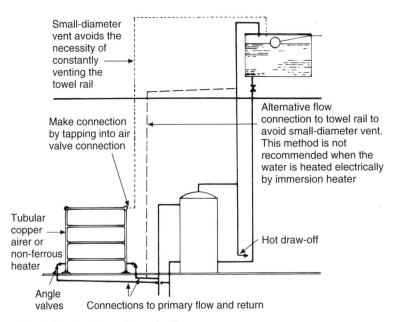

Fig. 4.27 Single towel and airer fitted to a domestic direct system.

on any of the schemes shown as many cheaper valves have a high head loss, which means they would restrict the flow of water to such an extent that a gravity circulation would not be effective.

Where fully pumped heating systems are employed it is considered undesirable to correct any form of space heating on the pipes serving the hot storage vessel. With a system of this type the bathroom radiator must be fitted on the pipework serving the other space-heating equipment in the building. When the bathroom only needs to be heated, all the other radiators on the system must be shut off.

Fuels used for hot water and central heating

The choice of a fuel is often dependent on its availability, as not all parts of the country are within reach, for example, of a natural gas supply. There are many other considerations such as cleanliness, convenience and both initial and running costs. Standing charges, as well as maintenance costs, apply to both gas and electrical supplies. Apart from this, the actual heat energy possessed by a fuel must be taken into account when a comparison is made between the cost of fuels. For example, certain types of solid fuel may be cheaper weight for weight, but the cheaper fuel may contain less effective heat content, which is usually expressed as its *calorific value* and is stated as the number of heat units (in joules or kilojoules) per mass or weight (in grams or kilograms). By comparing this value in different fuels some idea of the true cost will be ascertained.

While the average householder can maintain a solid fuel appliance, a specialist is required, usually once per annum, to clean out and check for good working order of both gas and oil appliances. The cost of new components for burners using these fuels can also be expensive and must be considered along with the running costs. Electricity has many advantages in that it is considered to be 100 per cent efficient, it is clean, it requires no storage and little maintenance is necessary. It is, unfortunately, expensive although savings can be made if off-peak electricity is used. This is usually referred to as Economy Seven heating, where during periods when the demand for electricity is low, e.g. overnight, its cost per unit is reduced considerably. Its use for space heating is limited because it is expensive at the normal tariff rates, and night storage heaters, which build up a store of heat during off-peak periods, are difficult to control effectively. Whether or not heating will be needed on the following day has to be decided the night before, and some reliance on the weather forecast, which may not be entirely accurate, is essential.

Water heaters

Instantaneous water heaters

These appliances are fuelled by either gas or electricity and, as the name implies, produce an instant supply of hot water but have no storage capacity. Instantaneous gas heaters have been used for many years and are well developed. Those using electricity as a fuel are comparatively new and have become very popular due to their use in instantaneous showers.

The advantages of instantaneous water heaters are as follows: they only function when a supply of hot water is required; a supply of hot water is immediately available; there is no limit to the supply of heated water; and, having no storage, heaters of this type have little heat loss. The main disadvantage with all instantaneous heaters is that the quantity of water delivered at a suitable temperature is considerably reduced if more than one draw-off is in use at the same time. They are, however, especially useful where a supply of hot water is desired in an isolated position which might need an excessively long draw-off if fed from a central source. A typical instance might be a supply of water for hand washing in a toilet situated a long distance from the main hot water services.

Storage water heaters

These are also fuelled by gas or electricity, possibly the most common being an electric immersion heater which is fitted into the main hot water storage vessel to heat the water when one does not want to operate the boiler. They are especially useful when the normal method of heating the water is by a solid fuel boiler, as the latter would tend to overheat the room in which it is situated, especially in hot summer weather.

Storage vessels are purpose-made for both gas and electric heaters, many of which are very economical to run due to the high standards of thermal insulation used. The main disadvantage of storage heating equipment is the limitation imposed by its capacity on the quantity of hot water available, which, when exhausted, takes a period of time to replenish. The temperature of the water is controlled by a thermostat, unlike that of instantaneous heaters where the temperature is governed by the volume of water passing through it. There are two other terms which the reader should understand relating to water heaters whether they are of the storage or the instantaneous type. *Single point* heaters are those serving only one point and are usually fitted with a swivel spout, while *multi points* serve several draw-offs.

It must be stressed that when gas or electric water heating appliances are to be installed the relevant regulations must be complied with. All manufacturers provide detailed fitting and fixing instructions with their products and these should be studied carefully before work is commenced. In all cases where multi-point heaters are installed long dead legs should be avoided, which means careful planning to ensure grouping of the appliances to which hot water is to be supplied.

One important point should be considered when the running and installation costs are compared between gas and electrical appliances. Although the running costs of gas are less than those of electricity, the installation costs are usually higher due to the need for both flueing and ventilation of the space in which the appliance is fitted.

Gas instantaneous water heaters

Although the market for this type of heater has diminished, mainly due to the increase of combined space and hot water systems in private dwellings, they do have many applications. Typical examples are small dwellings with alternative space-heating systems, and small industrial and commercial premises for ablutionary purposes. Most modern heaters of this type are multi-point, e.g. serving more than one outlet. They are very economical in use as there is little or no heat loss and any pipe runs should be as short as possible. If possible they should be fitted near to the draw-off most used, usually the sink. They do have the disadvantage of being capable of supplying water to only one tap at a time, e.g. if another tap on the system is opened the volume of water delivered will be halved. These heaters are normally designed to be connected to the main cold water supply, but some with modified pressure ratings can be used with a low-pressure cistern-fed supply. Most modern heaters are room-sealed, although open-flue types are still available. Room-sealed appliances must be used if installed in

bathrooms, bedrooms and garages. Figure 4.28 shows the basic working principles of a typical heater of this type. The essential controls they embody are very similar to those used when they were first produced.

Gas storage heaters

This term includes both boilers and circulators which are fitted independently of the water storage vessel and those which are purpose-made vessels with a small gas burner fitted directly underneath. Gas circulators are really small boilers and are quite often fitted in the airing cupboard close to the hot storage vessel. They can be used to augment another source of water heating, although they are often employed as the sole source. Although they are very effective and economical to run, the main disadvantage with this form of heater is the relatively high cost of both the heater and its installation. Due to this and the availability of alternative and more economic methods of achieving the same purpose, the demand for these heaters is very small.

A purpose-made gas storage heater is shown in Fig. 4.29. The burner is incorporated in the base of the storage vessel and is thermostatically controlled. These units are made with a great variation in water capacity, ranging from 75 to 285 litres. They can be used as the sole water-heating appliance in both domestic and industrial buildings where gas is used in preference to electricity for water heating. They are available for both unvented and open vented systems and are very economic in use. Many modern heaters of this type are room-sealed.

Electric water heating

The main features of instantaneous electric water heaters are discussed and illustrated in Chapter x where their application to shower heating is dealt with. Apart from this, their use is at present confined to small single-point heaters serving spray taps for hand washing. Storage heating by electricity has been well established for a long period of time. The methods used range from the installation of electric immersion heaters in a normal hot storage vessel to purpose-made units of varying capacities. The main advantage of these latter vessels is the

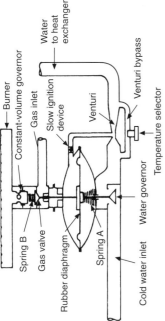

Burner

Constant-volume governor

Gas inlet

Slow ignition device

Water to heat exchanger

Venturi

Venturi bypass

Temperature selector

Spring B

Gas valve

Rubber diaphragm

Spring A

Cold water inlet

Water governor

(b) Detail of water valve

The temperature selector simply controls the volume of water passing through the heat exchanger and has the effect of raising a large volume of water to a relatively low temperature or a smaller quantity to a higher temperature. A constant-volume governor is built into the heater and automatically controls the volume of gas consumed. A detail of this type of governor is shown in (c).
Note that all valves are shown in the closed position and the water pressure both above and below the diaphragm is the same. When a water valve is turned on water flows through the water governor and venturi. The function of the venturi is to convert pressure head to velocity head which in this case has the effect of lowering the water pressure above the diaphragm. The higher pressure below the diaphragm pushes it upward, lifting the main gas valve so that gas can be admitted to the burner. The slow ignition device causes the main burner to light up in two stages which reduces condensation problems in both combustion chamber and flue.

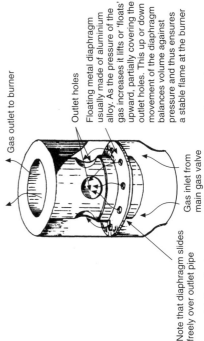

Outlet holes

Floating metal diaphragm usually made of aluminium alloy. As the pressure of the gas increases it lifts or 'floats' upward, partially covering the outlet holes. This up or down movement of the diaphragm balances volume against pressure and thus ensures a stable flame at the burner

Gas outlet to burner

Gas inlet from main gas valve

Note that diaphragm slides freely over outlet pipe

(c) Detail of constant volume governor used water heaters

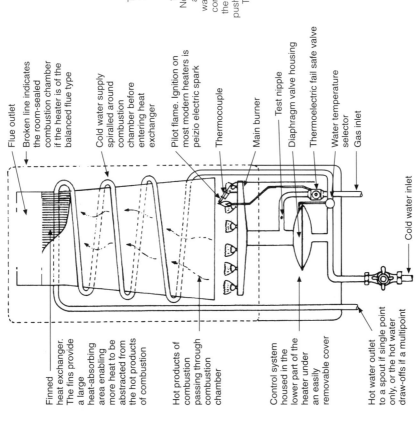

Flue outlet

Broken line indicates the room-sealed combustion chamber if the heater is of the balanced flue type

Cold water supply spiralled around combustion chamber before entering heat exchanger

Pilot flame. Ignition on most modern heaters is peizio electric spark

Thermocouple

Main burner

Test nipple

Diaphragm valve housing

Thermoelectric fail safe valve

Water temperature selector

Gas inlet

Cold water inlet

Finned heat exchanger. The fins provide a large heat-absorbing area enabling more heat to be abstracted from the hot products of combustion

Hot products of combustion passing through combustion chamber

Control system housed in the lower part of the heater under an easily removable cover

Hot water outlet to a spout if single point only, or the hot water draw-offs if a multipoint

(a) Basic working principles of a typical instantaneous gas water heater

This diagram does not illustrate any particular model; its purpose is to indicate the general principles of operation. These heaters are designed in such a way that a pilot flame is established and water is flowing before the main burner will light thus avoiding the risk of explosion or damage to the heat exchanger.

Fig. 4.28 Instantaneous gas water heater

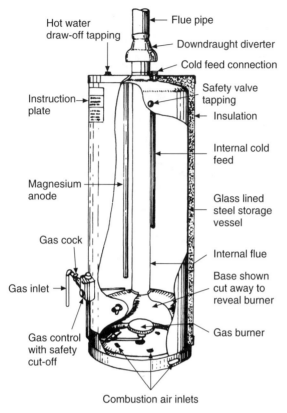

Hot water draw-off tapping
Flue pipe
Downdraught diverter
Cold feed connection
Instruction plate
Safety valve tapping
Insulation
Internal cold feed
Magnesium anode
Glass lined steel storage vessel
Internal flue
Gas cock
Base shown cut away to reveal burner
Gas inlet
Gas burner
Gas control with safety cut-off
Combustion air inlets

Fig. 4.29   Gas-fired hot water storage heater.

high-quality insulating jacket with which they are provided permitting only minimal heat loss.

Under-sink electric water heaters
These are small, single-point storage headers, usually of approximately 10 litre capacity with options of 3 or 1.2 kW element ratings. The 3 kW provides very fast recovery from cold. They must be fitted with a special tap which allows for expansion of the water during heating, but isolates the inlet until the tap is turned on (see Fig. 4.30). These heaters are a very convenient alternative to a wall-hung heater with a swinging arm outlet.

Electric immersion heaters
All methods of using electricity for heating stored water employ an immersion heater, one type of which is illustrated in Fig. 4.31. The type shown is available in various lengths and ratings and is suitable for installation in an existing storage vessel. Immersion heaters are often described as 100 per cent efficient as all the heat they generate must pass into the water, but as electricity is expensive it is important that any vessel into which they are fitted

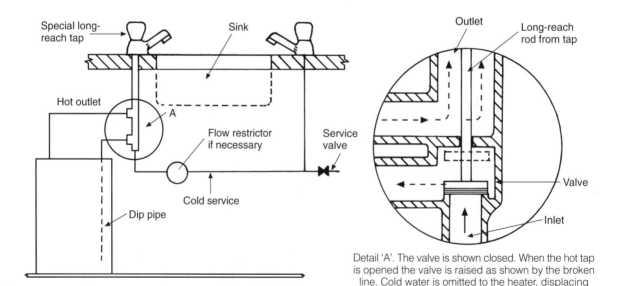

Special long-reach tap
Sink
Hot outlet
A
Flow restrictor if necessary
Service valve
Cold service
Dip pipe

Outlet
Long-reach rod from tap
Valve
Inlet

Detail 'A'. The valve is shown closed. When the hot tap is opened the valve is raised as shown by the broken line. Cold water is omitted to the heater, displacing the stored hot water which issues from the tap.

Fig. 4.30   Under-sink electric water heater.

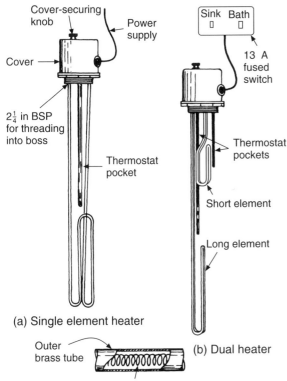

(a) Single element heater

(b) Dual heater

(c) Section of heater element

Coiled element in mineral insulation becomes heated due to resistance to the flow of current

Dual heaters have two elements and better types have two thermostats. Each element is separately switched in the heater cover or, as shown, a wall-mounted switch, when the 'sink' switch is thrown the short element is energised and will heat only the upper part of the hot-store vessel. The 'bath' switch controls the long element which heats the entire contents of the cylinder.

Fig. 4.31 Electric immersion heaters.

is well insulated. Some of these heaters are fitted with two elements, a short one heating the top third of the vessel's contents, which is normally sufficient for general use, and a long one used only when the entire contents of the storage vessel are required, each element being separately switched. The better types of heater have separate thermostatic controls for each element, others use one to control the temperature of both. While the latter are cheaper they are often not so effective in controlling the longer element, often causing it to switch off when only about half the contents of the storage vessel have achieved the desired temperature.

Thermostats controlling electric immersion heaters usually work on the invar rod principle. A diagrammic illustration is shown in Fig. 4.32. In hard water areas the thermostat setting should not be more than 60–65 °C, otherwise scaling of the element will take place causing it to overheat and burn out.

Most modern cylinders are supplied with a boss tapped with a $2\frac{1}{4}$ in BSP thread into which the immersion heater is fitted. If the storage vessel has no boss or if a bottom entry heater is used. Special flanges can be used which are made and fitted in such a way that they can be installed without access being gained to the inside of the vessel. They are suitable for both copper and galvanised storage vessels but, when ordering, the surface on which they are to be fitted must be specified, i.e. whether it is flat or circular. A boss of this type is illustrated in Fig. 4.33.

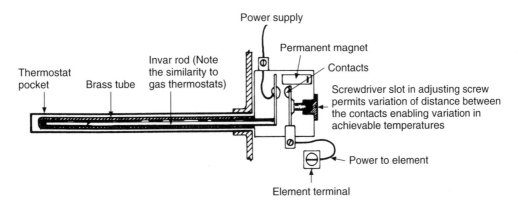

Fig. 4.32 Diagrammatic section of immersion heater thermostat. The permanent magnet ensures positive making and breaking of the contacts and avoids arcing with consequent burned contacts and possible fire risk.

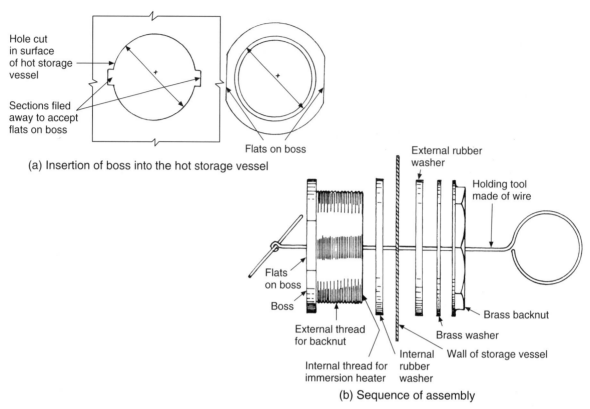

Hole cut in surface of hot storage vessel

Sections filed away to accept flats on boss

(a) Insertion of boss into the hot storage vessel

Flats on boss

External rubber washer

Holding tool made of wire

Flats on boss

Boss

External thread for backnut

Internal thread for immersion heater

Internal rubber washer

Brass washer

Wall of storage vessel

Brass backnut

(b) Sequence of assembly

Fig. 4.33 Patent boss for fitting immersion heaters. This is one of several types of boss manufactured to enable connections to be made in vessels accessible from one side only. A hole is first cut in the vessel to dimension x then two sections are filed away to admit the flats on the boss which is then turned through 90° so that the flats do not coincide with the filed area. A rubber washer is then passed through the hole and over the external thread of the boss which is now held firmly in position with the wire holding tool. The external rubber and brass washers are then fitted over the threaded end of the boss protruding from the vessel, prior to tightening the backnut which secures the complete assembly. The holding tool is then removed and the immersion heater can be fitted in the usual way, but care must be taken not to overtighten.

Immersion heater arrangements   Figure 4.34 shows arrangements for fitting immersion heaters for various purposes. That shown in Fig. 4.34(a) is a single, top entry type which will heat the entire contents of the vessel. It is useful as a supplementary form of heating for summer use, but is not as economical as the dual heater previously described. Figure 4.34(b) shows an arrangement using two short heaters and is very useful when electricity is the only source of heating as, by increasing the size of the storage vessel, full advantage can be taken of Economy Seven rates of electricity. The lower heater which heats all the water in the vessel is connected to the cheap rate supply, while the one at the top is used only for top-up purposes using electricity at normal tariff rates.

Fitting and removing immersion heaters   Specially made ring spanners are available for the installation or removal of immersion heaters. They form a good fit on the octagonal flange of the heater and cause less damage to the brasswork than large stilson wrenches. When installing an immersion heater make sure the element is not in contact with any part of the surface of the storage vessel (or the coil or annulus if an indirect cylinder is used) as this may cause the element to overheat and burn out.

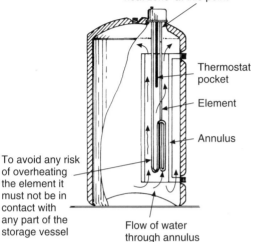

The unavoidable air space at this point is sometimes responsible for the element burning out. Some heaters are available with a 'no heat zone' at this point

Thermostat pocket

Element

Annulus

To avoid any risk of overheating the element it must not be in contact with any part of the storage vessel

Flow of water through annulus

(a) Single, top entry immersion heater

In vessels of this type the annulus acts as a circulator and provides a small quantity of hot water very quickly.

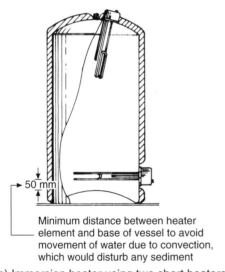

50 mm

Minimum distance between heater element and base of vessel to avoid movement of water due to convection, which would disturb any sediment

(b) Immersion heater using two short heaters

This arrangement is convenient and economical when electricity is the only source of water heating. The heater in the top of the vessel operates on electricity at the normal rate. The lower one heats the entire contents of the vessel and is normally supplied by cheap-rate electricity.

Fig. 4.34 Immersion heater arrangements.

If a heater is fitted at low level in the storage vessel it should be at least 50 mm above the base to prevent movement of the circulating water disturbing any sediment that may be deposited there. It is sometimes difficult to unscrew a defective heater from its boss, and damage to the vessel will be avoided if the heater is eased or unscrewed slightly before draining down when the vessel is more rigid due to its water content. This applies especially to vessels made of copper.

Introduction to solar heating

This form of heating is not new and has been commonly employed in countries nearer to the equator for many years. With certain exceptions it has not become popular in the UK, mainly due to the climate and the fact that the pay-back period on the initial cost is said to be 10–15 years. However, the economic cost of solar heating is becoming more realistic due to (a) the apparent increase in global warming, (b) the incidence of long, hot summer periods and (c) the gradual increase in the price of fuel. To encourage energy saving, VAT is reduced to 5 per cent on labour and parts for the installation of solar heating systems.

The basic principles are very simple. A collector panel or series of panels are fixed to the roof of the building in the most favourable position for maximum exposure to the sun. These panels are designed to absorb as much of the sun's heat as possible and transfer it to a fluid which is pumped through a coil in a storage vessel. The storage vessel may be of the vented or the unvented type and is often interconnected to a sealed or open system of space heating. Three types of solar heat collection are employed. In some instances the collecting panel may be fitted below the storage vessel using gravity convection currents to heat the water. This is not always convenient as the most advantageous point for solar heat collection is usually above the hot store vessel. The two most common systems used are shown in Fig. 4.35(a) and (b), both of which require a pump to circulate the heat transfer fluid, usually water, through the collector. Figure 4.35(a) is a sealed system, which as it is permanently charged requires protection against frost damage. This is achieved using a

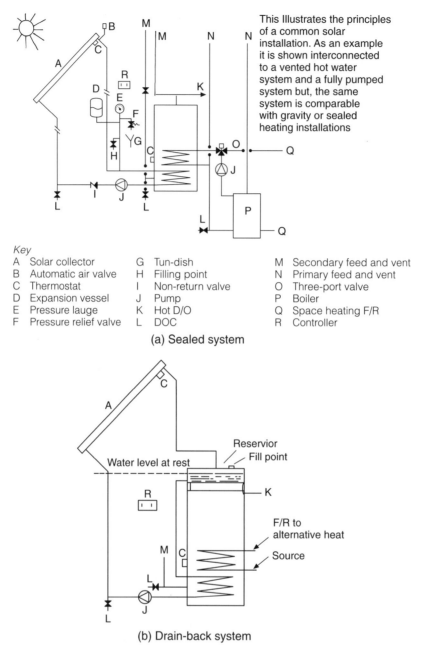

This Illustrates the principles of a common solar installation. As an example it is shown interconnected to a vented hot water system and a fully pumped system but, the same system is comparable with gravity or sealed heating installations

Key

A	Solar collector	G	Tun-dish	M	Secondary feed and vent
B	Automatic air valve	H	Filling point	N	Primary feed and vent
C	Thermostat	I	Non-return valve	O	Three-port valve
D	Expansion vessel	J	Pump	P	Boiler
E	Pressure lauge	K	Hot D/O	Q	Space heating F/R
F	Pressure relief valve	L	DOC	R	Controller

(a) Sealed system

Reservior
Fill point
Water level at rest

F/R to
alternative heat

Source

(b) Drain-back system

This is a very simple system requiring no frost protection as
the solar collector is only charged with water when the pump starts.

Fig. 4.35   Solar heating systems.

suitable anti-freeze fluid in the circulation system. The drain-back system (Fig. 4.35(b)) is designed so that when the pump is inoperative the collector is not charged and freezing is not a problem. To avoid the possibility of overheating the water in the hot storage vessel (outside temperatures can exceed 100 °C on very hot days) a thermostat is fitted in the storage vessel to stop the pump when the water achieves the temperature of approximately 60–65 °C. However, it is usual to match the panel size to the capacity of the storage vessel, largely overcoming the problem.

Controls

The principle function in a basic system is to activate the pump when there is sufficient heat available for collection. This information is obtained by the two temperature sensors, one on the solar panel, the other on the storage vessel. For example, if the storage water is at a higher temperature than that in the panel, heat will be lost — not gained — if the pump is operating. These two thermostats are interwired with the controller to ensure the pump only operates when heat can be gained. It will also be seen that the hot store vessel could act as a boiler, heating the solar panel by gravity when the pump is not operating. This reversed circulation on smaller schemes is prevented by a non-return valve or a two-port motorised valve governed by the controls.

Solar panels

These are the source of heat collection and vary considerably in cost and efficiency. The simplest type is a plate collector but there are other more complex types with increased efficiency; for details of these the solar heating manufacturers association should be contacted.

This section on solar heating deals mainly with water heating for domestic use. It can however be extended to heating systems, especially those of the underfloor radiant type, and can also be used for heating (or partially heating) the water in small swimming pools. Many manufacturers of solar heating market their products in package form and will give advice and supply all the necessary components required for any installation, providing they are given all the relevant details. This avoids time wasted in obtaining them from a variety of suppliers who are unlikely to keep them as stock items. As the traditional sources of heat and power dry up or become too expensive, solar heating and solar power will certainly become more frequently used as a source of renewable energy.

Legionnaires' disease

This is a potentially fatal form of pneumonia which was first identified following an outbreak of the disease among people who attended an American Legion convention in America in 1976.

It is normally contracted by inhalation. Legionella bacteria are common and low numbers are found naturally in water sources such as lakes, rivers and reservoirs. They can survive in water temperatures of between 6 °C and 60 °C, but water temperatures of between 25 °C and 45 °C appear to favour growth. The presence of sediment, sludge, scale and bio-films (a thin layer of microorganisms forming a slime on the surface of stagnant water) all provide favourable conditions in which Legionella bacteria can multiply. From a practical point of view, all crafts working in the Mechanical Service Industry should be aware of the following, especially those working on the maintenance of commercial and industrial systems.

Cisterns and storage vessels connected with cooling and air conditioning plants are known sources of infection, bearing in mind the critical temperatures of 25–45 °C. Careful installation of cold water storage cisterns is necessary to avoid stagnation (see Fig. 3.10). Cold water services running adjacent to hot water or heating pipes must be sufficiently well insulated to avoid any appreciable temperature rise. Pipes carrying mixed hot and cold supplies at temperatures of approximately 40 °C to showers or other sanitary fittings should be as short as possible; the Ministry of Health recommends a maximum of 2 m. From the foregoing it will be seen that some degree of thought must be given to any situation in the working environment in which Legionella bacteria could proliferate. As the Health and Safety at Work Act may be invoked here, it may be necessary for employers to carry out a COSHH risk assessment and provide suitable measures, including information and instruction, to protect their employees. Where considered necessary the building owner and HSE should be notified of any possible danger to health.

Further reading

BS 6283: 1991 Parts 2 and 3 Safety devices for use in hot water systems.

BS 6144 and 6920 Specification for expansion vessels using an internal diaphragm for unvented hot water systems.

Unvented hot water systems BRS Digest No. 308 from BRS Bookshop, Building Research Establishment, Garston, Watford, WD2 7JR.

Heating appliances and hot water systems

Solid Fuel   Details from local offices.

Gas Storage Heaters   Andrews Industrial
   Equipment Ltd, Dudley Road, Wolverhampton,
   WV2 3DB, Tel. 0121 506 7440.
Instantaneous water heaters   Baxi (Mains)
   Brownedge Road, Bamber Bridge, Preston,
   Lancs, Tel. 0870 606 0623.
Johnson and Starley Ltd, Rhosili Road,
   Bracknell, Northampton, NN47 0LZ.
   Tel. 01604 762881.
Santon Electric Water Heating, Hurricane Way,
   Norwich, Norfolk, Tel. 01603 420149.
Gledhall Water Storage Ltd, Sycamore Trading
   Estate, Squires Gate Lane, Blackpool, Lancs,
   FY4 3RL, Tel. 01253 693304.
Reliance Water Contracts Ltd, Worcester Road,
   Evesham, Worcestshire, WR11 4RA,
   Tel. 01386 47148.

Combination boilers   Worcester Heat Systems Ltd,
   Cotswold Way, Warnedon, WR4 9SW,
   Tel. 01905 754624.

Pumped systems   Harton Heating Appliances Ltd,
   Thameshead Industrial Estate, Erith, Kent,
   DA18 4AN, Tel. 020 8310 0421.
*Reliability and performance of solar heating
   systems*   BRS Digest 254.
Legionnaires' Disease. Advice to employers.
   HSE Books, PO Box 1999 Sudbury, Suffolk,
   CO10 6FS.

Dunsley Baker Neutralizer System   Dunsley
   Heat Ltd, Fearnough, Huddersfield Road,
   Holm Forth, West Yorkshire, HD7 2TU,
   Tel. 01481 682 635.
Andrews Boilers 2 Water Heaters   Wednesbury
   One, Black Country New Rd, Wednesbury,
   West Midlands, WS10 7NZ.

Self-testing questions

1. In what circumstances would it be necessary to
   install a supplementary system of hot water
   supply?

2. (a) Make a sketch of an indirect cylindrical hot
       water storage vessel fitted in the horizontal
       position showing all the connections
       including those for a secondary circulation.
   (b) State why it is sometimes necessary to
       install a cylinder in this position.
3. (a) List three advantages of using
       instantaneous water heaters for domestic
       supplies.
   (b) State the main disadvantages of these
       heaters.
4. List and describe the essential operating and
   safety controls necessary for the safe and
   efficient functioning of unvented hot water
   systems.
5. (a) Explain what is meant by the term
       stratification in a hot water storage vessel.
   (b) Assuming the temperature of the water at
       the top of the vessel is 65 °C, and at the
       bottom 45 °C, state the mean water
       temperature in the vessel.
6. State the most likely cause of persistent
   discharge of an expansion valve in an unvented
   hot water installation.
7. Describe three appliances or methods of
   providing mains-fed hot water systems that
   can be fitted by a non-BBA-approved installer.
8. State the reason for fitting a single check valve
   on the mains inlet to an unvented hot water
   system.
9. State the advantages and disadvantages of
   combination boilers.
10. (a) State the possible causes of burning out the
       element on an electric immersion heater.
   (b) Sketch the fixing position of the heater for
       (i) a small quantity of water heated in a
       short period of time, (ii) a larger quantity
       heated over a period of 2 or 3 hours.
11. List the essential safety requirements for
   discharge pipes from pressure and temperature
   relief valves on unvented hot water storage
   vessels.
12. State the purpose of sacrificial anodes.
13. Specify the type of towel rail for use on a
   secondary circulation.
14. State how the relative height between a boiler
   and hot water storage vessel affects gravity
   circulation.

5 Hot Water Heating Systems

After completing this chapter the reader should be able to:

1. Recognise and evaluate heat emission equipment.
2. Recognise the basic pipework systems common to wet central heating.
3. Understand the purpose and working principles of equipment and components used in hot water heating schemes.
4. Identify the main types of heating systems and their advantages for specific applications.
5. Identify simple control systems essential for economy.
6. Identify methods of energy conservation in housing.
7. Understand the basic principles of designing domestic heating systems.

Space-heating systems

Some form of space heating has become a prerequisite for modern comfort and it has become a recognised part of the work of a plumber to install and maintain hot water heating systems. This subject is complex and is impossible to cover completely in the limited space available. Modern schemes include smallbore heating, schemes using very small pipes having diameters of only 8 or 10 mm, called mini-bore, and fully pumped schemes where the circulation of both the domestic hot water and heating water depend on the use of a pump. The type of boiler and the fuel used also influence the type of system employed.

The term wet *central heating* is derived from the provision of heat by hot water from a central source as distinct from a series of unconnected heaters. Central heating systems have been in use for many years, some installed with little concern for design or the conservation of fuel. Most of these early systems worked on a gravity basis, e.g. a cold column of water displacing a hot column.

Due to the low circulating pressures available when used to heat radiators long distances from the boiler, gravity circulation has some limitations. The lower circulating pressures require larger pipe diameters and large-radius bends to ensure that minimum resistance is offered to the flow of water. Larger pipes are often very difficult to conceal and installation costs are higher.

For these reasons, during the early 1950s, the British Coal Utilisation Research Association introduced a system of heating using silent, sealed rotor pumps to circulate the water, thereby enabling smaller pipes to be used, the pump head or pressure easily overcoming the frictional resistance of the smaller pipes. The system was designed to enable owners of small- and medium-sized dwellings to enjoy the benefits of low-cost central heating. Although this system was designed initially for solid fuel installations, it is readily adaptable for use with gas- or oil-fired boilers. A further advantage of using pumps and small-diameter pipes is the degree of flexibility and control that can be obtained. The small volume of water in the system circulates quickly and less time is necessary for it to heat up

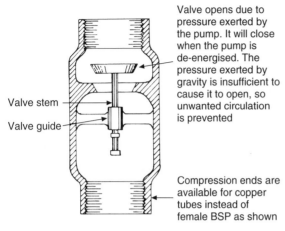

Valve opens due to pressure exerted by the pump. It will close when the pump is de-energised. The pressure exerted by gravity is insufficient to cause it to open, so unwanted circulation is prevented

Valve stem

Valve guide

Compression ends are available for copper tubes instead of female BSP as shown

Fig. 5.1 Anti-gravity valve.

or cool down. When the pump is stopped, in a well-designed system, circulation to the heat emitters should also cease.

Despite the resistance of the small pipes used, some systems tend to work on gravity when the pump is not functioning, especially radiators fitted on upper floors. This is very unlikely to occur with fully pumped systems — it is one of their advantages. It is likely to happen when the domestic hot water is heated by a gravity flow and return, as may be the case if a solid fuel boiler is used. To avoid this an anti-gravity valve may be fitted in a vertical position on the space heating flow pipe. Figure 5.1 illustrates such a valve, which is normally closed and will only open due to pressure exerted by the pump.

Heat emitters

There are two main types of heater used in hot water heating systems to heat the rooms in which they are situated. One is called a *radiator*, the other a *convector*.

The object of a radiator is to expose a hot surface to the air in a room. The air passes over the radiator, being warmed as it does so, and a convection current is created, causing a circulation of air in the room. Despite its name, approximately 90 per cent of the heat emission from a radiator is by convection; the remaining 10 per cent only is by direct radiation although this does vary according to the type of radiator.

A convector heater relies for its heat output on a relatively small area of heating surface over which air flows through a series of fins. It is essential that the water temperature in this type of heater is in excess of 80 °C if this form of heating is to be effective, and for this reason high-pressure sealed systems are sometimes employed with heating schemes using convector heaters.

Radiators
Most modern radiators are made of pressed steel sheets of welded construction, although originally they were made of cast iron. Aluminium alloy, although expensive is also commonly used as a material for modern column-type radiators. Although they are lighter and easier to handle, steel radiators are more prone to corrosion than those made of cast iron or aluminium alloy. It is recommended that heating systems are treated with an inhibitor, not only to prevent corrosion but also to avoid pump seizure. Radiators may be of the column, panel or hospital type (see Fig. 5.2).

Column radiators are the most compact for any given heating surface. The number of columns in each section varies and a radiator of this type is named by the number in each section. Hospital radiators are very similar, but each section consists of only one column which reduces their overall heating surface. The name 'hospital' derives from the fact that these radiators are easier to clean than column radiators, an important detail where cleanliness is essential. The most popular and effective radiator in use is the panel radiator, despite the fact that it occupies more space than other types. Its popularity is mainly due to its higher heat output in relation to its area in comparison with other radiators. These, too, were originally made of cast iron, but they are now easily and quickly produced by passing a sheet of steel through a set of rollers which press out the waterways. A continuous welding process is used to join the top and bottom of the two sheets, after which they are cut to the required length prior to sealing the ends and welding in the connections. Sheet steel panel radiators can be curved or angled to order, enabling a radiator to be fitted into a shaped bay window. The radiators shown in Fig. 5.2 are basic types, but there are many variations of

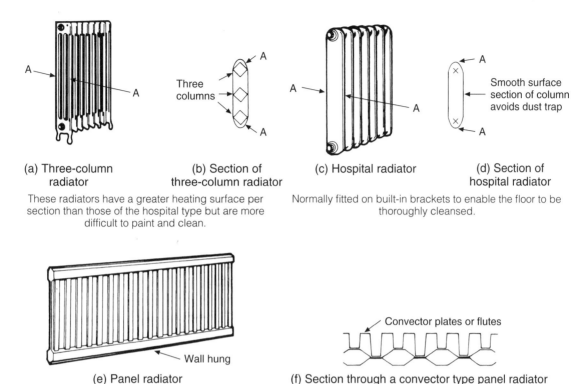

(a) Three-column
radiator

(b) Section of
three-column radiator

(c) Hospital radiator

(d) Section of
hospital radiator

Three
columns

Smooth surface
section of column
avoids dust trap

These radiators have a greater heating surface per section than those of the hospital type but are more difficult to paint and clean.

Normally fitted on built-in brackets to enable the floor to be thoroughly cleansed.

(e) Panel radiator

Wall hung

Convector plates or flutes

(f) Section through a convector type panel radiator

Fig. 5.2   Radiators.

each which are designed to increase the heat output and improve the appearance. It is possible to buy a top and side casing for some panel radiators, the top having a grilled outlet to permit air movement. Some panel radiators have a series of flutes or convector plates (see Fig. 5.2(f)) welded on to the back which improves the convected effect thus increasing the output. When fitted on the inside surface of double panel radiators these flutes help to reduce the radiant heat loss from one panel to the other. The heat output of a radiator is given as watts per square metre at a temperature difference of 60 °C between the water and the air temperature. This means that if the air temperature is, for example, 20 °C and the water temperature in a radiator is is 80 °C, its heat emission in W will be as shown in the manufacturer's catalogue. This will vary depending upon the type of radiator used — those of the panel type are generally more efficient than those of the column type for example, due to the slightly lower level of radiant heat they emit.

It is recommended that heat emitters are fixed under windows for the following reasons: much of the incoming cold air enters the room via a window; condensation that forms on the glass is evaporated; and as furnishings are seldom situated under windows, useful wall space is not taken up.

Pattern staining
Where it is not possible or desirable to fit a radiator under a window (this is obviously impossible in a room with patio doors), it is important that a suitable shelf is provided over the radiator. This prevents what is called 'pattern staining' which is caused by dust in the air, due to its movement by convection, being deposited on the decorations immediately above the radiator. The shelf has the effect of diverting the air out into the room. Most radiator manufacturers produce shelves of varying lengths made of metal and it is also possible to make or purchase them in polished hardwood. Any shelf should be fitted with a minimum clearance of

70–75 mm above the radiator to allow the convected air to circulate freely.

Convector heaters

These fall into two main groups, those which rely on natural air movement and those which are fan-assisted (Fig. 5.3). A typical example of the former type is called skirting heating. The heating unit is encased in a moulded steel case and is fitted instead of a normal skirting. Specially shaped angles are produced, enabling changes of direction to be made, and also valve boxes with hinged doors allowing the valves to be concealed. Those heaters which incorporate a fan are much more compact and often occupy less space than a radiator. They are not completely silent in operation due to the rapid movement of the air caused by the fan. Most are provided with two-speed fans, the greater speed being used for a quick heat-up of the room, while the lesser (and quieter) speed can be used for normal running.

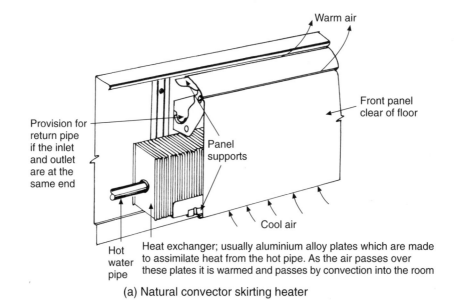

Warm air

Front panel clear of floor

Provision for return pipe if the inlet and outlet are at the same end

Panel supports

Cool air

Hot water pipe

Heat exchanger; usually aluminium alloy plates which are made to assimilate heat from the hot pipe. As the air passes over these plates it is warmed and passes by convection into the room

(a) Natural convector skirting heater

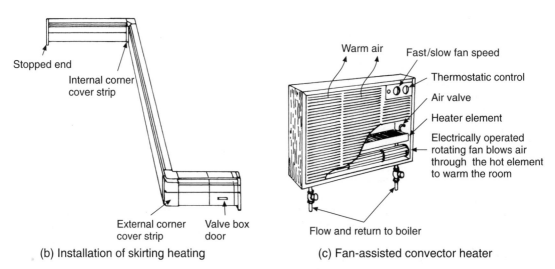

Stopped end

Internal corner cover strip

External corner cover strip

Valve box door

(b) Installation of skirting heating

Warm air

Fast/slow fan speed

Thermostatic control

Air valve

Heater element

Electrically operated rotating fan blows air through the hot element to warm the room

Flow and return to boiler

(c) Fan-assisted convector heater

Fig. 5.3 Convector heaters.

Central heating components

Radiator valves
Both angle and straight patterns are made,
but the angle type is the most popular due to
its convenience of connection and neatness in
appearance. All straight valves are of the gate pattern
and are therefore full way, but some angle valves
cause considerable head or pressure loss to the
flow of water. This has little effect on a two-pipe
scheme, but if for any reason a one-pipe system is
used where the water gravitates to the radiator from
a pumped circuit only, the use of a better-quality,
more expensive angle valve is essential. In most
cases they are specified as full way.

The glands of radiator valves are prone to
leakage, especially when they are continually being
turned on and off. The leak is often very slight and
often imperceptible, especially in carpeted rooms
where the carpet soaks up the water. It is important
that when servicing heating systems the valves are
inspected for leakage as it may avoid serious
damage to valuable household furnishings. The
glands of some types of valves may be resealed
without draining the system or freezing the pipes
locally. Both valves shown in Fig. 5.4 are of this
type, although it is still necessary to take extreme
care when carrying out this operation. It is
recommended that a sheet of plastic material is laid
under the valve, and always have an absorbent cloth
handy for mopping up if necessary.

Each radiator must be fitted with two valves, one
for on/off operation, the other to control the flow of
water. The latter must be of the lockshield pattern.
This prevents short-circuiting, to which two-pipe
systems are especially prone, and ensures an equal
distribution of water to each radiator. The lockshield
valve is regulated by the plumber when the system
is commissioned. This is called 'balancing' and is
carried out by adjusting the lockshield valve and
noting the temperature at both the inlet and outlet
of each radiator, until a difference of approximately

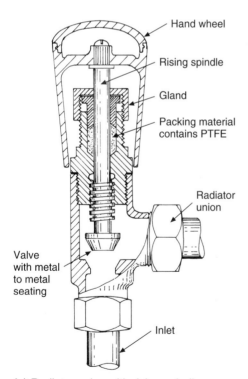

(a) Radiator valve with rising spindle

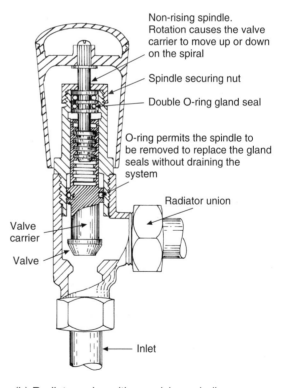

(b) Radiator valve with non-rising spindle

Fig. 5.4 Radiator valves.

11 °C is achieved. Providing the pipes have been correctly sized this should not be too difficult, but if it is found that the difference is more than 11 °C it may be necessary to increase the pump pressure. If the difference is less than 11 °C the pump pressure can be reduced. Electronic methods of balancing heating systems are available and the results they give are very accurate. An alternative is to use two rotary clip-on thermometers.

Pumps

These provide the pressure causing the water to circulate through the system. Most modern pumps are made so that their output can be regulated to accommodate a variety of requirements. It is obvious that the frictional resistance on a run of pipe 100 m long will be less than that of a run of, say, 150 m, and the pressure necessary to overcome this resistance will be greater. By regulating the speed, a pump can be made to cover a wide range of pressures. If the pressure is too high it will result in high water speed (high velocity) causing noise to occur in the system. The recommended velocity for smallbore schemes should not exceed 1 m per second. For mini-bore schemes 1.5 m per second is recommended to overcome the frictional resistance of the smaller pipes used.

Pumps should be valved so that they can be removed without draining down the system in the event of a breakdown. They are normally fitted in such a way that the central bearing on which the impeller rotates is in a horizontal position in order to avoid undue wear (see Fig. 5.5(b)). Some modern pumps may be fitted which allow the bearing to be at an angle, as recommended by the manufacturer, but this is not the general case. To ensure a long working life, the pump must be fitted to comply with the manufacturer's instructions.

The system must be washed out to remove any debris such as traces of wire wool or grit before the pump is fitted. It must also be purged of air before it is commissioned, as if the impeller runs dry, the bearing will be damaged. Normal lubrication is provided by the water in the system only, and no oil or grease should be used on the bearing at any time.

The pump must be fitted into the system in what is termed the *neutral point* to avoid water being

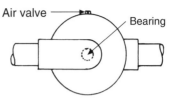

(a) Pump in a horizontal position

Pumps fitted in this position require the air in the top part of the pump to be purged when the system is filled

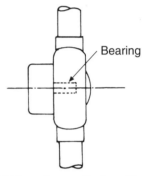

(b) Pump in a vertical position

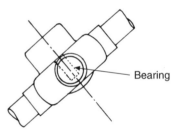

(c) In this case the bearing is not horizontal and such a position is undesirable

Fig. 5.5 Positioning central heating pumps. To avoid excessive wear on the bearing on which the impellor rotates it must always be in a horizontal position as shown in (a) and (b). If fitted as (c) the pump is likely to fail after a very short period of time. Note that pumps should be fitted with isolating valves (not shown here) so that they can be removed for maintenance.

pumped into the expansion cistern. Figure 5.6 shows in a simple way how this can happen. It is much less effort for the pump to draw water from the expansion cistern than to overcome the frictional resistance of the radiator pipework.

In the early period of forced circulation systems of heating it was recommended that the pump

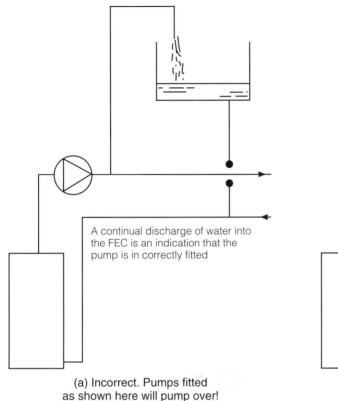

A continual discharge of water into the FEC is an indication that the pump is in correctly fitted

(a) Incorrect. Pumps fitted as shown here will pump over!

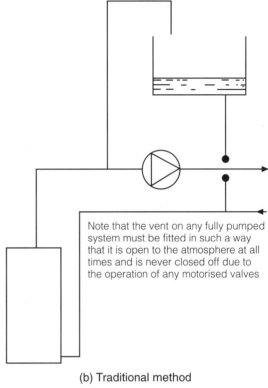

Note that the vent on any fully pumped system must be fitted in such a way that it is open to the atmosphere at all times and is never closed off due to the operation of any motorised valves

(b) Traditional method

Fig. 5.6 Position of heating pump in the system.

should be fitted on the return. It has been found, however, that this causes a negative pressure on most systems which allows air to be drawn in through leaks which are normally unseen as they are too small to admit the passage of water. This is due to the fact that the cohesion between the molecules of a gas is less than between those of a liquid. This can be proved quite simply with a fine-meshed wire strainer. It will be found that water will not pass through it. However, if a candle is lit and air is blown through the strainer, it will be found possible to extinguish the flame. Air can often be drawn into a heating system via valve glands and defective joints, and also through an open vent and, due to its oxygen content, it can be the source of many corrosion problems. It is now common practice, for the pump to be fitted on the flow pipe as shown in Figs 5.14 and 5.20, thus pressurising most of the system, which avoids the problems mentioned.

Air eliminator
Figure 5.7(a) shows a simple device for ejecting any air or other gas contained or picked up by water in low-pressure heating systems. Water containing air can cause many problems such as air locks, corrosion due to oxidation and noise, and their use prevents water being pumped through the vent where the static head on the system is very low. The associated illustration Fig. 5.7(b) shows its application in a heating installation. It is equally effective both with fully pumped or pumped heating/gravity hot water schemes.

Boilers

General details of boilers are dealt with in Book 1 of this series. More detailed information on gas and oil appliances is covered in Chapters 4 and 6 of this volume. Low-water-content and condensing boilers are dealt with in the following text.

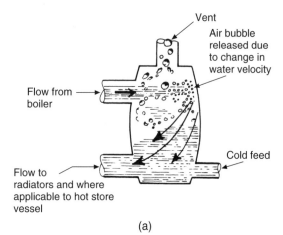

(a)

This very simple device is designed to disperse and eliminate air bubbles in a low-pressure heating system. As the water from the boiler impinges on the back of the chamber its velocity is suddenly arrested causing it to release any air it contains, the air passing into the vent. These fittings are not essential in boilers having a high water capacity, as any air contained in the water can easily escape via the vent. However, they can be used with advantage on schemes using boilers having a high-pressure drop across the flow and return.

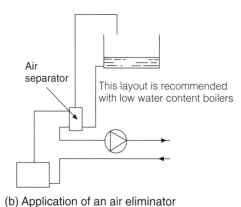

(b) Application of an air eliminator

Fig. 5.7   Air eliminator.

Low-water-content boilers

These are mainly gas fired but some using oil as a fuel are available. Care must be exercised when low-water-content boilers are used, as failure to follow any instructions in fitting and connecting such a boiler can result in serious damage to the heat exchanger. These boilers must only be used on fully pumped schemes. They rely for their efficiency upon circulating a small amount of water very quickly. It is essential, however, in order to avoid overheating, that a movement of water occurs at all times though the heat exchanger. To this end these boilers will only operate when the pump is functioning; a delay switch is incorporated which ensures that the pump continues to run for a short period of time after the main boiler shuts down. A bypass valve must be built into the system to enable any build-up of heat in the boiler to dissipate quickly.

Condensing boilers

As the cost of fuel increases and the limitations of sources of energy become more apparent, fuel suppliers and boiler manufacturers are diverting more time and thought to improving the efficiency of their appliances. One of the more recent products of this research is the condensing boiler. The reader will appreciate that 100 per cent efficiency is not possible with any burning appliance. This is mainly due to the fact that products of combustion carry away some of the heat given off by the burning fuel up the chimney and into the atmosphere. The actual efficiency of a boiler varies considerably depending on a number of factors, some of which may be listed as follows: the age of the appliance, whether it is oversized in relation to the scheme it is heating and whether or not it has a permanent pilot or is fitted with electronic (spark) ignition. A modern boiler of the conventional type is said to be approximately 75 per cent efficient, the remaining 25 per cent of the heat passing up the flue. To achieve greater efficiency some of this heat must be reclaimed, but this poses a serious problem. There must be a considerable difference between the ambient air temperature and that of the products of combustion in order that convection takes place. The second difficulty is that water vapour is formed during the process of combustion, and if the flue temperature is lowered significantly the vapour will condense and turn to water. The first problem can be overcome by using a flue extractor, indeed some non-condensing boilers already use this method to reduce flue sizes. The difficulty of disposing of the condensate (water) has to some degree been overcome by allowing it to collect in the base of the boiler where it can be disposed of into the drainage system. Although the condensate is acidic in nature containing traces of nitric and sulphurous acids, tests have been conducted on materials from which domestic drains are constructed and the following conclusions reached. Drains made of plastic material and clayware showed insignificant damage while cast iron is likely to be affected in the long term and gives rise to staining. Cement and concrete products appeared to be affected more seriously than other materials. This could lead to problems in older properties having salt-glazed drain pipes with cement joints. In practice, however, the adverse effects due to condensation are unlikely to be serious as it will be appreciated that it will be diluted very quickly by the discharges from sanitary appliances.

There are two basic types of condensing boilers using either wet or dry heat-transfer principles. The wet type requires a purpose-made system which at present is only produced for commercial boiler plant and is still being developed. Dry condensing boilers are more suitable for domestic work. Unlike the wet system where the circulating water is in direct contact with the combustion gases, they operate

on the more conventional principles of traditional boilers where the products of combustion are separated from the circulating water via a heat exchanger. A diagrammatic illustration of a condensing boiler is shown in Fig. 5.8. It will be seen that the combustion gases pass through the primary heat exchanger in a similar way to boilers of a traditional pattern, but at this point the similarity ends. Instead of passing the combustion products directly into the flue, bearing in mind they are at a temperature of approximately 200–250 °C, they are circulated around the secondary heat exchanger where more heat is given up to the cooler return water entering the boiler. Heat is extracted from the combustion products in two ways.

(a) In the form of sensible heat, i.e. the transmittance of heat from a hotter medium (in this case the combustion products) to the cooler return.

(b) By the latent heat of evaporation.

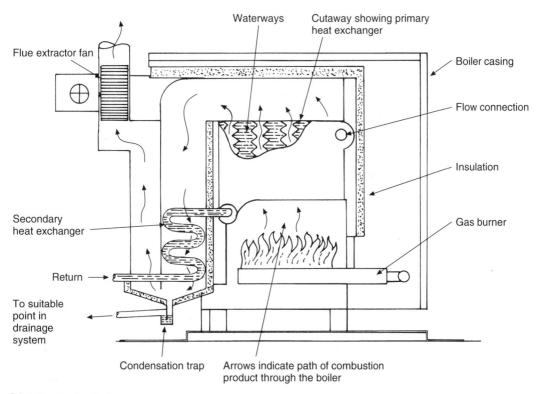

Fig. 5.8   Condensing boiler.

The reader should understand that the water vapour in the combustion products reverts to a liquid when condensed, giving up the heat that caused it to become a vapour. These two sources of heat can be usefully employed in properly designed boilers to increase dramatically their efficiency, even when the electrical energy necessary to operate the extractor is taken into account. Obviously such a component is necessary on all boilers of this type as the temperature of the outgoing products of combustion are so low that the convective effect of natural draught is impossible. The water produced by condensation in the boiler is collected in the base and discharged into a drain, preferably via a gully where the acidic nature of the condensate will be diluted. One further point is worth noting, the relatively small quantity of water produced when the boiler is in operation would, in frosty weather, cause the discharge pipe to freeze. To prevent this a device similar to that employed in automatic flushing cisterns is used which allows a body of condensed water to build up in the base of the boiler until it operates what the plumber would call an automatic siphon. The relatively large body of water thus discharged is then unlikely to freeze.

A further study of the illustration will show that the lower the temperature of the return water in relation to that of the products of combustion, the greater will be the heat transference to the water. Some authorities have suggested that by using much larger radiators to permit lower operating temperatures with consequently lower boiler return temperatures, greater efficiency can be achieved. In practice, however, such a design concept would only fractionally increase operating efficiency and any saving would be offset when compared with the initial cost of installing much larger heat-emission equipment such as radiators. A more realistic approach would be to increase the heating flow and return differential from the recommended 11 °C to approximately 20 °C. This would have the effect of reducing the average or mean temperature of the heat emitters, but if the radiators are accurately sized this might not be a practical proposition. It is, however, worth checking on a refurbishment job where a condensing boiler is replacing an older

model, as it is quite common for existing radiators to be oversized.

The initial cost of condensing boilers is more than that of conventional types, and although the 92–94 per cent efficiency claimed by manufacturers is attractive, as already explained this depends largely on operating temperatures as illustrated by the graph in Fig. 5.9. Dewpoint is the term used to signify the temperature at which the water vapour condenses and reverts to water. The ideal temperature at which this occurs is 59 °C when the air-to-gas ratio is just sufficient to cause complete combustion of the gas. In practice, however, to ensure safe working conditions, a quantity of air in excess of the ideal requirements must be provided which has the effect of lowering the combustion temperature and the dewpoint of the combustion products. Despite this it will be seen that with a dewpoint of approximately 53–54 °C very high efficiency can be obtained with low return-water temperatures. With regard to the viability of changing an existing boiler for a condensing type, several factors must be considered. If the existing boiler is of the old pattern with an open burner, giving efficiencies of only 60–65 per cent, and taking into account its remaining service life, it will be an economic proposition. If, however, it is of

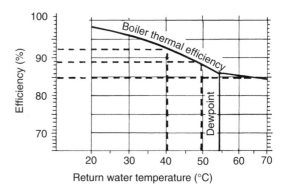

Return water temp (°C)	Efficiency (%)
70	85.9
48	89
40	91

Fig. 5.9 The relationship between boiler efficiency and the return-water temperature.

the new pattern with a closed combustion chamber giving an average efficiency of 75 per cent, despite the savings on operating costs, the initial cost of the condensing boiler would not be viable, bearing in mind that a modern conventional boiler is likely to be operating a heating scheme where the return temperature is relatively high. Condensing boilers are, however, an economic alternative to a conventional boiler in terms of running costs when fitted with new installations designed to give a 20–22 °C drop between the flow and return temperatures. Radiant heating schemes employing a system of pipework embedded in walls, floors or ceilings, operating at lower temperatures and heated by a condensing boiler, would show significant savings on heating costs.

When upgrading an existing boiler one of the practical aspects that must be considered is the position of the flue outlet. Due to their efficiency the flue gases from condensing boilers tend to produce a greater degree of 'pluming' than those of the traditional type. Flue termination under windows or adjacent to doors, car ports or opposing walls, must be avoided.

Pipework systems used for heating

There are basically only two main pipework installations used for heating, the one-pipe and two-pipe systems. There are many variations of these, but they can all be classified under one of the two main headings. Both systems are well established for heating by gravity, and are just as effective for use with forced circulations. The reader should note that the following illustrations are diagrammatic only and do not include details of valves or pumps, etc.

The one-pipe system
This consists of a single pipe to which both the radiator connections are made and it is illustrated in Fig. 5.10. The main advantage of this system is that only one pipe is necessary to convey hot water to the radiators; the main disadvantage is that hot water passing through radiator no. 1 is cooled and so, when it returns to the main pipe and then supplies the next radiator, it has the effect of producing a lower temperature in radiator no. 2 than in radiator no. 1. As this process is repeated at each radiator, the water becomes progressively cooler and the temperature of the last radiator on the circuit is noticeably lower than that of the first. This can be overcome, to some extent, by careful regulation of the lockshield valves on each radiator and by limiting the number of radiators on each single pipe circuit to three, or at most four. There is a further factor to take into account: with the one-pipe system, hot water is pumped round the main circuits only, the radiators being heated by the convection currents which occur between the cool water in the radiator and the hot water in the supply pipe beneath it. As most modern radiators have only $\frac{1}{2}$ in BSP tappings, these may not be large enough to convey

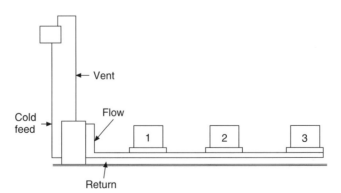

Fig. 5.10 One-pipe system. The main feature of this system is that both flow and return connections to each radiator are connected to a single loop of pipe.

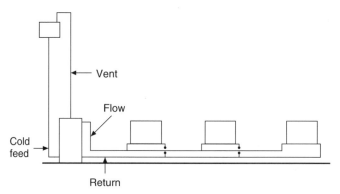

Fig. 5.11 Two-pipe system. The inlet of each radiator in this installation is connected directly to the boiler flow which ensures an even distribution of hot water to each radiator.

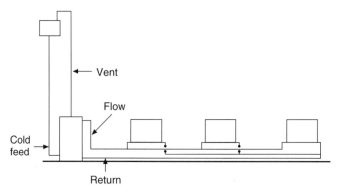

Fig. 5.12 Two-pipe reversed return system, sometimes called a three-pipe system. This is very similar to a two-pipe scheme but, due to the fact that the pipe run to and from each radiator is of the same length, this system is easier to balance.

sufficient water to the radiators at the low flow rates produced by gravity resulting in an unacceptable temperature difference between the top and bottom of the radiators. This and the fact that full-way radiator valves are required limit the application of this scheme when modern heating components are used.

The two-pipe system
This is shown in Fig. 5.11. It has quite a number of advantages over the one-pipe system and, in most cases, should be used in preference to it. Its main advantage is that the hot water is pumped through each heat exchanger or radiator, which gives a quick heat up and a more positive flow. It is for this reason that when convector heaters are used, this system of pipework is essential. The main

disadvantage is that the first radiator on the circuit tends to short-circuit the system, as water will take the easiest path and it will be seen from the diagram that there will be less resistance to the flow of water through the first radiator on the system than to that through the last. This can be overcome in most cases by careful regulation of the lockshield valves. Reference should be made to balancing on page 155.

A modification of the two-pipe system is the three-pipe system, or reversed return system, shown in Fig. 5.12. It is expensive to install due to the extra quantity of pipe necessary, but close inspection of the diagram will show that the length of pipe to each radiator is the same so that the flow of water to each is subjected to the same frictional resistance. This makes the balancing of the system a very simple operation.

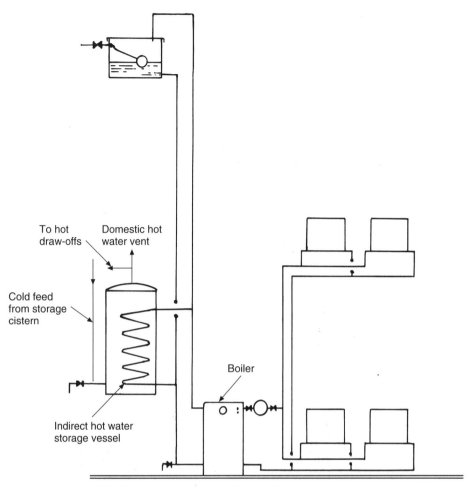

Fig. 5.13 Small bore system. This is a traditional system with the domestic hot water heated by a gravity circulation. Note that no electrical controls are shown.

Heating systems

Pumped heating, gravity hot water systems
Heating systems using small-bore pipes and a pump to circulate the water round the system, have been used for many years. The original arrangement was for the heating circuit to be pumped, the domestic hot water operating on a gravity circulation. Such a system is shown in Fig. 5.12. Although a fully pumped system has some advantages over this system, it is still commonly used and is recommended for schemes employing solid fuel boilers, where a heat leak

is essential. It should be noted that under no circumstances should any valve or temperature control be fitted on a gravity circulation that provides a heat leak. The one possible exception to this are boilers fitted with fan assisted draught. If, however, a gas or oil boiler is used, temperature control of the domestic hot water is mandatory, and on a scheme of this type a two-port valve controlled by the cylinder thermostat would be used. The purpose of controlling the hot water circuit in this way is to avoid what is called 'cycling'. If no control is used, every time a small quantity of water is drawn off, the boiler will fire. This increases fuel

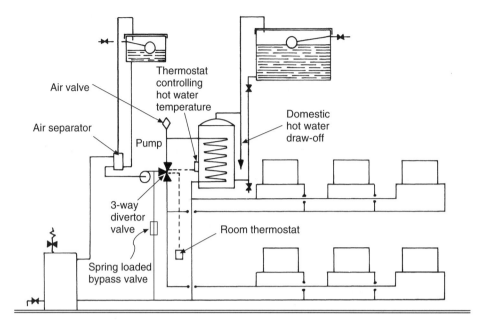

Fig. 5.14 Fully-pumped system. General layout for a fully pumped small-bore system. The room and hot water thermostats control the three-port valve to divert hot water from the boiler to the radiators or heaters in the domestic hot water storage vessel. If the valve is closed against the hot water circulation and all the radiators are shut off water can still circulate via the bypass valve which opens the bypass circuit due to pump pressure.

costs and wastes energy. The main disadvantages of gravity hot water, pumped heating systems are the slow recovery rate of the domestic hot water, and also it is sometimes difficult to arrange a gravity circulation between the boiler and hot storage vessel, especially in bungalows, and for this reason fully pumped systems have a greater degree of flexibility.

Fully pumped systems
This arrangement of heating is a development of the small-bore principle where not only the water to the radiators is pumped, but also the primary flow and return to the heating element in the hot store vessel. Figure 5.14 illustrates the installation details and it will be seen that the heart of the system is the three-way motorised valve shown in Fig. 5.15 which is controlled by a room thermostat and a cylinder thermostat. The original concept of this system was to give priority to the hot water storage vessel should the thermostat be calling for heat. This has the effect of causing

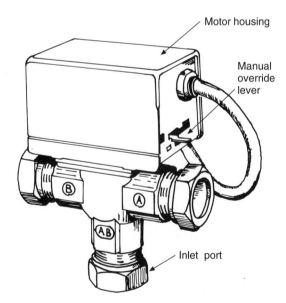

Fig. 5.15 Three-port motorised valve used with fully pumped heating systems. They must be fitted in such a way that the vent and feed pipe to the system are not closed off.

the heating port in the motorised valve to be closed, allowing all the hot water produced by the boiler to be diverted to the coil or annulus in the storage vessel. This continues until the thermostat signals that the water is hot enough and only then will the water be diverted to the space-heating system. This ensures a constant hot water supply, and because the recovery rate is so rapid (approximately 120 litres of water can be raised 40–50 °C in 15–20 minutes) any slight cooling of the radiators during this period is unnoticed. Only when both requirements of domestic hot water and space heating are met will the boiler shut down. Three-way valves of this type are provided with a lever, which when set in the manual position, have the effect of allowing both outlet ports to be held in the open position while the system is being filled, and should the motor be defective, enable both hot water and space-heating demands to be met. This is useful, as although it cancels the control system, both hot water and heating will function until the defective valve is serviced. It is perhaps worth noting that manufacturers of these valves make them in such a way that the motor can be replaced without removing the valve from the system. Similar valves called 'mid' position valves are also available which permit both space heating and domestic hot water to be heated simultaneously. A similar system of control can also be achieved by using two two-port motorised valves, one serving the heating circuit, the other the primary circuit to the hot storage vessel. With all systems of this type the recommendations of the boiler manufacturers relating to the pipework details and the electrical wiring system must be observed. The advantages of these systems can be summarised as follows:

(a) Rapid recovery period for hot water supply even after a heavy draw-off such as running a bath.

(b) Greater circulating pressure by the pump avoids the limitations of a gravity primary circulation to the hot storage vessel in single-storey buildings.

(c) It reduces the effects of stratification; this means the temperature difference between the water at the top and bottom of the storage vessel is much less, resulting in a greater quantity of water at a higher temperature.

(d) Much more positive control is exerted over a system which is electrically operated. Unwanted circulation by gravity is non-existent in a well-designed scheme.

(e) The circulating pipes to the hot store vessel can be reduced in diameter enabling costs of pipework and fittings to be reduced. This must of course be offset against the cost of the motorised valves.

Mini-bore systems

This system shown in Fig. 5.16 is really a development of the small-bore principle using smaller diameter soft temper copper pipes which are obtainable with a PVC sleeve which serves to both protect and insulate the pipes. The main difference is that the flow and return to each heat emitter is connected to the main flow and return through a manifold, two typical examples being shown in Fig. 5.17. If the manifold can be situated in such a way that the branch flow and return to each radiator are approximately the same length, the frictional resistance will also be approximately the same, making the system self-balancing. Some thought must also be given to the flow rates or delivery of hot water to the radiators. It will be obvious that if the pipes are too small in diameter it will be impossible for sufficient water to be delivered to enable the mean output of the heat emitter to be achieved without the use of excessive pump pressure. Of course these facts must be considered in all types of installations. In the case of mini-bore systems the soft temper copper pipes used are made having 8, 10 and 12 mm bores. It is found in practice, however, that for domestic work 8 and 10 mm diameters are satisfactory for most purposes and the use of only two sizes reduces the number of adaptors for valves and manifolds

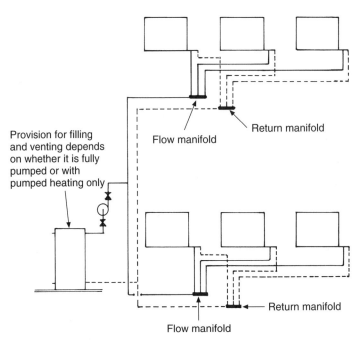

Provision for filling and venting depends on whether it is fully pumped or with pumped heating only

Flow manifold

Return manifold

Return manifold

Flow manifold

Fig. 5.16 Mini-bore system. This system differs from that of small bore in that each radiator has its own flow and return pipes which are connected through a manifold to the main flow and return. The system shown can be used in conjunction with unvented, vented and fully pumped systems heating water for domestic use.

that have to be carried. All diameters of this pipe may be obtained in coils of 7.5, 15 and 30 m in length. Being soft temper they are prone to damage where they are exposed, especially at floor level and Fig. 5.18 shows a method of protecting them at this point. These small pipes are easily concealed and can be run behind skirtings with a minimum of trouble, or in shallow ducts in a floor screed. Where they are passed through or over joists in suspended floors the holes or notches necessary are smaller than required for larger pipes, therefore there is less possibility of weakening the joists. The soft non-rigid nature of the pipe, being dead soft temper, reduces the noise factor sometimes associated with more rigid pipes. The longer lengths require fewer joints and the lower water content in the system makes it very responsive to temperature change and therefore more efficient. Because of the smaller pipe diameters the pump pressure must be raised to overcome the increase

in frictional resistance, the recommended flow rate being 1.5 m per second. The use of soft copper pipe may lead to traps of air in the pipe runs, but the higher pump pressure used (unlike the traditional small-bore scheme) is capable of shifting the air. Because these systems may not be naturally self-venting, they present a good case for the use of the automatic air valves and eliminators described on pages 158 and 169 of this chapter. Mini-bore schemes are suitable for use with all types of system layouts, including solid fuel, providing provision for a heat leak is made. The radiator valves are usually of the annular type (see Fig. 5.19) which enables the flow and return connections to be made on one end of the radiator only. No provision for balancing is made on the valve shown, but some do have this facility. In a well-designed system balancing should not be necessary as the aim is to keep all pipe runs to heat emitters at, as near as possible, the same lengths.

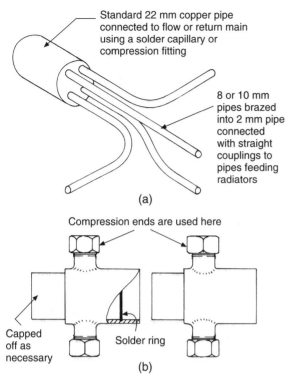

Standard 22 mm copper pipe connected to flow or return main using a solder capillary or compression fitting

8 or 10 mm pipes brazed into 2 mm pipe connected with straight couplings to pipes feeding radiators

(a)

Compression ends are used here

Capped off as necessary

Solder ring

(b)

These four-way tees are used together to build up manifolds as required. Reducers are used in the branches to accommodate the appropriate pipe size.

Fig. 5.17 Manifolds for mini-bore heating.

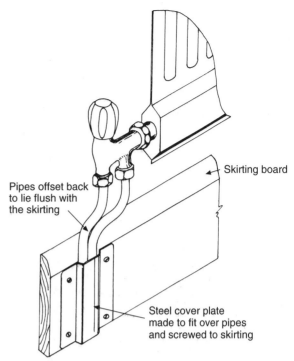

Skirting board

Pipes offset back to lie flush with the skirting

Steel cover plate made to fit over pipes and screwed to skirting

Fig. 5.18 Protecting soft copper or plastic pipes at skirting level.

Sealed heating systems

The basic principles of sealed heating systems are similar to those of unvented hot water systems, but while these have only recently been permitted by water authorities, a sealed system of heating in its present form has been used for a long period of time and historically it is almost as old as low-pressure heating. They have many advantages, possibly the most important are as follows: pipework and cisterns at high level or in a roof space are unnecessary and the system can be operated at higher working temperatures if required. If radiators are used the maximum temperature recommended is the same as for low-pressure schemes i.e. 82 °C. For the really effective use of convector heaters higher temperatures are necessary, and if this type of heating is specified serious consideration should be given to a sealed system

of heating. Figure 5.20 illustrates a typical scheme combined with an unvented hot water system. As with unvented hot water systems they are usually provided by the manufacturer in the form of a package. It should be noted that both ventilated and unventilated hot water systems can be used with all types of heating schemes, whether they are small-bore, mini-bore or fully pumped systems.

System components
The system is filled by one of two methods, either manually via a small top-up vessel at the highest point of the system or by direct connection to the mains, providing the following requirements of the water by-laws are met. *The connection between the mains supply and the heating system must be temporary only, to be used when filling or topping up; after use it must be disconnected.* A backflow device such as a check

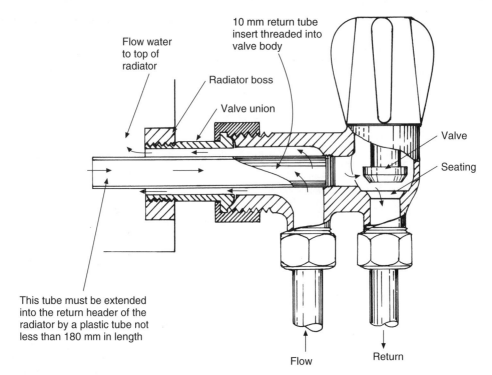

Flow water to top of radiator

Radiator boss

Valve union

10 mm return tube insert threaded into valve body

Valve

Seating

This tube must be extended into the return header of the radiator by a plastic tube not less than 180 mm in length

Flow

Return

Fig. 5.19 Double entry radiator valve. They may be manually or thermostatically controlled and are specially designed for use with small pipes and permit both the flow and return to the heater to be connected at one end. When used with double panel radiators a special flexible insert is required.

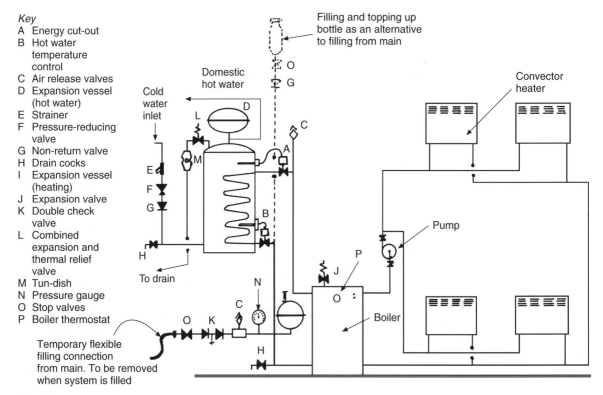

Key
A Energy cut-out
B Hot water temperature control
C Air release valves
D Expansion vessel (hot water)
E Strainer
F Pressure-reducing valve
G Non-return valve
H Drain cocks
I Expansion vessel (heating)
J Expansion valve
K Double check valve
L Combined expansion and thermal relief valve
M Tun-dish
N Pressure gauge
O Stop valves
P Boiler thermostat

Filling and topping up bottle as an alternative to filling from main

Domestic hot water

Convector heater

Cold water inlet

Pump

Boiler

To drain

Temporary flexible filling connection from main. To be removed when system is filled

Fig. 5.20 Sealed system of heating with an unvented domestic hot water system.

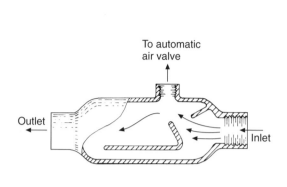

(a) Air purger: sealed heating systems

This works in a similar way to that of an air eliminator for low-pressure heating. The turbulence caused by the internal plates cause the minute air bubbles to merge together where they are easily dispersed through the automatic air valve.

(b) Automatic air valve or eliminator

These valves are used in combination with an air purger in sealed systems of heating, but they have many other applications in low-pressure systems where the use of an open vent is undesirable, i.e. high points on a heating system or on radiators that persistently collect air. It must be stressed, however, that the use of these valves in low-pressure systems does not mean that an open vent at some point in the system is unnecessary.

Fig. 5.21 Air elimination in sealed heating systems.

and anti-vacuum valve or double check valve must be permanently fitted at the inlet point. This ensures there is no possibility of contamination of the mains supply by water in the heating system, which may contain a corrosion inhibitor or other unpleasant additives.

Air purger This is necessary only when filling takes place from the mains. Its purpose is to remove the air bubbles from the water as it enters the system. It is expelled through an automatic air valve which is sometimes made as an integral part of the air purger. Both these components are illustrated in Fig. 5.21.

Expansion vessel This takes the place of the feed and expansion cistern used in a low-pressure heating system to accommodate the expansion of the water when it is heated. Its capacity is important and it must be capable of absorbing the expansion of water when it is raised through 100 °C. Manufacturers of these vessels will advise on its

capacity for a specific installation. The factors involved are the static pressure on the system, the maximum working pressure which will be the same as the relief valve setting and the volume of the water in the system. An approximation of the water content can be made working on a basis of 14 litres per kW of boiler power. The static pressure is calculated by measuring the highest point in the system to the expansion vessel. The situation of the expansion vessel in the system is not critical, but it is likely that the life of the flexible membrane will be extended if it is not subjected to water at very high temperatures. For this reason it may be fitted to the system return via a non-circulating pipe as shown in the illustration. It is also suggested that the pump exerts a negative pressure rather than positive pressure on the membrane, as this reduces the pressure to which it is subjected. The gas charge in the vessel can be air or nitrogen, but the letter is considered to be the better, as if, due to a fractured membrane the gas escapes into the system, air, containing oxygen, is likely to cause corrosion.

Figure 5.22(a) shows an expansion vessel when the system is cold, Fig. 5.22(c) when the normal operating temperature is achieved.

Safety devices

As with unvented hot water systems there are three safety devices to prevent the build-up of excess air pressure. The first is the thermostat permitting

variation of temperature of the appliance. Should this fail, a temperature cut-out comes into operation in the same way as an unvented hot water system. Thermal relief valves are not normally fitted, but a pressure relief valve is of course essential. They are very similar to the expansion relief valves used with unvented hot water systems. The safety valves traditionally used for low-pressure heating are unsuitable for this type of work as they are often made of poor-quality brass which is affected by dezincification, and the valves often become sealed on to the seating due to corrosion, or in some cases, lime scale. Safety devices for sealed systems should be made of bronze and any seals, washers or O-rings made of heat-resistant synthetic rubber or plastic. A pressure gauge, often combined with a rotary-type thermostat, should be provided to enable the user and service engineer to monitor the working characteristics of the system. Working pressures of these systems should not exceed 3 bar and the water temperature should not exceed 99 °C.

Under-floor systems of heating

Both installation and design procedures are dealt with later in this chapter.

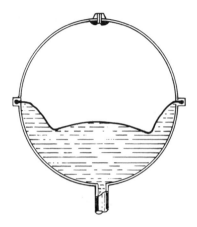

(a) The system cold with the vessel charged with gas

The gas charge is equal to the static head exerted on the membrane + a margin. Normally this is 3 bar.

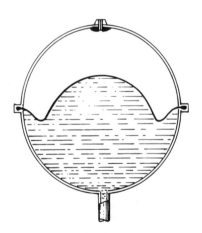

(b) The water is heated and the gas is compressed due to expansion of the water

Note that expansion vessels used for sealed heating systems are not suitable for unvented hot water.

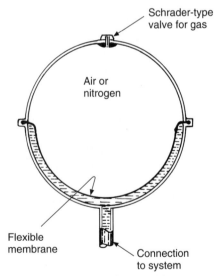

Schrader-type valve for gas

Air or nitrogen

Flexible membrane

Connection to system

(c) Operating temperature is achieved

Fig. 5.22 Expansion vessel. The gas, air or nitrogen in the vessel is precharged to a predetermined pressure related to the static head on the system. A gas charge of 1 bar will support approximately a 10 m head of water. If there were no charge, water would fill the vessel leaving no space for the expansion of the water when it is heated.

Heating controls

Controls for automatic shut-down of the system and temperature are absolutely necessary for both hot water and heating installations.

The Building Regulations are currently under review, which may mean even more stringent controls to avoid energy wastage and reduce CO_2 emissions. The 1995 Regulations currently in force specify the following.

(a) Time control for all wet central heating systems except natural draught solid-fuel boilers.
(b) Zone control to provide separate temperature control in areas with different requirements.
(c) Cylinder heat exchangers (the internal coil) must comply to at least BS 1566. This is to prevent wastage of fuel due to boiler cycling. The cylinder must also be insulated to a minimum standard and the first metre length of any pipe connected to it must also be insulated.
(d) Temperature control of the stored water is also mandatory.
(e) The thermostats on gas and oil-fired boilers used for wet central heating systems must be interlocked with the thermostats controlling space and hot water storage heating. This ensures that the boiler will only fire when there is a call for heat, not simply when the water in the boiler cools.

Boiler thermostats
These are usually an integral part of the boiler. Those used for modern gas- and oil-fired boilers work on the principle of expansion, either of heat-sensitive fluids or metals which operate an electric switch. Typical examples are shown in Chapter 2. The boiler thermostat controls the temperature of both the space heating and domestic hot water equipment.

Time switches
Some provision for automatically switching on the apparatus is the next essential, and this is achieved by means of a time switch, or its modern counterpart, a programmer. This is basically an electric time switch which not only determines the times when the system is operational but also controls the boiler and pump independently. If the programmer is set for hot water, the boiler controls are energised causing it to heat the domestic hot water only. If it is set for both heating and hot water the pump is also energised causing the space-heating equipment to operate as well. A typical programmer is shown in Fig. 5.23.

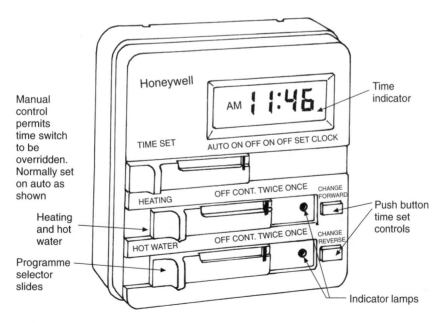

Fig. 5.23 Programmer.

Cylinder thermostats

The average or mean design temperature on which most domestic heating schemes operate is approximately 77 °C, which would overheat the domestic hot water, the maximum temperature of which should not exceed approximately 65 °C. To control the hot water temperature a cylinder thermostat controlling a motorised valve is the usual method employed. A cylinder thermostat and its application is shown in Fig. 5.24.

The position of the thermostat on the vessel is important, and if it is fitted as shown, it will prevent the boiler 'cycling', which has been mentioned previously. The motorised valve will remain closed preventing wasteful circulation until a call for heat is made by the cylinder thermostat.

Space heating control

Room thermostats   The boiler is normally set for the temperature required for hot water, usually about 80 °C, and unless some form of control is used, this can lead to unnecessarily high space heating temperatures. The heating system seldom operates at its designed temperature, except for a few weeks in the year when the weather is very cold. Overall control of space heating is achieved by a room thermostat which is normally set at a temperature of 20–22 °C. When the temperature falls the contacts close and start the pump. On modern heating systems fitted with TRVs, they provide overall control of the space heating system and should therefore be fitted in an area of the building requiring a lower temperature than that of living rooms and bedrooms. The room thermostat shown in Fig. 5.25 is a basic type. Some now available can be programmed to provide both timing and variations of temperature at different periods in a 24 hour or 7 day cycle. The latter is very useful when the pattern of building occupation varies during the period of a week. A further development in control systems is the use of radio waves, which makes conventional wiring unnecessary.

Zone control   In some types of commercial and larger dwellings, energy savings may be

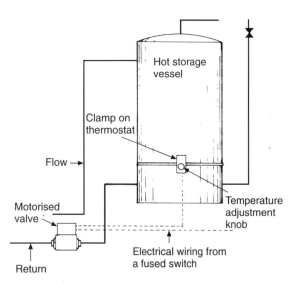

Fig. 5.24   Temperature control of domestic hot water with a motorised valve. This is an electrical system of control which is suitable for both domestic and industrial use. The thermostat controls the motorised valve causing it to open when the water temperature in the vessel drops, and close when the predetermined temperature is achieved.

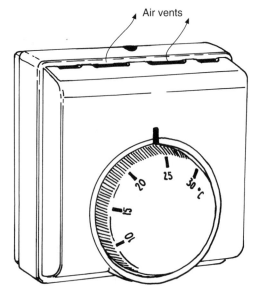

Fig. 5.25   Room thermostat. These operate by a flow of air passing over a bimetallic coil causing it to expand or contract making or breaking a microswitch.

made by breaking down the space heating system into zones, e.g. living spaces and bedrooms. This allows different parts of the building to be either isolated when not in use or heated to different temperatures. The use of thermostatic radiator valves accomplishes this and is considered satisfactory for small dwellings. In the case of larger buildings considerable savings can be made by controlling the various zones using two-port motorised valves and programmable room thermostats. These control not only the zone temperature but also the on/off periods when the heating is required.

Temperature control interlock   To ensure maximum economy and efficiency, heating control systems must be fully interlocked, which means the boiler will not fire until the room or cylinder thermostat calls for heat. Figure 5.26 shows in schematic form how the controls are wired to achieve this. With older systems the boiler thermostat was controlled only by the programmer which allowed the boiler to fire every time its water content cooled. This resulted in wasteful cycling when there was no call for heat from the room or boiler thermostat.

Weather-compensating controls   These have been fitted for many years in commercial buildings and may be used with advantage in larger domestic properties. They ensure that energy is not wasted

in milder weather when a design flow temperature of, for example, 82 °C for space heating is unnecessary. A simple valve of this type is shown diagrammatically in Fig. 5.27. Some of the larger types used in industrial and commercial buildings are electrically operated, but the basic principles are the same as shown.

Thermostatic radiator valves (TRVs)   These are now mandatory for space heating control and work in conjunction with the room thermostat. Details of these valves are shown in Fig. 5.28. They are fitted to each radiator and control the temperature of each room independently of any other. The fitting instructions supplied by the manufacturers of these valves must be carefully observed to ensure their full effectiveness. They should not be positioned where they can be affected by heat from the sun as this can result in their closure before the desired room temperature is achieved.

TRVs embody a sensor containing a heat-sensitive fluid or wax which expands when the required air temperature in the room is achieved, closing the valve, and contracts when the temperature in the room falls below that required, thus opening the valve.

TRVs have a low temperature setting so that if the room temperature falls below approximately 4 °C, they will automatically open. This means that if in very cold weather the heating system is turned on in empty premises, these valves provide some degree of protection in the event of a frost. In a property where all the heat emitters are fitted with TRVs, it is essential that one of the circuits is open, so that in the event of all valves being in the automatically closed position and the pump starts, it is not operating against a static pressure. While it is unlikely that all the manually operated valves would have been turned off, in the case of thermostatic valves this could easily happen causing possible damage to a low-water-content boiler and the pump if the room thermostat is calling for heat. This problem has hitherto been overcome by connecting the flow and return together after the last radiator connection on the system. This can cause unnecessary heat losses which can be avoided by the use of an automatic spring-loaded bypass valve.

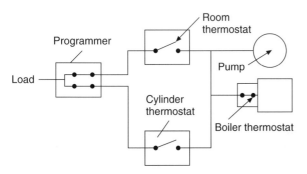

Fig. 5.26   Interlocking of controls. Although the programmer switches and the boiler thermostat are closed, the boiler will not fire until the room or cylinder thermostats close on a call for heat.

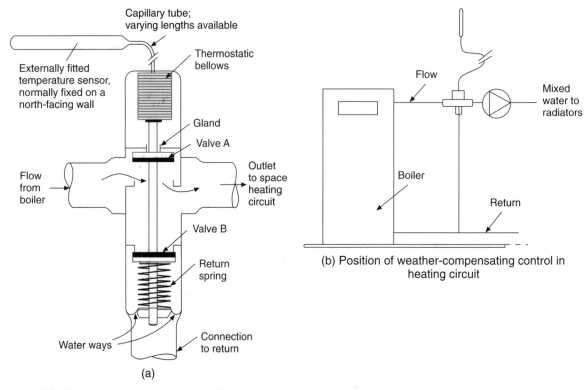

Fig. 5.27 Weather-compensating control. The valve is shown with valve A open, and in this position water at boiler temperature is flowing to the radiators. An outside temperature rise will be transmitted via the sensor to the bellows, partially closing valve A and opening valve B. This allows water from the heating return to mix with the water from the boiler, effectively lowering the temperature of the water circulated to the radiators.

It must be fitted on a bypass pipe between the outlet side of the pump and the main heating return. It is designed to open when there is a build-up of pressure in the system, e.g. if all valves are closed. When a heating circuit opens the valve closes, allowing the full output from the pump to circulate through the system. It must be manually adjusted to meet the requirements of the system.

Frost thermostats   As with room thermostats, frost thermostats sense the temperature of the air and, except for working over a lower temperature scale, they are almost identical. They are essential if the premises are left empty for a long time or during periods of frost or if the boiler is fitted outside the building. They are wired into the control scheme in such a way that the time switch or programmer, when in the off position, is overridden when there

is a sharp drop in temperature. They are sometimes installed with a limit thermostat as shown in Fig. 5.29. The foregoing controls are generally applicable to oil- and gas-fired schemes. Due to the necessity of providing a heat leak with solid fuel appliances, it is not recommended that any automatic control is fitted to the domestic hot water circuit.

Wiring electrical control systems   Most manufacturers of control systems publish suitable information for installers who undertake this work, and also run short courses on this subject, which are available for a modest sum. It must be stressed, however, that it is essential that all electrical work complies with the IEE wiring regulations and no short cuts should be taken. Failure to comply could not only be dangerous but also make it very difficult

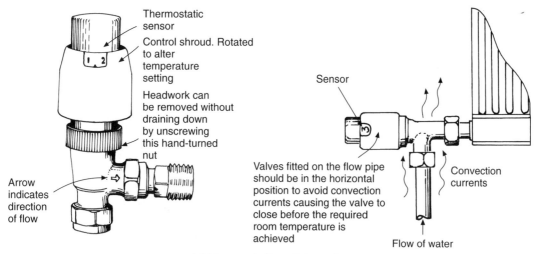

(a) Thermostatic radiator valve

Compliance with the manufacturer's instructions is essential if the valve is to operate effectively. Generally, the best position is on the flow with the sensor in the horizontal plane.

The following labels appear in figure (a):

- Thermostatic sensor
- Control shroud. Rotated to alter temperature setting
- Headwork can be removed without draining down by unscrewing this hand-turned nut
- Arrow indicates direction of flow
- Sensor
- Valves fitted on the flow pipe should be in the horizontal position to avoid convection currents causing the valve to close before the required room temperature is achieved
- Convection currents
- Flow of water

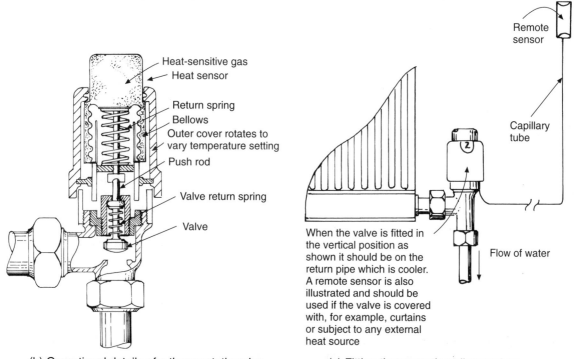

(b) Operational details of a thermostatic valve

The following labels appear in figure (b):

- Heat-sensitive gas
- Heat sensor
- Return spring
- Bellows
- Outer cover rotates to vary temperature setting
- Push rod
- Valve return spring
- Valve

(c) Fitting thermostatic radiator valves

The following labels appear in figure (c):

- Remote sensor
- Capillary tube
- When the valve is fitted in the vertical position as shown it should be on the return pipe which is cooler. A remote sensor is also illustrated and should be used if the valve is covered with, for example, curtains or subject to any external heat source
- Flow of water

On a rise in room temperature the heat-sensitive gas expands causing the bellows to push the valve on to its seating. Both the valve and bellows are spring loaded to ensure a positive opening of the valve on a drop in temperature.

Fig. 5.28   Thermostatic radiator valves.

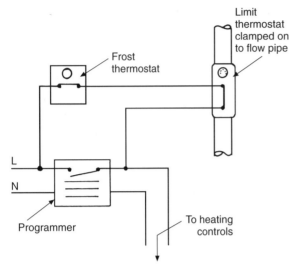

Fig. 5.29 Wiring to frost thermostats. The illustration shows a schematic wiring diagram for frost protection. The time switch is open and the system is shut down. If the air temperature falls below the thermostat setting (2–3 °C) the contacts close, overriding the time switch. The limit thermostat, shown strapped to a flow pipe adjacent to the boiler, will normally be closed when the system is cold. Under these conditions the system will operate until the limit thermostat opens. It is generally set at 10–15 °C which prevents the system operating at design temperatures when, but for very cold weather, it would normally be shut down.

to trace a fault should one occur at a later date. In short, the electrical installation must be carried out strictly according to the diagrams supplied with the components, and the wiring colour code used throughout, if applicable.

Central heating design

Heat loss

Modern, efficient central heating systems are designed on the basis of calculated heat loss, the main sources being loss of heat through the building fabric by conduction, and ventilation or air change. While bearing in mind that some ventilation of rooms is necessary for fresh air, odour removal, prevention of condensation and combustion air for heating appliances, excessive air changes are both unnecessary and costly.

Prior to designing and installing a heating system, a simple survey should be made to investigate and rectify unnecessary heat loss, especially in older buildings. Heat loss can usually be minimised quite simply without too much cost, being identified as follows:

(a) Is the roof space sufficiently well insulated? A minimum thickness of 100 mm of insulation blanket is essential. Heat loss in this area can be further minimised by a thicker blanket and a layer of aluminium foil.

(b) Non-insulated cavity walls can be filled with insulating material by companies who specialise in this work. New buildings must comply with the requirements of the Building Regulations, and external walls must be constructed to specific standards to reduce thermal transmittance. Prior to 1965 many buildings were constructed with solid brickwork 225 mm thick, the heat loss through this form of structure being very high by today's standards. The cost of dry lining external walls in such buildings is worth consideration.

(c) Double glazing does reduce heat loss through windows by approximately 50 per cent, but is very expensive, and because window areas are relatively small, any savings on heating costs would take a long time to pay for its installation. The main advantages of double glazing are the reduction of condensation on inside windows, and the minimisation of noise levels from external sources.

(d) The closure of all apertures causing draughts is also worth investigating. Gaps around doors and windows can easily be weatherstripped. Shrinkage of floors and skirting board on suspended floors can also present a problem unless the room is carpeted. In rooms without carpet, gaps can be sealed using a suitable filler, or fixing a small wooden moulding around the base of the skirting. Open fires in a room should be sealed if no longer in use, as they are very wasteful due to the convective effect of the flue, but should a client wish to retain a fire, a sealed solid fuel or gas appliance will result in lower heat losses than an open fire.

Table 5.1 Room temperature and ventilation rates.

Room	Temperature (°C)	Air changes per hour
Living	21	1.5
Dining	21	1.5
Bed/sitting	21	1.5
Bedroom	18	1.0
Hall/landing	18	1.5
Bathroom	22	2.0
Toilet	18	2.0
Kitchen	18	2.0

Heating requirements

Calculated heat loss is normally based on an outside temperature of −1 °C. Table 5.1 gives the inside temperature recommended by BS 5449:1990.

Design flow temperatures for systems where the heat-emitting surfaces are exposed, e.g. radiators, should not exceed 82 °C to avoid the possibility of burns. The recommended return temperature is 71 °C for pumped schemes giving an 11 °C temperature drop. By circulating the water more quickly the temperature drop could be lowered, but the higher water velocity necessary could result in pipework noise. Where heating systems operate intermittently, i.e. where they shut down overnight, an allowance should be added to the calculated heat loss of between 10 and 20 per cent. This has the effect of oversizing the heat emitters and allows for a quicker heat-up. The higher percentage is normally used for very well insulated buildings only, because of the low design heat requirements.

Boiler output rating

The boiler should be capable of meeting the sum total of the design heat requirements and the emission of the system pipework. An allowance of 2–3 kW is normally made for boilers supplying both hot water and heating.

Thermal transmittance (conduction)

This term relates to the flow of heat through the building fabric. The first step in heating design is to determine the heat loss. There are three main factors to consider here:

(a) The building fabric (some types resisting the rate of flow better than others).
(b) The superficial area of walls, ceilings and floors.
(c) The temperature difference between the inside and outside of the building. The greater the difference, the greater will be the heat flow.

These factors are all taken into account by the use of U-values, which have been devised by experiment and calculation and are published in tables covering most forms of construction. A U-value may be defined as the rate of heat flow through a structure in watts/m^2 per hour, per 1 °C temperature difference and is illustrated in Fig. 5.30(a). Window glass has a very high U-value of 5 W/m^2 because of its comparatively thin section. To carry this one stage further, assuming a 22 °C temperature rise is required as shown in Fig. 5.30(b) the heat loss will be very much greater, and is calculated as follows: Area of glass m^2 × temperature rise × U-value, thus:

$$1 \times 22 \times 5 = 110 \text{ W}$$

The result is a heat loss of 110 W through a single-glazed window.

It will be seen that this is a considerable loss of heat, and fortunately most other construction materials have a lower conductivity rate. In order to save energy, it should be noted that current Building Regulations specify that the maximum permitted U-value for domestic buildings is

0.25 W/m^2 °C for roofs
0.45 W/m^2 °C for external walls and floors.

The reader should bear in mind that U-values relate to the heat transference factor of various combinations of building materials, the lower its numerical value, the less will be the heat loss.

The following text shows how heat loss is calculated (a) through the building fabric and (b) due to air changes. Figure 5.31 illustrates a typical room of which the heat loss is to be determined, and shows the relevant U-value of its components the heat loss of these being tabulated in Table 5.2.

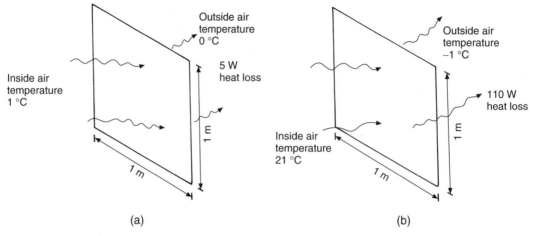

Fig. 5.30   Heat losses through glass.

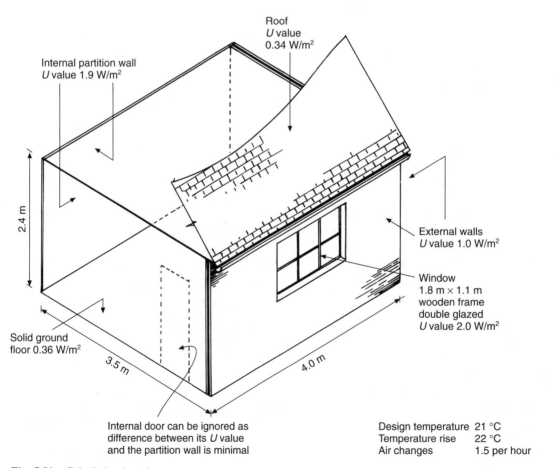

Fig. 5.31   Calculating heat losses.

Table 5.2 Tabulating heat requirements.

Component	Area (m²)	U-value	Temperature rise	Watts
Floor	14.0	0.36	22	110.88
External walls	16.02	1.0	22	352.44
Internal walls	18.0	1.9	5	171.00
Window	1.98	2.9	22	126.32
Roof	14.0	0.34	22	104.72
Heat loss through the building fabric (total)				865.36
+ Heat loss due to air change 1.5 watts per hour				365.90
Total heat loss in room				1231.26
+ 15% for intermittent firing				184.69
Heat requirements (watts)				1415.95

To illustrate how this is done the floor is dealt with first thus. Measurements:

$$3.5 \times 4.0 = 14 \text{ m}^2 \text{ total floor area}$$

If the loss through 1 m² of floor area is 0.36, the loss through the whole floor will be

$$14.0 \times 0.36 = 5.04 \text{ W}$$

This will be the heat loss for 1 °C temperature difference. A temperature difference of 22 °C is, however, required:

$$5.04 \times 22 = 110.88 \text{ W}$$

This will be the heat loss per hour through the floor. The other components are calculated in the same way: Total area of two external walls:

$$7.5 \times 2.4 = 18 \text{ m}^2$$

Less window area:

$$1.98 \text{ m}^2 \qquad = 16.02 \text{ m}^2$$
$$16.02 \times 1 \times 22 = 352.44 \text{ W}$$

The remaining components have been calculated and entered into the Table 5.2 using the same method.

To ascertain the heat loss due to air change, the cubic contents of the room are calculated and multiplied by the temperature rise, the number of air changes and 0.33, this figure being a constant relating to the specific heat of air:

$$2.4 \times 3.5 \times 4.0 = 33.6 \text{ m}^3$$
$$33.6 \times 1.5 \qquad = 50.4$$
$$50.4 \times 0.33 \qquad = 16.63$$
$$16.63 \times 22 \qquad = 365.9$$

Thus 365.9 W must be added to the fabric losses in the example, and by adding 15 per cent for intermittent firing, the total heat loss in watts can be seen.

Domestic heating calculators

A typical calculator is shown in Fig. 5.32 and has been used by heating installers for many years to determine the design requirements of domestic properties without resorting to lengthy calculations. It should be noted that their use is generally confined to domestic buildings with standard methods of construction, using traditional materials. It has been found that their use may slightly oversize heat emission surfaces, but this can be ignored for all practical purposes. A calculator is basically a series of rotating scales, which, when lined up, provide the information necessary to design the complete system. The calculator used here is for general purposes. For buildings complying to the latest requirements of the Building Regulations, a calculator should be used which takes into account the permissible heat transference rate of energy-saving buildings. This is necessary, as the heat emission surfaces in such buildings will be smaller. All calculators take into account any

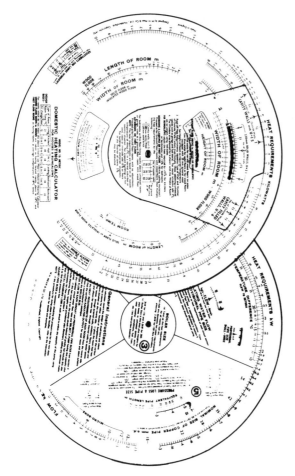

Fig. 5.32 Illustrates the two sides of a domestic central heating calculator. (Courtesy of M.H. Mear & Company Limited)

necessary allowances, e.g. the percentage increase of heat emission where the heating system is used intermittently.

Figure 5.33(a) and (b) shows the principle dimensions of a small two-bedroomed house, and Figure 5.34(a) and (b) shows how the calculator is used to determine the heat requirements of the lounge area. This is demonstrated by illustrations of the relevant scales using a step-by-step approach.

Set the width of the room by its length — in this case for solid floors (a separate scale is provided for suspended floors).

Set the height of the room to the required temperature rise, e.g. 22 °C.

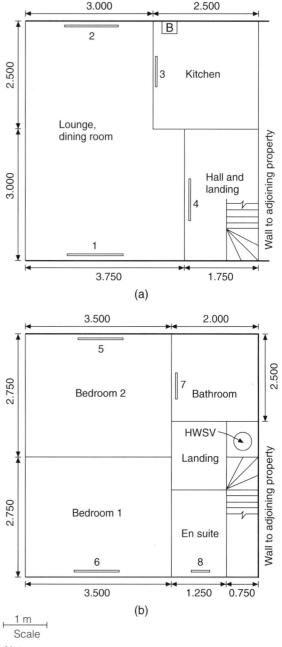

(a)

(b)

1 m
Scale

Notes
1   Solid ground floor
2   Wood floor on joists first floor
3   280 mm external cavity walls
4   All room heights 2.400

Fig. 5.33 Plan of small two-bedroomed semi-detached house: (a) ground floor showing measurements of rooms, radiator and boiler positions; (b) first floor plan.

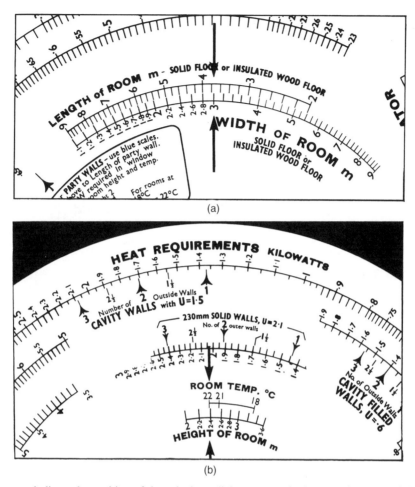

Fig. 5.34 The arrows indicate the position of the calculator dials to assess the heat requirements of the lounge which are shown as 1.8 kW.

Read off the heat requirements against the number of outside walls and its construction. In this case they will be kilowatts.

The calculator can be used to indicate the area of heat emitter surface required, but it is often more convenient to read this information directly from the manufacturer's information sheets, a section of which is shown in Table 5.3. All radiators are of the single panel convector type, 600 m high. A radiator schedule should be prepared at this stage as shown in Table 5.4. The boiler capacity may now be selected by totalling the space heating requirements of each room and making an allowance 2.5 kW for domestic hot water as shown in Table 5.5.

An increase of 15 per cent to allow for a quick heat-up period is shown for systems operating intermittently.

The next step is to determine the pipe sizes that will provide sufficient hot water to supply the radiators and the main circuits. It can generally be assumed that for small-bore schemes 15 mm is of sufficient size to supply each individual radiator. If, for example, there is any doubt about a very long circuit serving a large radiator, the pipe size can be ascertained in a similar way to that of sizing a circuit. To determine the circuit loadings, a drawing illustrating the circuit layout for the building under consideration is given in Fig. 5.35.

Table 5.3 Radiator sizing chart for a 600 mm high radiator.

Nominal length		Single convector type 1P		
mm	in	Order no.	Watts	BTUs
400	15.7	6004/1P	536	1830
500	19.7	6005/1P	664	2265
600	23.6	6006/1P	790	2696
700	27.6	6007/1P	915	3123
800	31.5	6008/1P	1040	3548
900	35.4	6009/1P	1164	3971
1000	39.4	6010/1P	1287	4392
1100	43.3	6011/1P	1410	4810
1200	47.2	6012/1P	1532	5227
1300	51.2	6013/1P	1654	5643
1400	55.1	6014/1P	1775	6057
1600	63.0	6016/1P	2017	6881
1800	70.9	6018/1P	2257	7700
2000	78.7	6020/1P	2496	8516
2200	86.6	6022/1P	2734	9328
2400	94.5	6024/1P	2791	10136

Table 5.5 Boiler sizing.

	kW
Total space heating requirements	9.06
+ 15%	1.36
Domestic hot water	2.5
Boiler capacity = Total	12.02

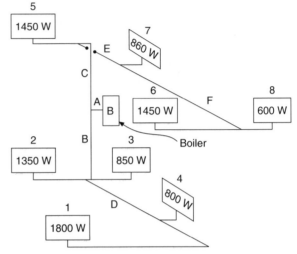

Fig. 5.35 Diagrammatic layout of pipe runs showing how they are broken down into individual circuits feeding groups of radiators, so that their heating load and pipe diameters can be established. Each radiator is identified by its number on the radiator schedule. It will be seen that circuit A, although very short, is carrying the whole circuit load. Circuit B is feeding the ground floor; circuit C the first floor; E, F, D, are sub-circuits of C and B.

Figure 5.36(a) shows how the calculator is used to determine the flow rates and pipe diameters. The method of determining the flow pressure drop for section B is shown in Fig. 5.36(b). The circuit lengths are taken from the plans (Fig. 5.33(a) and (b)), the usual practice being to take the measurements from the centre of each radiator. Note also that the circuit length must be doubled for two pipe schemes. All the foregoing details for

Table 5.4 Radiator schedule.

Radiator No.	Location	Design temperature (°C)	Heat emission (kW)	Radiator size (mm)
1	Lounge area	21	1.70	600 × 1400
2	Dining area	21	1.35	600 × 1100
3	Kitchen	18	0.85	600 × 700
4	Hall/landing	18	0.80	600 × 700
5	Bedroom 1	18	1.45	600 × 1200
6	Bedroom 2	18	1.45	600 × 1200
7	Bathroom	22	0.86	600 × 700
8	En suite	22	0.60	600 × 500

Note: Although the design temperature is shown, the actual temperature rise is 1 °C in each case. This must be taken into account when using the calculator, e.g. for a design temperature of 21 °C the actual temperature rise will be 22 °C. The radiator sizes are read from Table 5.3, which is an extract from a manufacturer's list.

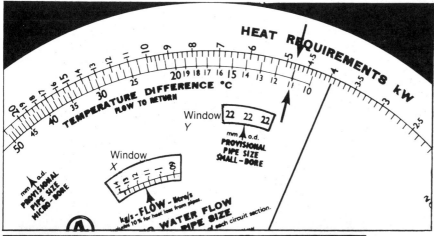

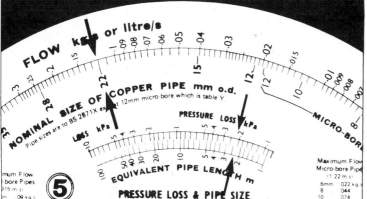

(a) Determining the water flow and provisional pipe size

Section B has a total heating load of 470 kW, and the scale is rotated until the load lines up with the temperature drop (11 °C) shown arrowed. The flow rate can then be read in window X as just under 0.12 kg/s (litres per second), and a provisional pipe size of 22 mm is shown in window Y.

(b) Determining the pressure drop on circuit B

From Fig. 5.34 it has been established that the circuit length is 2.4 m, with a flow rate of 0.12 litre/sec. The pipe size (22 mm) is set against the water flow scale (top arrows). The pressure loss can now be read off through the window and will be seen to be approximately 2.5 mb (bottom arrows).

Fig. 5.36 Determining flow rates and pipe diameters.

each pipe section must be recorded as shown in Table 5.6.

The final step is to determine the 'index circuit'. This is the circuit which is the most difficult to feed by virtue of its heating load and pressure drop. It will be seen that the scheme has two main circuits serving the ground and first floors. Dealing with the ground floor first, it will be seen it comprises of pipe sections A, B, D. The total water flow for these sections is obtained from Table 5.6 and found to be 0.42 litres per second. Also from Table 5.6 it will be seen that the total pressure loss in 85 mb, which divided by 10 gives 8.5 kN (kPa). The first floor circuit is dealt with in the same way, and from Table 5.6 it will be seen the volume of water to be circulated is 0.48 litres per second against 95 mb. This will be the index circuit upon which the pump

loading is determined. Reference should be made to Fig. 5.37 which shows a typical graph for setting the pump to give the required outputs. The broken lines show the correlation of the water volume and the pressure. It will be seen they intersect in the area where a pump setting of (2) is shown to be adequate.

Under-floor central heating

This type of system employs, in effect, a heating pipe coil embedded or laid in a floor, which heats the space above it mainly by radiant heat. As radiant heat requires no medium through which to travel, floor coverings of whatever type used have little effect on the heat output. It is not entirely new: radiant heating using copper tubes has been

Table 5.6   Circuit sizing.

Col. 1	2	3	4	5	6
Pipe section	Heat emission per section k/W	Flow rate litres/sec	Pipe diameter (mm)	Circuit length (m)	Pressure drop (mb)
A	9.06	0.23	22	0.5	20.0
B	4.70	0.12	22	2.4	2.5
C	4.36	0.11	22	2.4	2.0
D	2.50	0.07	15	18	60.0
E	2.91	0.08	15	16	20.0
F	2.05	0.06	15	15	35.0

Notes: Column 1 lists the pipe section taken from Fig. 6.34. Column 2 is the total heating load carried by each section, e.g. pipe section B is carrying the load for all the ground floor, section D is carrying the load for the lounge and hall radiator only. Columns 3–6 list the values obtained from the calculator, for example, Fig. 6.35 shows calculations for section B. The calculator gives the maximum flow rates for 15 mm pipe as 0.13 litre and for 22 mm pipe as 0.29 litre at a velocity of 1 m/s. It can be seen above that these velocities are not exceeded and that the original pipe sizes selected are suitable.

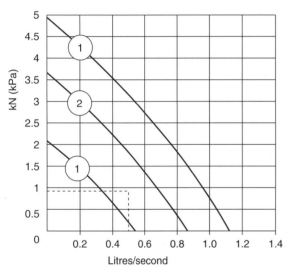

Fig. 5.37   Pump duty selection graph. The pressure 0.85 kN and volume of water to be pumped, 0.48 litres per second are identified by the horizontal and vertical broken lines. It will be seen they fall into area 2 on the graph which corresponds to setting no. 2 on the pump.

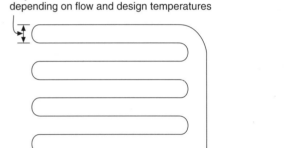

Fig. 5.38   Pipe coil for under-floor heating.

fitted for many years but has never been used extensively for domestic work. The use of modern flexible plastic pipes, e.g. made from cross-linked polythene, and more sophisticated methods of control have made this system more viable, certainly for new-build work. It has many

advantages, not the least being that the design temperature is much less than that for radiators, and at 45–60 °C is ideal for the use of condensing, energy-saving boilers. While most people do not object to the appearance of radiators, they do occupy wall space and might in some cases limit the interior design, and in very old properties radiators are hardly authentic.

Figure 5.38 illustrates the layout of the pipe coil for each room or space and Fig. 5.39 shows a

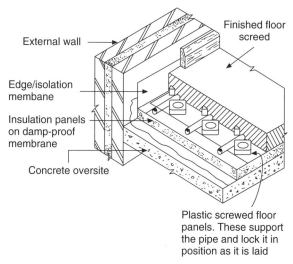

External wall

Edge/isolation membane

Insulation panels on damp-proof membrane

Concrete oversite

Finished floor screed

Plastic screwed floor panels. These support the pipe and lock it in position as it is laid

Fig. 5.39 Section of under-floor heating in solid floors. Refer to manufacturers for details of suspended and floating floors.

section of a solid floor installation. The methods of pipe support vary if suspended or floating floors are heated (the term 'floating floor' relates to an installation fitted over an existing floor). The flow and return from each coil are taken to a distribution panel housed in a box and connected to the main manifolds from the boiler. This box must be situated in a convenient position in the building and also houses the on/off and lockshield valves to each circuit. As with most modern systems a fully pumped scheme is recommended. Individual control

of each room or area is more difficult as the use of TRVs is not practical in most cases, and for smaller schemes a normal room thermostat is used. The revised Building Regulations may make individual room temperature control mandatory, and if this is the case the use of two-port motorised valves controlled by a room thermostat for each circuit will be necessary.

Because the flow temperature is lower than that necessary for hot water and radiator schemes, where a single boiler is used for both space heating and hot water, the flow temperature to the under-floor circuits must be reduced. This is accomplished by a blending valve very similar to a three-way thermostatically controlled mixing valve, which allows water from the return manifold to mix with that from the boiler flow, thereby lowering the water temperature in the under-floor coils. The blending valve is normally manually operated to give the prescribed temperatures. Figure 5.40 shows this valve and its position in relation to the distribution panel.

System design
The following procedures are common to most methods of installation and the following text will give the reader a broad idea of the design of under-floor heating. Most manufacturers supply on request sufficient information for this type of heating to be carried out successfully. The heat emission will depend upon the pipe length per m² of floor area. There are some variations depending upon the type

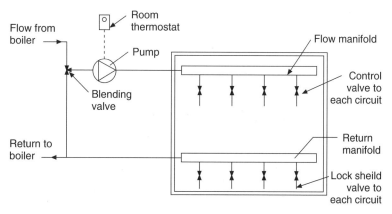

Flow from boiler

Room thermostat

Pump

Flow manifold

Control valve to each circuit

Blending valve

Return to boiler

Return manifold

Lock sheild valve to each circuit

Fig. 5.40 Detail of manifolds and control box.

Table 5.7   Design criteria for under-floor heating (solid floors).

Design room temperature	Pipe centres (mm)	Watts/m²		
		Flow temperature at 45 °C	Flow temperature at 50 °C	Flow temperature at 55 °C
18 °C	100	75	92	109
	200	62	75	89
21 °C	100	65	82	99
	200	53	67	90
22 °C	100	62	79	95
	200	51	64	78

Reproduced Courtesy of Polypipe Ltd.

of floor construction — the following relates to solid floors.

It is first necessary to calculate the heat losses for each room in the usual way. If a calculator is used it should be of a type specifically for under-floor heating. Heat loss through the floor on which the coil is fixed can be ignored as it is laid on a well-insulated base. The next step is to divide the calculated heat losses by the floor area, which will give the heat requirements in W/m². To give an example, assume the heating requirements of a room with a floor area of 14 m² are 1.120 watts and the design temperature is 21 °C.

$$\therefore \frac{1{,}120}{14} = 80$$

Thus, each m² must produce 80 watts.

Table 5.7 shows output per m² at various flow temperatures and pipe centres. It will be seen that for pipes fitted at 100 mm centres, the higher the flow temperature, the greater will be the output per m² of floor area. From the table it will be seen that to give 80 W/m² there are two alternatives. Pipes fixed at 100 mm centres at 50 °C flow temperature gives 82 W/m². If the flow temperature is raised to 55 °C the pipe centres can be 200 mm which gives 90 W/m². Generally the temperature drop over the flow and return on the system is approximately 10 °C. The following points must always be taken into consideration:

(a) The maximum heat output from a solid-floor installation with a design temperature of 21 °C and a floor temperature of approximately 30 °C is 99 W/m². If this is insufficient it may be necessary to fix two separate coils or provide heat backup using a radiator or other heat source.

(b) The maximum recommended pipe length in one coil is 100 metres. Pipes fixed at 100 mm centres will require 8.2 metres of pipe per m² of floor area, giving a total coverage of floor of approximately 11 metres. If the pipes are fixed at 200 mm centres, each m² will require 4.5 metres of pipe with a maximum floor coverage of 21 m².

Commissioning and testing under-floor heating systems

The system should be pressure tested to 6 bar before the floor screed is laid. A constant pressure of 3 bar must be maintained throughout the period of screeding and curing. The optimum thickness of screeds is 50 mm from the top of the pipes as this gives the best performance. When the under-floor heating is initially turned on it must be allowed to warm up gently for a few days. Adjust the flow temperature at the manifold and each floor area to the design temperatures using a suitable thermometer in a similar way to when balancing radiators.

Commissioning and maintaining central heating systems

On completing the installation it should be washed out prior to refilling, venting the radiators, pump and all high points on the system. Check for leaks, fire the boiler and run until operating temperature is achieved — then examine the system for leaks again. Switch off the boiler — pump and drain the system while it is still hot. This will flush out any remaining debris or chemical deposits. Refill the system, and if specified add a suitable inhibitor to the proportions recommended by the manufacturer. A label stating the type of inhibitor used, and the date of the application, should be exhibited in a prominent position on the system. Check and adjust the float-operated valve in the feed and expansion cistern to the correct level. After a final examination for leaks any insulation necessary should be fitted. The boiler should be commissioned to the manufacturer's instructions, and the pump adjusted to the required setting to provide the designed flow rate. The radiators should be balanced to ensure an even distribution of hot water. To maintain the system in good working order the following checks should be made on an annual basis.

1. Service boiler as specified by the boiler manufacturer.
2. Examine the system for leaks, paying special attention to the radiator valve glands. Ensure the spindle of any stop- or gate valves are not seized and can be operated if necessary.
3. Examine all pipe fixings and insulation for damage and rectify if necessary.
4. Check the water level in the feed and expansion cistern (it can evaporate over a period of time). Also ensure the float-operated valve is in working order.
5. Vent the radiators and any high points on the system.
6. Fire the system and verify the correct operation of the controls. If the service is carried out at the beginning of the heating season, ensure the pump is operational.
7. Notify the client in writing that the service has been carried out, carefully noting any factors which may require attention but are not covered under servicing arrangements.

Useful maintenance tools for central heating systems
It is often necessary to replace radiator valves if they are beyond repair. This would normally

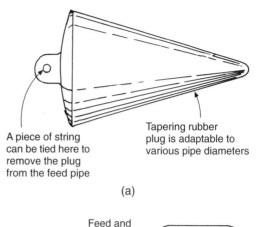

A piece of string can be tied here to remove the plug from the feed pipe

Tapering rubber plug is adaptable to various pipe diameters

(a)

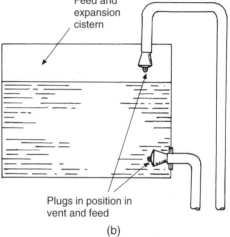

Feed and expansion cistern

Plugs in position in vent and feed

(b)

Fig. 5.41 Use of plugs to avoid draining heating systems for maintenance.

necessitate draining and refilling, possibly admitting air to the system in the process, causing an air lock. The freezing equipment shown in Book 1 is one method of overcoming this problem, but if this is not available, the tapering plugs shown in Fig. 5.41 can be used temporarily to plug the feed and vent pipes connected to the primary part of the system.

The theory is that if no air can obtain access to the system, little or no water will escape if a radiator is disconnected. This method is not advocated where there is any possible chance that extensive damage may occur should something go wrong, but apart from the loss of a little water, which is easily mopped up, the use of these plugs can save time. This method of avoiding drain-downs

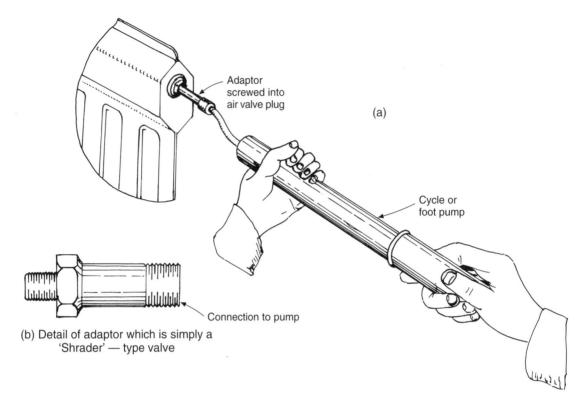

Adaptor
screwed into
air valve plug

(a)

Cycle or
foot pump

Connection to pump

(b) Detail of adaptor which is simply a
    'Shrader' — type valve

Fig. 5.42   Emptying radiators by air pressure. By pumping air into the radiator the water is pressurized and forced back into the feed and expansion cistern. Note that if the cycle pump is inadequate, a foot pump would exert greater pressure.

should not be used for the domestic hot water system — its use has been known to result in a collapsed cylinder!

A useful adaptor shown in Fig. 5.42 can be used to avoid draining a radiator, thus preventing the loss of water which may have been treated with an inhibitor. Using this adaptor enables the water to be pumped back into the feed and expansion system. To use, first close both radiator valves and remove the air valve, substituting it with the adaptor, which is connected to a cycle or foot pump. Open one of the radiator valves and commence pumping until a gurgling sound is heard, indicating the radiator is emptied. Reclose the valve and the radiator may be removed.

If a radiator has to be drained, Fig. 5.43 shows a radiator union incorporating a drain cock, enabling it to be drained easily if the valves are closed and the air valve is open.

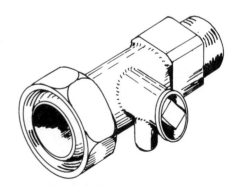

Fig. 5.43   Radiator union with integral drain cock. The type shown has a miniature drain cock moulded into an extended union. Some have a screw which if removed serves the same purpose.

Noise in hot water supply and heating systems

Most noise in hot water and heating systems is caused by expansion and contraction due to temperature changes in pipes and components such as radiators and boilers. One of the most objectionable noises encountered with hot water systems occurs when a pipe has insufficient freedom of movement where it is notched into a joist or passes through a floor board. While it is not necessary to cut massive notches in the joists, or overlarge holes in flooring, pipework must have freedom of movement in such situations. The foregoing also applies to clips and brackets securing hot water pipes, especially if long straight runs are considered. Air entrapped in a system can also give rise to noise, especially if it is present in circulating pipes. Hot water pipes should always be fitted in such a way that any air or gases in the system are automatically passed into the atmosphere via a vent or air valve at the highest point on the system. The remarks relating to pumps in the cold water supply relate also to those used with hot water if pump noise is encountered. It is rare, however, as the velocities are usually lower, and the pumps used for hot water supply and heating are seldom as powerful as those used for cold water.

Boiler noise is not uncommon and may be due to a restriction of the waterways, scale formation or the movement of rust deposits and debris in the base of the boiler when a violent circulation of water takes place. A singing noise is sometimes caused by the explosion of small bubbles of steam on the interior surface of the boiler when heat is transferred quickly through the boiler walls causing the water in immediate contact with them to boil. This results in the rapid formation of minute bubbles of steam which burst on contact with the cooler water. Noise due to this cause may cease as the water temperature increases in excess of 80 °C. If the boiler is firing a combined hot water and heating scheme, an increase in pump velocity may improve the situation. Where the boiler is fired by gas, a check must be made to make sure the blue oxygen core of the flame is not impinging on the boiler as local overheating may be the cause of noise. The addition of a suitable chemical inhibitor usually solves the problem, but it cannot be guaranteed to do so.

Corrosion in heating systems and its prevention

There were very few instances of corrosion problems in heating systems where the radiators were made of cast iron and the pipework of low-carbon steel. Both these materials are ferrous metals and any electrolytic corrosion between them would be minimal, and because of their physical thickness any chemical attack would have little effect. The use of systems with radiators made of thin sheet steel and copper pipe coupled with the use of acidic fluxes have brought about very serious problems, one of the most common being 'pin-holing' in the radiators. The effect produced is a series of very small holes on the external surface of the radiator with subsequent leakage of water. The adverse effects of air containing oxygen in the system have been mentioned in several instances in this chapter and modern systems should be designed to exclude air where possible. Electrolysis takes place between dissimilar metals, and in simple terms, minute local corrosion cells are formed giving off hydrogen, which, like air, collects at the top of the radiator making continual venting necessary. The worst effects of corrosion can be overcome, however, by the use of a chemical inhibitor, the word *inhibit* meaning 'to slow down or prevent' the process of chemical change that produces corrosion. When a new system is commissioned it should be thoroughly flushed through to remove any debris or residue in the system. Inhibitor manufacturers recommend a cleansing additive to assist the process. The system is heated for approximately 1 hour and then flushed out prior to adding the inhibitor, usually via the feed and expansion cistern, or in the case of sealed systems through the filler bottle. Always follow the instructions on the vessel containing the inhibitor regarding the quantity to be used, and in the case of cleansing additives always observe the rules on health and safety, as these additives often contain strong acids. For existing systems which may be severely scaled and corroded, special treatment may be necessary and advice should be sought on treatment procedures from a suitable builders' or plumber's merchant.

Further reading

BS 5449:1990 Specification for forced circulation in hot water central heating systems for domestic premises.

BS 2767:1991 Specification for valves and unions for hot water radiators.

BS 4814:1990 Specification for expansion vessels using an internal diaphragm for sealed hot water heating systems.

BRE Digest 108 *U values*. Building Research Establishment, Garston, Watford, Herts, WD2 CJR.

Heating controls and systems

The Association of Control Manufacturers (TACMA), Westminster Tower, 3 Albert Embankment, London, SE1 7SL, Tel. 020 7193 3007/8.

Energy Savings Trust, 11–12 Buckingham Gate, London, SW1E 6LB, Tel. 020 7931 8401.

Poly Plumb (Underfloor Heating), Broomhouse Lane, Edlington, Doncaster, DN12 1ES, Tel. 01709 710000.

Honeywell Ltd, Charles Square, Bracknell, Berkshire, RG12 1EB, Tel. 0800 521121.

Danfoss Randell Ltd, Amphill Road, Bedford, MK42 9ER, Tel. 01234 364621.

Radiators

Stelrad Ideal, Caradon Heating Ltd, PO Box 103 National Avenue, Hull, North Humberside, HU5 4JN, Tel. 0800 975 8790.

Barlow Ltd, Barlow House, Spinning Jenny Lane, Leigh, Lancashire, WN7 4PE.

Grandfos Pumps Ltd, Grovesbury Rd, Leighton Buzzard, LU7 8TL, Tel. 01525 850000.

Clyde Combustion Ltd, Cox Lane, Chessington, Surrey, KT9 1SL, Tel. 020 8391 2020.

Self-testing questions

1. Name three types of heat emission devices for space heating.
2. State the differences between, and advantages and disadvantages of one-pipe and two-pipe systems of heating.
3. List three advantages of using an air eliminator in a low-pressure heating system.
4. Explain the working principles of a gas condensing boiler.
5. Explain the need for balancing a system and state how it is carried out.
6. List the advantages of fully pumped systems.
7. Explain the main difference between mini-bore and small-bore heating systems.
8. Describe the method and equipment used for filling a sealed heating system directly from the main in compliance with the Water Regulation.
9. Make a list of the essential controls necessary for a domestic heating system and state their purpose.
10. Describe how *U*-values are used to ascertain heat losses through the fabric of a building.
11. List the factors that must be taken into account when determining a boiler output rating.
12. Explain the term boiler cycling, its effects and how to prevent it.
13. List any observations that you would take into account relating to heat loss in an existing building prior to installing a central heating system.
14. During a survey of an existing building prior to installing a central heating system, list the points you would make to the client relating to improvements in the structure to reduce heat loss.

6 Oil Firing

After completing this chapter the reader should be able to:

1. Identify the components used in connection with oil storage and know their requirements.
2. Understand the principles of burning oil.
3. Know the functions and working principles of the components of an atomising burner.
4. Understand testing, commissioning and service procedures.

Fuel oils

The concept of heating using oil as a fuel goes back for many years and was pioneered in America c. 1920 due mainly to the fact that a plentiful supply of cheaply produced oil was available. Oil-fired boilers were also available in this country during the 1930s, but were mainly confined to large commercial installations. It was not until the 1950s, when large quantities of oil began to be refined in this country, that it became a serious competitor to solid fuel and gas for domestic use. Very briefly, crude oil as it is called when abstracted from the earth, is a mixture of gases, light and heavy oils and bituminous residues. Before it can be used commercially it must be refined or broken down into its constituent parts. The two basic methods used to refine oil are a chemical process called cracking, and distillation, where the lighter constituents are separated by heating the oil. After these have been extracted, a mixture of fuel and heavy (thick) bituminous oils remain which must be further refined to separate them for commercial use.

Oil is generally classified by its viscosity or fluidity, which is determined by the length of time in seconds it takes to measure a given quantity as it flows through a standard orifice, see Fig. 6.1. The main oils used for domestic and some industrial burners are C2 (28) second oil and class D (35) second oil. The heavier or thicker oils 200, 960 and 3,500 seconds viscosity respectively, although much cheaper, are unsuitable for domestic use as the preheating equipment necessary to keep the oil fluid, and the special burners required, would make it uneconomic to both purchase and maintain. Even 35-second oil can thicken and become waxy at temperatures just below freezing point. The use of these heavy oils is therefore confined to industrial premises where its low cost makes it economic to use. Table 6.1 shows the main characteristics of light fuel oils.

Oil with a viscosity of 35 second has been and still is used for domestic work as it is cheaper than 28-second oil. The difference in price is now minimal and most modern atomising/pressure jet burners for non-industrial use burn 28-second oil. Its advantages are its very low sulphur content, also its solidification temperature is well below that normally experienced in the United Kingdom.

Fuel storage

A suitable tank must be provided in which to store the oil, and for domestic use this usually presents no problems if suitable accommodation can be found to site the tank. The regulations relating to liquid fuel storage are, however, more stringent in public buildings and are usually subject to local authority inspection.

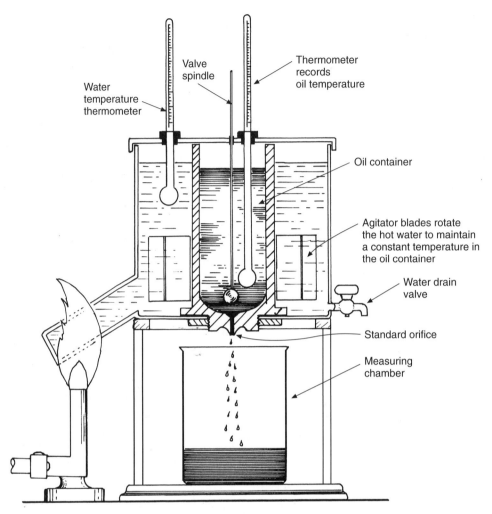

Fig. 6.1 Diagrammatic viscosometer. This apparatus is used for testing the viscosity of light oils.

Table 6.1 Approximate characteristics of light fuel oils.

Characteristic	Class	
	C2.28s	D.35s
Specific gravity at 15 °C	0.79	0.83
Viscosity. Redwood No. 1 scale		
(oil at 37.8 °C) (seconds)	28	35
Calorific value (mJ per kg)	46.6	45.5
Flash point	38 °C	55 °C
Sediment	—	0.01
Solidification	–40 °C	10 °C
Ash (% by mass)	—	0.01
Sulphur (% by mass)	0.2	1.0
Water content (% by mass)	—	0.05

In the event of a leak in the storage vessel, oil might spread over a large area and vaporise more readily, constituting a serious fire risk. To prevent this the tank must be 'bunded', which means that any leak will be confined to the area in which the storage vessel is situated. The confinement area must be of sufficient capacity to contain its contents plus 10 per cent. In large public and commercial installations the storage vessels are made of steel, and to confine any leakage are usually surrounded by a brick-built tank. These are often constructed as a completely roofed-in structure adjacent to the boiler room. Figure 6.2 illustrates an example of an externally fitted tank where a pump must be fitted to remove

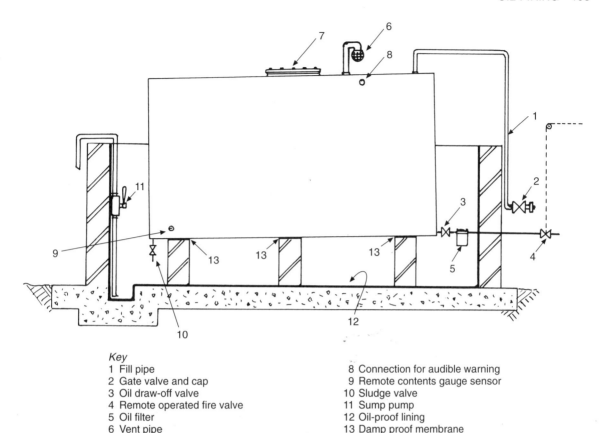

Key
1 Fill pipe
2 Gate valve and cap
3 Oil draw-off valve
4 Remote operated fire valve
5 Oil filter
6 Vent pipe
7 Manhole-normally only required for vessels
   of more than 4,500 litres capacity
8 Connection for audible warning
9 Remote contents gauge sensor
10 Sludge valve
11 Sump pump
12 Oil-proof lining
13 Damp proof membrane

Fig. 6.2 Typical oil storage requirements for a public or commercial building.

any rainwater that may collect in the container. Brick piers are shown, which is the normal method of supporting large steel fuel storage vessels. A damp-proof membrane must always be used between any supports and the vessel to avoid corrosion.

Bunding
Until recently bunding of vessels for domestic use has not been obligatory, but new regulations make this necessary in situations where spillage or leakage of oil could gain access to water courses, cause land pollution or a nuisance to neighboring premises. Some vessels made of plastic are obtainable with built-in bunding, they are in effect a tank within a tank. Because any oil leakage from the tank cannot be seen, a sensing device is available which is fitted in the bunded area so that periodic checks can be made to ensure there are no leaks in the main storage vessel.

Fuel storage inside buildings Bulk storage of fuel oil is not considered a fire risk itself, but it is necessary to protect contents of the storage vessel from a fire which may originate externally to it. The room or chamber in which the vessel is housed should have at least a one-hour fire resistance and be provided with adequate high- and low-level ventilation. The access door must also have a one-hour fire resistance rating and should operate outwards from inside the chamber without a key. When oil is stored in a garage the foregoing applies, the chamber being inside the garage. Any lighting should be of the bulkhead or wall glass types with switches located externally to the chamber.

Fuel tanks
These are traditionally made of low-carbon steel, but more recently polythene has become commonly used, certainly for domestic purposes due to

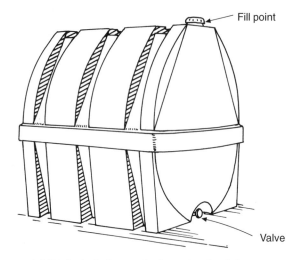

(a) Typical oil storage tank made of polythene

These tanks are manufactured from a special grade of polythene which is unaffected by oil fuels. Capacities vary from approximately 1200 litres (260 gals) to 2500 litres (550 gals) depending on the manufacturer. Provision is made for all the ancillary components in the moulding. As with water cisterns made of plastic materials the base must be fully supported. A smooth finished concrete base or carefully levelled paving slabs meet the necessary requirements.

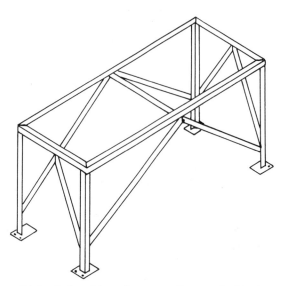

(b) Angle iron frame for supporting small oil tanks

Suitable for both tanks made of polythene and steel, bearing in mind a steel plate must be provided to give full support to tanks made of polythene. It is important that this frame is well painted with a suitable corrosion-resistant paint before the tank is fitted. The feet must be supported on a level concrete base.

Fig. 6.3 Fuel storage tanks and their support.

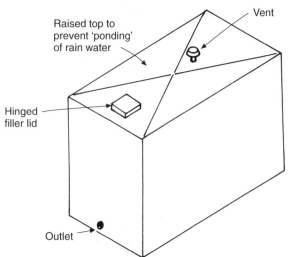

(c) Steel tank fabricated with steel plates with welded joints

its maintenance-free characteristics. Figure 6.3 illustrates typical types of fuel storage tanks and their method of support. The fact that polythene tanks are permissible for oil storage may be surprising as plumbers are usually taught that oil and its derivatives are harmful to plastic materials. Certain grades of polythene are, however, not adversely affected by fuel oil and can be safely used for this purpose.

Some of the main points to bear in mind when installing oil storage tanks are as follows:

1. They must have a slight slope towards the drain or sludge cock to enable any water or debris to be drawn off — the general recommendation is 6 mm in every 300 mm.
2. The capacity of the storage vessel should be as large as possible as most suppliers reduce the cost for large deliveries. For domestic properties capacities of not less than 1,250 litres is recommended. To take advantage of cheaper rates for larger oil deliveries, a tank capacity of not less than 2,700 litres is normally specified. Most oil suppliers these days provide a top-up service, and even if only a small quantity of oil is required its cost usually depends on the capacity of the tank.
3. Some consideration must be given to the distance of the tank from the road or hard

standing where the delivery vehicle can park. Most tankers carry a length of flexible hose 32 mm diameter × 36 m long so that any tank within this measurement can be filled without the necessity of fixed extension pipes to the boundary of the building. The foregoing relates to domestic premises. Extended fill pipes are often necessary in large commercial or public buildings where the storage vessel may not be accessible to the tanker. The termination point depends on such variables as the type of oil used and the length and diameter of the pipe. The advice of the oil supplier should be sought in such cases.

4. Adequate space must be provided for the maintainance of storage vessels especially those made of steel, which require periodic painting. Any loose scale or rust must be removed and treated with an inhibitor prior to repainting. Red oxide or bituminous paints are recommended.

Low temperature effects on fuel oil

In very cold weather fuel oil may be less viscous and become very 'waxy', i.e. it begins to solidify. Normally 28-second oil is not affected but heavier oils are, and it is necessary to maintain their fluidity in very cold weather by raising their temperature. Figure 6.4 shows some of the methods used.

Oil storage tank accessories

Figure 6.2 illustrates the components associated with oil storage vessels. A main control valve, usually a gate valve must be fitted as close as possible to the tank outlet to isolate the supply when necessary. A sludge or drain cock must also be provided to allow debris, rust particles and water to be periodically drained off, the main source of water being due to the formation of condensation. It is suggested the outlet of this valve is plugged to avoid oil loss due to unauthorised operation of the valve. An open vent is necessary to maintain atmospheric conditions in the tank.

Where easy access to the top of the storage vessel is provided, for example suitable ladders or steps, filling is achieved by opening a hinged lid, oil being delivered through a trigger-operated nozzle similar

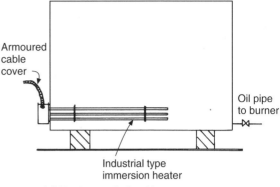

(a) During periods of low temperature

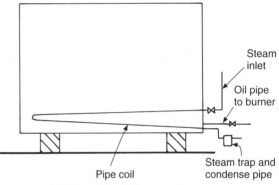

(b) Oils with a high viscosity require constant heating

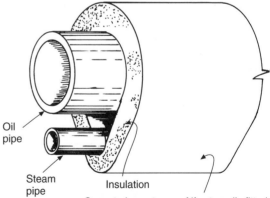

(c) Tracing an oil supply pipe where a steam supply is available. An alternative to steam is the use of an electrical tracing tape

Fig. 6.4 Methods of keeping oil fluid. Thermostatic control of temperatures will depend on the viscosity of the oil.

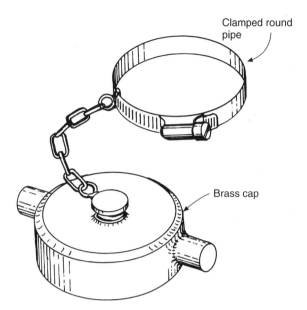

Fig. 6.5   Cap for sealing the end of an oil fill pipe.

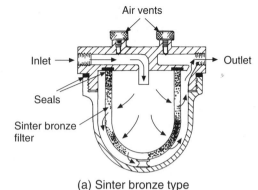

(a) Sinter bronze type

All oil filters are clearly marked inlet and outlet. They must be fitted correctly to avoid air locking.

to those used on petrol pumps. The alternative is to provide a fill pipe complete with valve and plug to avoid oil spillage when the filler hose is disconnected (see Fig. 6.5), some types of which are provided with a lock to prevent tampering by unauthorised persons.

Oil filters

The amount of solids in the oil causing sediment is negligible, but after a period of time deposits on the base of the tank may be disturbed, for example, particles of rust in the case of steel tanks, possibly after an oil delivery. To avoid such deposits causing trouble in the burner, a main filter must be provided adjacent to the outlet valve. Figure 6.6 illustrates some of the types used. In the case of atomising burners, additional filters are also component parts of the oil pump and nozzles.

Although oil fuels are 99.9 per cent free of solids when delivered, a filter must be used on the fuel pipe outlet to prevent any debris or sludge deposited in the base of the tank contaminating the fuel pipe to the burner. They must be cleaned at regular intervals, preferably with clean oil of the type in use. Under no circumstances should petrol be used.

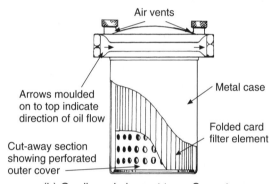

(b) Cardboard element type. Some types may be reused after cleansing; others must be renewed

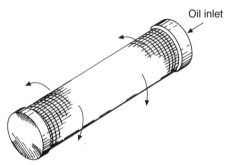

(c) Fine wire mesh strainer usually used with vaporising burners. The strainer only is shown as it is normally housed in the burner control box

Fig. 6.6   Oil filters.

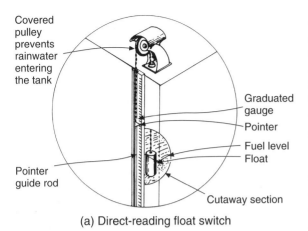

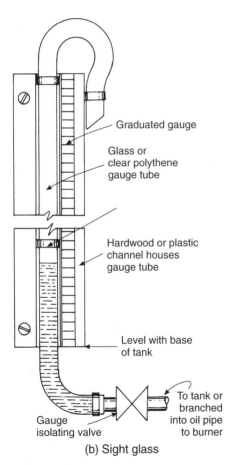

(a) Direct-reading float switch

(b) Sight glass

Fig. 6.7 Direct-acting fuel gauges.

Audible alarms

An audible filler alarm is often necessary if the contents gauge is out of sight when filling the tank. An automatic switch permanently fitted in the tank is temporarily connected by a cable to the delivery vehicle. This triggers an audible warning when the tank is full, allowing the driver to stop the flow of oil immediately.

Oil contents gauges

The two most simple of these are the direct-reading float-operated gauge and sight gauges, the latter usually being employed for domestic work. Both these are illustrated in Fig. 6.7. A simple indirect system, primarily for domestic use and operated by radio signals, is shown in Fig. 6.8. The read-out does not show the contents in litres but simply displays the oil level on a scale of 1–10. A reading of 5 indicates the vessel is half full. The transmitter, via a probe fitted in the vessel, sends radio signals which are picked up by the receiver, which is fitted into a 13 amp socket outlet. In large commercial installations it is necessary to be able to read the tank contents in the control or boiler room itself, and direct-reading gauges would be unsuitable for this purpose. There are many types of indirect gauges, some operate electrically by means of sensors fitted in the tank, others are self-powered and operate on the pressure exerted by the oil in the

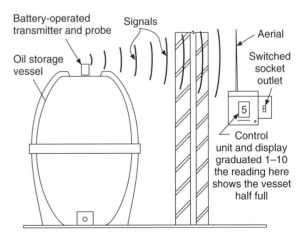

Fig. 6.8 Radio-operated fuel contents gauge.

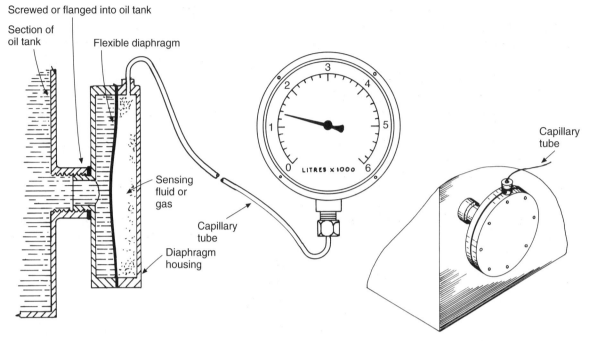

Fig. 6.9 Pressure-operated remote reading fuel contents gauge. Pictorial illustration of the diaphragm housing fitted to the tank. Pressure of the oil in the tank is exerted on the flexible diaphragm pressurising a sensing fluid or gas which is converted into a dial gauge reading. The diaphragm housing is shown here in the base of the tank, but by using a purpose-made extension it can be completely immersed inside the tank via a manhole cover.

tank. Figure 6.9 illustrates a typical example of this type. Due to the differences in tank dimensions and capacities, the manufacturers should be consulted as to the suitability of the gauge reading.

Fire valves

These automatically shut off the oil supply in the event of a fire in the boiler room and should comply with BS 5839:Part 1. Some of the main points relating to the foregoing are as follows (others are illustrated). Shut-off valves not being automatic can only be used in addition to, and not in place of, fire valves. Fire valves must be capable of withstanding the same pressure tests as the pipes to which they are fitted, and must be manufactured in such a manner that they cannot be rendered inoperative by over-tightening a sealing device or gland. They must be capable of a positive shut-down, even if debris is deposited on the valve seating. Heat-sensitive devices such as fusible links, should normally operate at temperatures of 68 °C to

74 °C. In cases where the ambient (surrounding) air temperature exceeds 49 °C, the operative temperature of the heat sensor, e.g. fusible link, may be increased to 93 °C. A fire valve must be fitted preferably as near as possible to the entry of the building. If this is not practical it should be fitted at the point where the fuel pipe enters the boiler room. All fire valves must be capable of being manually reset in the event of accidental operation. Details of these valves and their associated components are shown in Figs 6.10, 6.11 and 6.12. Figure 6.13 shows details of some components used with free fall fire valves.

Oil supply pipes

The most common and convenient material for this purpose is soft copper tube — 8 or 10 mm diameter is suitable for most domestic installations where comparatively short runs are necessary. The main points to remember when using soft copper tube are as follows:

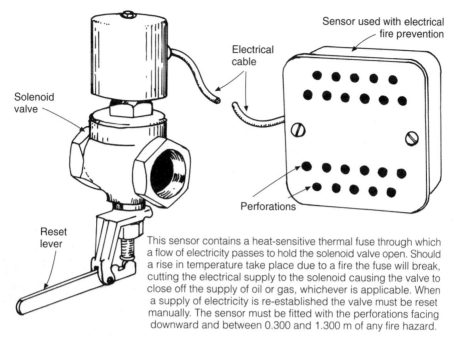

This sensor contains a heat-sensitive thermal fuse through which a flow of electricity passes to hold the solenoid valve open. Should a rise in temperature take place due to a fire the fuse will break, cutting the electrical supply to the solenoid causing the valve to close off the supply of oil or gas, whichever is applicable. When a supply of electricity is re-established the valve must be reset manually. The sensor must be fitted with the perforations facing downward and between 0.300 and 1.300 m of any fire hazard.

Fig. 6.10   Electrical system of fire safety control for oil or gas.

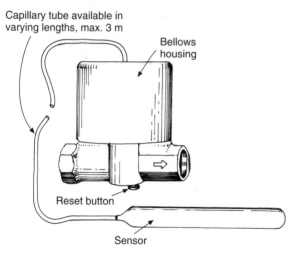

Fig. 6.11   Heat-sensitive fluid fire valve. These are suitable for small, mainly domestic installations and work on the expansion of a heat-sensitive fluid which closes the valve at a predetermined temperature. If the valve has closed it may be reset manually by pressing the reset button. The sensor must be situated inside the boiler casing at least 300 mm from hot surfaces. The valve should preferably be fitted externally and suitably protected from mechanical damage.

1. Make sure it is laid in such a way that there are no high points resulting in air locks giving persistent trouble.
2. If it is buried it must be deep enough to prevent it becoming damaged by gardening activities — or alternatively suitably ducted.
3. Manipulative compression joints or brazed capillary fittings are the approved method of jointing. Soft soldered joints are no longer permissible.
4. Joints underground should be avoided. Where this is not possible, they should be tested prior to back-filling.

Pipework systems for oil supply

The one-pipe system is usually fitted in domestic properties where the boiler is on the ground floor, and oil can be delivered to the burner by gravity. The base of the tank should be approximately 600–700 mm above the burner, as this normally provides sufficient pressure to overcome the frictional resistance of the pipeline. In exceptional circumstances where the oil store is a long way

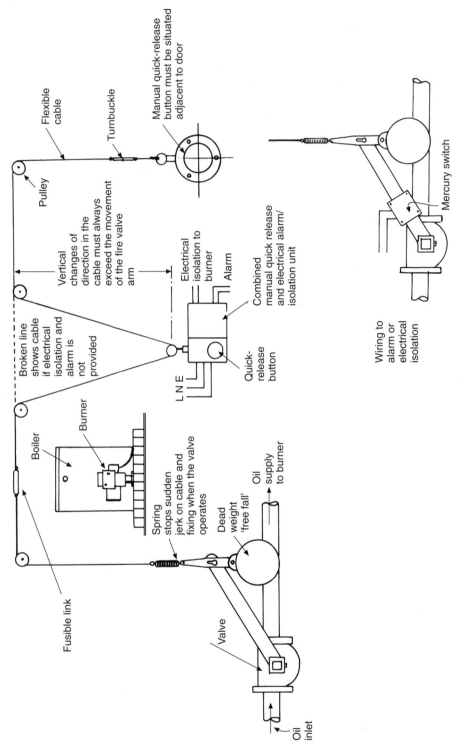

Fig. 6.12 Free-fall fire valve.

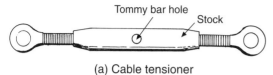

(a) Cable tensioner

By rotating the stock the right- and left-handed threaded ends permit the cable to be correctly tensioned.

Brass plates soldered together with low-melting-point solder, lead/tin/bismuth alloys. Melting points vary between 160 °F (72 °C) and 356 °F (180 °C)

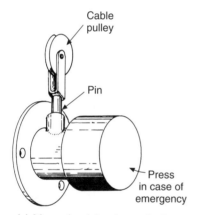

198 °F 92 °C

(b) Fusible link

The fire valve will close if the solder melts due to an excessive rise in temperature adjacent to the boiler.

Cable pulley

Pin

Press in case of emergency

(c) Manual quick-release button

Should be fitted in such a way that it is not necessary to actually enter the boiler room to operate. Combination quick-release and electrical isolation types are available. To test the fire valve to ensure it closes this component may be activated and the pin will be released. It can be re-engaged by pressing the button and simultaneously replacing the pin in its housing.

Fig. 6.13 Accessories used with free-fall fire valves.

from the burner, it may be necessary to raise the level of the tank or enlarge the diameter of the supply pipe. It should be noted that the oil supply to vaporising burners, having no integral oil pump,

must always be gravity fed, see Fig. 6.14(a). In situations where the burner is at a higher level than the storage vessel, there are several alternatives.

(a) A two-pipe system may be employed as shown in Fig. 6.14(b).
(b) Some manufacturers produce specially adapted pumps which make a two-pipe system unnecessary.
(c) The use of a component marketed as the 'Tiger Loop System' shown in Fig. 6.15. Its use overcomes the problem previously mentioned, but also enables the release of air bubbles and minute particles of debris, adding to the pump's efficiency. As small quantities of flammable vapour may be released from the equipment when it's in use, it must always be fitted externally of the boiler room.

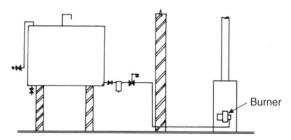

Burner

(a) Normal installation of oil tank where the oil flows to the burner by gravity

Both the height of the tank above the burner and the length of the supply pipe influences its diameter. The manufacturer's recommendations must always be complied with on this point.

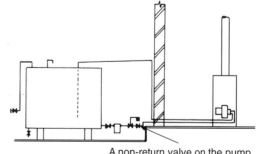

A non-return valve on the pump suction pipe may be specified by some burner manufacturers

(b) Two-pipe system of oil supply

Fig. 6.14 Oil pipe supplies to burners.

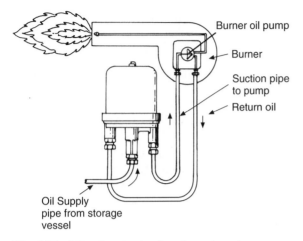

Burner oil pump

Burner

Suction pipe
to pump

Return oil

Oil Supply
pipe from storage
vessel

Fig. 6.15 'Tiger loop' valve for oil supply to burners.

The main reason for these three alternatives is to avoid a vacuum in the oil line.

Oil pumps are made to pump more oil than is necessary to supply the nozzle, the excess being recirculated in the pump itself. When the oil is supplied by gravity feed the pump is designed to undertake this work, but if it has to draw oil from a tank at a lower level, the extra work will cause it abnormal wear. Figure 6.16 illustrates an oil supply system in a large commercial building where the boiler room is at high level, possibly on the roof of the structure. The burners are gravity fed from a service tank to which oil is independently pumped from the main storage tank.

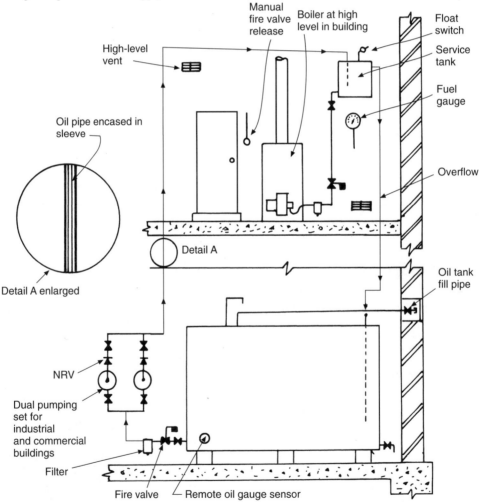

Fig. 6.16 Service tank system.

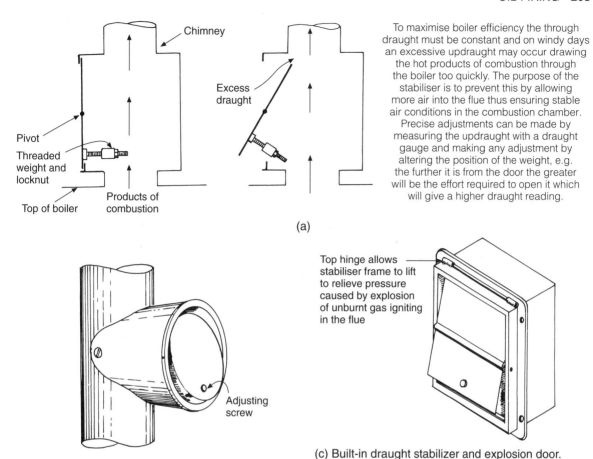

To maximise boiler efficiency the through draught must be constant and on windy days an excessive updraught may occur drawing the hot products of combustion through the boiler too quickly. The purpose of the stabiliser is to prevent this by allowing more air into the flue thus ensuring stable air conditions in the combustion chamber. Precise adjustments can be made by measuring the updraught with a draught gauge and making any adjustment by altering the position of the weight, e.g. the further it is from the door the greater will be the effort required to open it which will give a higher draught reading.

(a)

Top hinge allows stabiliser frame to lift to relieve pressure caused by explosion of unburnt gas igniting in the flue

(b) Fuel pipe adaptor for circular stabiliser

(c) Built-in draught stabilizer and explosion door. These are sometimes specified for solid fuel appliances

Fig. 6.17 Principles of draught sterilizers.

Draught stabilisers

Many types of burner are provided with a draught stabiliser, which, to obtain maximum efficiency, must be carefully regulated to ensure a constant draught in the flue (see Fig. 6.17(a), (b) and (c)). It should be noted that stabilisers are also used with solid fuel boilers. The type shown in Fig. 6.17(c) is designed to act as both a stabiliser and explosion door. This relieves pressure in the flue should any unburned gases ignite.

Types of oil burner

Vaporising burners

These boilers were once very popular for domestic work, but apart from a few exceptions they are no

longer in use, mainly due to their limited heating output and the fact that modern atomising burners, which can be controlled electrically, are much smaller and quieter than older models. Because of this they can now be accommodated in the boiler casing and fitted into modern kitchens. Some types are made that enable servicing to be carried out from outside the building. Of the main types of vaporising burner once commonly in use, only the sleeve type has any application in building services, where it is used in some types of cooking stove; because no electrical supply is necessary this does have some advantage in rural areas. For further information reference should be made to the companies listed at the end of the chapter. A small market exists for pot vaporising burners, although

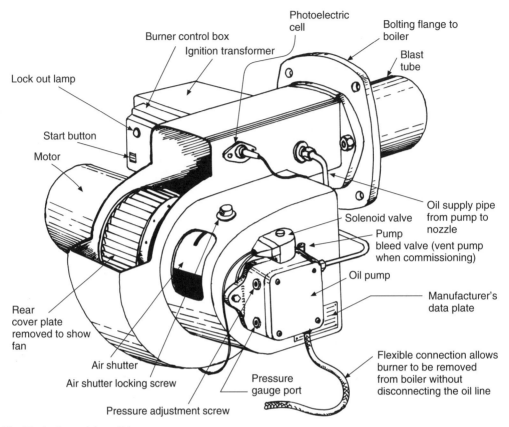

Fig. 6.18   Typical atomising oil burner.

they are not used now for water heating. Although they are very efficient, wall flame burners are no longer made for use in the UK due to their limited output.

Atomising burners

Sometimes called pressure jet burners, these are currently the most popular method of utilising oil for heating and hot water supply. They are available in both floor-standing or wall-hung mode with conventional or balanced flues. The burner is a self-contained unit, being bolted on to the front plate of the boiler by a flange. Some of the larger types used in industry are also provided with a leg or stand for additional support. A typical burner is shown in Fig. 6.18; whatever its heat output the same basic components are used. The rear end of the burner houses a fan which rotates on a spindle driven by an electric motor; the same spindle also drives the

oil pump. Some manufacturers make the spindle in such a way that the oil pump is connected to it by a hard rubber sleeve. In the event of oil-pump seizure the sleeve will break, preventing damage to the motor. The delivery side of the pump is connected to a solenoid valve, the purpose of which is to positively close the supply of oil on burner shutdown. From the solenoid valve the oil is delivered at high pressure to the nozzle, which breaks it down into a fine oil mist so that it can mix easily with the air from the fan enabling combustion to take place. Figure 6.19 shows the end of the blast pipe which is fitted with a swirl or diffuser plate. As the air passes through it, this plate imparts a rotary motion to the air, this and the fact that the nozzle is designed to give the oil mist a contrarotating motion, ensures thorough mixing of the oil and air to form a highly combustible compound. Ignition is achieved by a spark produced

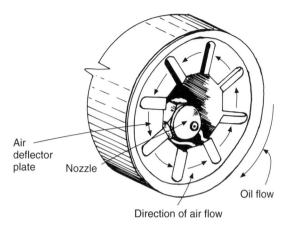

Fig. 6.19 Air deflector plate and its relationship with the nozzle. The rotative motion of the air flow through the deflector plate and the contrarotation of the oil through the nozzle produce a flammable oil mist which is ignited by a spark across the electrodes.

by a pair of electrodes shown in the nozzle assembly in Fig. 6.20. The position of the electrodes in relation to the nozzle is fairly critical and it is wise to check their position against the manufacturer's data.

All oil burners must be provided with a fail-safe device which prevents the combustible oil/air mixture entering the combusion chamber and flue without igniting. If this happened and the mixture accidentally ignited, the result could be a very serious explosion. In wall flame burners protection against this is provided by a flue thermostat. These were originally used in atomising burners, but a much more effective device called a photoelectric cell or resistor, which actually sees the flame, is now employed. Figure 6.21 shows this component. Reference should also be made to Fig. 2.44.

The control box

The control box is to the oil burner, what the brain is to man, its purpose being to coordinate the signals relayed by the remote controls to ensure that the burner fires and shuts down when necessary. In the event of failure to fire, it will automatically stop the fan and oil pump momentarily until the unburnt oil/air mixture has vacated the combustion chamber; this is called the 'purge' period. The control box will then again attempt to start the burner. If this meets with failure, the control box is able to recognise there is a fault in the system, causing its integral fail-safe mechanism to operate. This is called 'lock out' and a red light showing on the control box will indicate this. When the

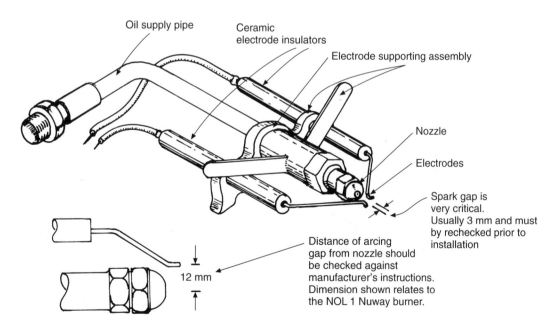

Fig. 6.20 Electrode assembly housed in blast tube.

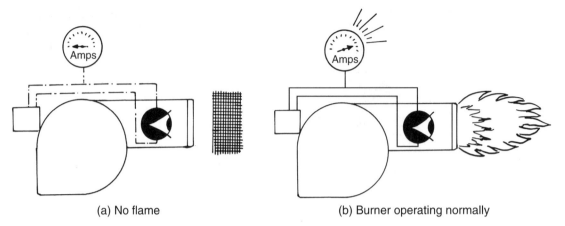

(a) No flame   (b) Burner operating normally

Fig. 6.21 Fail-safe equipment. The photoelectric resistor or cell is illustrated as an eye. If the burner does not fire it will not see a flame and will send a signal to the control box to go to lockout.

fault is located and rectified the burner can only be restarted manually by pressing the appropriate button; a typical control box is seen in Fig. 6.22.

Oil pumps
The reader may be aware that pumps vary considerably depending on the work they are required to do. There are three basic pump types:

(a) The reciprocating pump which has many applications, typical examples being raising water from a well, and pressure testing water installations.
(b) Centrifugal pumps used in heating systems, cold water and fire services in tall buildings.
(c) Gear pumps, shown diagrammatically in Fig. 6.23 which are widely used in situations where good suction and pressure characteristics

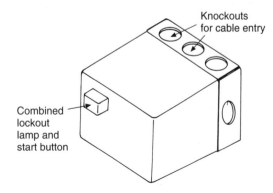

Knockouts for cable entry

Combined lockout lamp and start button

Fig. 6.22 Oil burner control box.

are required, for example to force a flow of oil through the burner nozzle.

Burner oil pumps also incorporate a filter and a regulating valve, which maintains both a stable oil pressure and directs the volume of oil not used at the nozzle back to the suction side of the pump, via a bypass, or if a two-pipe system is used, back to the storage tank. Figure 6.24 illustrates the working principles of a typical oil pump.

Modern oil pumps are also provided with an integrally operated cut-off, or solenoid valve. This permits oil to pass to the nozzle on start-up only when the fan has developed sufficient air pressure to ensure the oil/air mixture is correct. Conversely, when the oil burner stops, the motor speed falls, and both oil and air pressure will fall evenly as the motor slows down. Unless the flow of oil is positively stopped during this period, a sooty, pulsating flame will be momentarily produced; nozzle dribble may also occur, both of which are very undesirable.

Oil burner nozzles
A nozzle plays a very important part in the process of oil firing, and unless it is functioning correctly it will be impossible to achieve a clean, stable flame burning with maximum efficiency. Figure 6.25 shows all the components which make up the nozzle unit. The sintered bronze filter is provided to prevent any solid particles that may remain in the oil after passing through the main filter adjacent to the storage tank and pump filter. From the nozzle

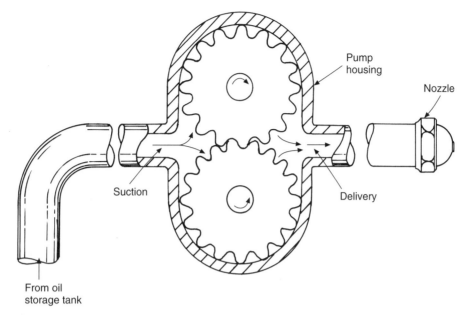

Fig. 6.23 Oil burner pumps. Basic principle of oil pump used on atomising oil burners. Gear pumps are used to develop the high oil pressure necessary to force it through the nozzle orifice. As the gear wheels rotate the disengagement of their teeth creates a negative pressure and oil is sucked into the pump housing, where it is carried round by the gear teeth to the delivery side of the pump and the nozzle.

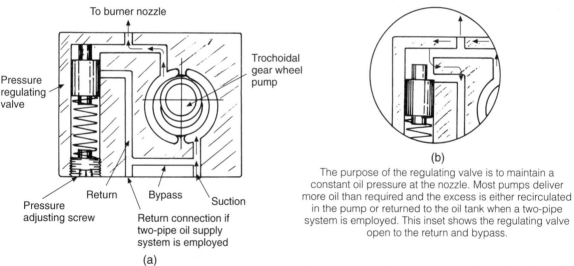

(a)

(b)

The purpose of the regulating valve is to maintain a constant oil pressure at the nozzle. Most pumps deliver more oil than required and the excess is either recirculated in the pump or returned to the oil tank when a two-pipe system is employed. This inset shows the regulating valve open to the return and bypass.

Fig. 6.24 Pressure regulation of oil pump.

filter the oil runs along the outside of the cone and through the slots which give it a rotative motion, ensuring complete mixing with the combustion air provided by the fan. In practice the oil is forced through the nozzle at such a high velocity that a 'tube' of oil is formed in the nozzle orifice. This tube expands on leaving the nozzle and breaks up into very fine droplets which easily mix with air. Nozzles are marketed to give a variety of spray angle patterns and their hourly flow rate is usually quoted in US gallons or litres at a specified pressure. The spray angles and patterns are designed to match the boiler combustion chamber, the burner manufacturers taking this into account when supplying a burner for a specific boiler. In view of

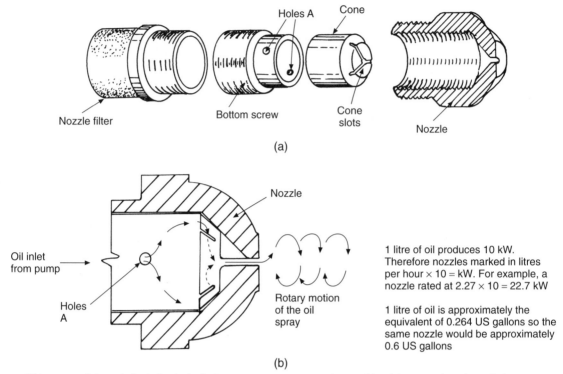

(a)

1 litre of oil produces 10 kW.
Therefore nozzles marked in litres
per hour × 10 = kW. For example, a
nozzle rated at 2.27 × 10 = 22.7 kW

1 litre of oil is approximately the
equivalent of 0.264 US gallons so the
same nozzle would be approximately
0.6 US gallons

(b)

Oil is pumped through the holes in the bottom screw, passes over the outside of the cone then through the cone slots which are made in such a way that they impart a rotary movement to the oil as it passes through the nozzle. The contrarotative flow of air from the fan mixes with the oil to produce a highly flammable oil mist.

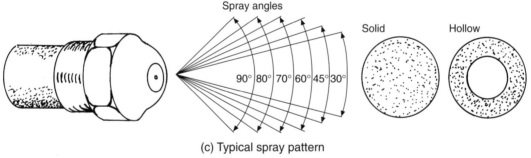

(c) Typical spray pattern

Nozzles are specified by the volume of oil they pass per hour, their spray angles and flame pattern. They are supplied with the burner and are matched with the boiler combustion chamber to give maximum firing efficiency.

Fig. 6.25 Burner nozzles.

this it is important to replace worn-out nozzles with another exactly the same, therefore they are stamped with the relevant data for easy identification. It is important to ensure that nozzles are treated with care and kept in their protective casing until they are fitted. Avoid touching the tip of the nozzle — the orifice is very easily blocked — they are best handled by holding them across the spanner flats.

Scratches on the face must be avoided at all costs. Nozzles do wear, the high oil pressure at which the oil passes through them can cause erosion in the orifice resulting in it becoming enlarged. This will upset the oil/air ratio causing poor combustion. It is generally accepted that a nozzle will remain serviceable provided good burning and flame characteristics can be maintained. This can be determined by using the standard tests described later in this chapter.

Preheaters

Maximum burner efficiency can only be achieved if the oil is of the correct viscosity at the nozzle. It has been stated that at low temperatures 35-second oil starts to solidify and become waxy, and even 28-second oil tends to thicken in cold weather. This means that burner efficiency will vary in differing weather conditions. If the viscosity increases due to low temperatures, the result will be larger oil drops which make its combustion more sluggish, and a higher oil flow through the nozzle, which, because there is no corresponding increase in the air for combustion, will result in a sooty flame. The presence of soot in any boiler, irrespective of the type of fuel used, means that combustion is incomplete, resulting in a loss of efficiency. To overcome this problem the oil must be preheated to approximately 65 °C prior to entering the nozzle assembly. This will result in precise atomisation and clean, stable, soot-free combustion. A typical preheater is shown in Fig. 6.26 and will normally improve efficiency of the burner. Preheaters incorporate a heater element controlled by a thermostat which maintains the supply of oil at the nozzle at a temperature of 65–70 °C. They are an optional extra on new oil burners and may also be fitted to existing burners subject to consultation with the manufacturer.

Transformers

Step-up transformers must be used to increase the normal 204 V supply to approximately 10,000 V — there are some variations, depending on the manufacturer. This voltage will 'jump' the gap between the electrodes, causing a spark which will ignite the oil/air mixture.

Installation of oil-fired boilers

Floor-standing boilers should be sited on a solid hearth similar to those required for solid fuel appliances. Provision must be made for water connections, maintenance and cleaning. A suitable flue must also be available, except for boilers having a balanced flue, in which case the terminal position will be similar to those required for gas appliances. Boilers situated in outbuildings or specially constructed compartments must be provided with adequate frost protection, e.g. frost thermostats and effective pipework insulation. As with all heating appliances, it is essential to ensure that any mechanical ventilation in the building does not adversely affect the operation of the flue.

Air requirements for combustion and ventilation

This is illustrated in Fig. 6.27. The ventilation requirements shown apply if the boiler is fitted in a

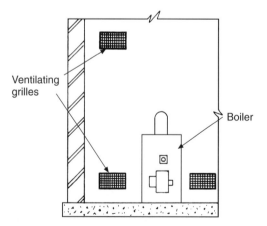

Fig. 6.27 Ventilation requirements. For open-flued appliances a permanent supply of air for combustion at the rate of 550 mm² per kW input minus the first 5 kW must be provided. Air for ventilation, if taken from inside the building, must have a free area of grille 11.00 mm² per kW. When taken from outside the free area of each grille must be 550 mm² per kW.

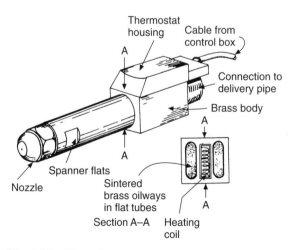

Fig. 6.26 Oil pre-heater.

small compartment. In larger rooms the ventilation requirements may be less. Manufacturer's data will give air supply requirements precisely, but as a guide an average domestic burner needs about 60 m³ of air per hour for good combustion. To this must be added sufficient air for ventilation of the compartment itself.

Commissioning and testing

A short length of flexible armoured pipe supplied with the boiler is used to connect the oil pump to the supply pipe so that the burner unit can be easily removed for servicing without disconnecting the oil pipe. A stop valve should be fitted at this point — especially in commercial installations. The oil pipework and the pump must be completely purged of air, the first step being to disconnect the delivery pipe at the pump. All valves on the fuel pipe should then be opened allowing oil to discharge into a container until a good flow, free of any air bubbles is observed. The delivery pipe can now be reconnected. Next the oil pump must be purged by starting the motor and opening the bleed screw until, as with the suction pipe, an air-free flow of oil is discharged. During the two foregoing operations have a supply of rag or cotton waste at hand to mop up any oil spillage to reduce fire risks.

The burner is next removed from the boiler and placed in a position so that the delivery pipe is vertical. The nozzle, if fitted, must be carefully removed and the leads from the transformer must also be temporarily disconnected at this stage. On some burners it may be necessary to remove the blast tube to achieve this. The burner should then be restarted so that oil is pumped into the delivery pipe. When it reaches the top, switch off the power supply and carefully refit the transformer leads and nozzle. A small quantity of oil should be seen to discharge through the nozzle orifice, proving that the discharge pipe and nozzle are completely purged of air. Some of these operations are illustrated in Fig. 6.28. Finally, before refitting the burner to the boiler, a final check must be made on the position of the nozzle in relation to the swirl plate, and that of the electrodes in relation to the nozzle. Figure 6.29 illustrates the effects of not carrying out this operation correctly.

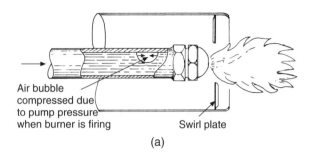

Air bubble compressed due to pump pressure when burner is firing

Swirl plate

(a)

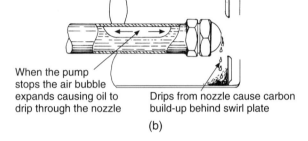

When the pump stops the air bubble expands causing oil to drip through the nozzle

Drips from nozzle cause carbon build-up behind swirl plate

(b)

(c)

To ensure any air is purged in the delivery pipe first remove the blast pipe and turn the burner so that the oil delivery pipe is vertical and full of oil. Screw in the nozzle by hand so that oil is seen to pass through the orifice prior to tightening with spanners.

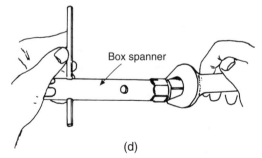

Box spanner

(d)

Finally tighten the nozzle on to the delivery pipe. Do not overtighten and use the correct size spanners.

Fig. 6.28 Effect of entrapping air in the nozzle delivery pipe.

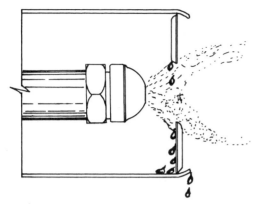

(a) In this case the nozzle assembly is set too far back into the blast pipe so that the atomised oil impinges on the diffuser plate

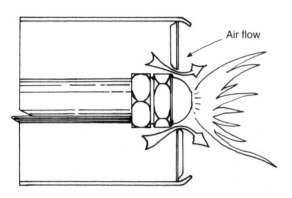

(b) If a nozzle is too far forward the air velocity will be too great. The burner may not fire under these circumstances, but if it does the flame will be very ragged as shown.

Fig. 6.29   Effects of incorrect positioning of nozzle assembly.

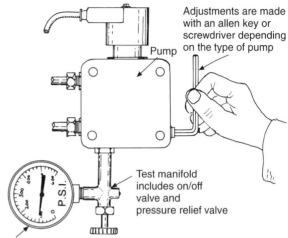

Pressure reading shown at 100 p.s.i.
Note that most gauges are also calibrated in bars

Fig. 6.30   Checking and adjusting the oil pressure at the pump.

Testing combustion and burner efficiency
Tests must now be conducted to ensure the burner is functioning correctly and efficiently. The oil pressure developed by the pump must be checked as shown in Fig. 6.30. Pressures usually vary between 100 p.s.i. (7 bar) and 140 p.s.i. (10 bar) depending on the pump used. The correct operating pressure for any unit will be found in the manufacturer's instructions.

Prior to conducting combustion and efficiency tests the burner should be allowed to run for 5–10 minutes to warm the flue. To avoid soot deposits in the boiler during this period, open the air shutter until the flame pulsates — then close it slightly. The flame can usually be seen, either through a small glass window in the front of the boiler, or a small hole in the top. When it is burning properly it should look similar to a blowlamp flame with yellowish tinges. No smoke should ever be seen from the chimney. All the following tests are made through a small hole below the stabiliser, if one is fitted. For boilers not having a stabiliser a small hole must be drilled in the flue pipe about 150 mm from the top of the boiler. This hole must be sealed when testing has been completed.

All the following tests must be conducted both when commissioning and at least once a year or when the boiler is serviced.

Chimney updraught   This test is made using a draught gauge shown in Fig. 6.31 to indicate the updraught or 'pull' on the flue. They are graduated in inches or millibars water gauge (WG). There are slight variations for different burners, but the usual requirements are 0.04 in WG. If a draught stabiliser is provided with the boiler, it can be adjusted to maintain a steady updraught — if not it may be necessary to fit one where excessive updraught readings are indicated.

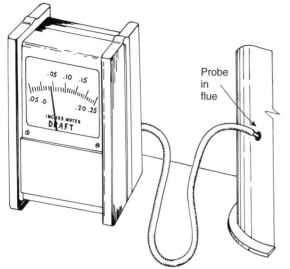

Fig. 6.31 Draught gauge shown reading .04 in WG which is acceptable for most burners. Note that a reading to the left of .0 indicates downdraught. Gauges reading in mm are available, but most manufacturers currently quote draft requirements in inches WG.

Table 6.2 Effect of smoke on burner performance (refer to Fig. 6.32(b)).

Bacharach smoke scale no.	Burner rating	Sooting produced
1	Excellent	Extremely light if at all
2	Good	Slight sooting which will not increase stack temperature appreciably
3	Fair	May be some sooting but will rarely require cleaning more than once a year
4	Poor	Borderline condition, some units will require cleaning more than once a year
5+	Very poor	Soot rapidly and heavily

Smoke testing (see Fig. 6.32)   This is done with a pump which extracts a sample of flue gas and draws it through a filter paper. The standard test needs ten strokes on the pump — more or less will give a false reading. The filter paper is then removed from the pump, and a spot, varying in shade between almost white to black, will be seen. This is compared with the shades on a standard card, known as the Bacharach scale, which is numbered from 1 to 10. If the spot on the filter paper is too dark, it is a sure indication that there is excessive soot in the flue gases and more air must be admitted to the burner by opening the air shutter. If the filter paper is unshaded it is an indication that too much air is being passed and the air shutter requires closing down. When it is properly adjusted the shade on the filter paper should match No. 1 or 2 on the scale. If adjusting the air shutter fails to give a good reading, the position of the nozzle assembly in the blast pipe should be rechecked, and in the case of an existing boiler the nozzle may need replacing.

If the previous two tests give good readings then the burner should be operating efficiently, but this must be checked by comparing the flue gas temperature against the CO_2 content. Figure 6.33 shows the type of dial thermometer used for flue gas temperature measurements. Net stack temperatures of approximately 230–240 °C are normally specified by manufacturers. The net temperature is that minus the ambient air temperature — thus a flue temperature reading of 260 °C when the air temperature is 20 °C will have a net stack temperature of 240 °C. Excessive flue temperatures may be the result of:

(a) Excessive draught through the boiler, in which case the draught stabiliser setting should be rechecked.
(b) Dirty, carboned or sooty surfaces, remedy — clean.
(c) Incorrect setting of air shutter.
(d) Overfiring — possibly due to incorrect or worn nozzle.

Figure 6.34 illustrates apparatus for testing the CO_2 content of the flue gas. The fluid contained in the gauge absorbs CO_2, and prior to use, the adjustable scale should be set at zero against the fluid level. A hand pump is used to sample the flue gases and the standard test requires 18 pumps, no more, no less. On removing the probe, the gauge is turned twice through 180° to ensure thorough mixing of the CO_2 prior to taking a reading. Shaking the tube is not recommended.

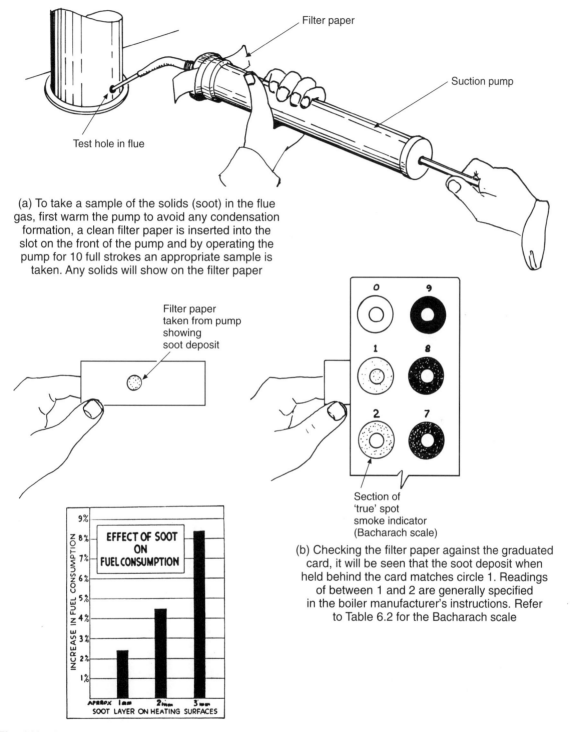

(a) To take a sample of the solids (soot) in the flue gas, first warm the pump to avoid any condensation formation, a clean filter paper is inserted into the slot on the front of the pump and by operating the pump for 10 full strokes an appropriate sample is taken. Any solids will show on the filter paper

Filter paper
Suction pump
Test hole in flue
Filter paper taken from pump showing soot deposit
Section of 'true' spot smoke indicator (Bacharach scale)

(b) Checking the filter paper against the graduated card, it will be seen that the soot deposit when held behind the card matches circle 1. Readings of between 1 and 2 are generally specified in the boiler manufacturer's instructions. Refer to Table 6.2 for the Bacharach scale

EFFECT OF SOOT ON FUEL CONSUMPTION

INCREASE IN FUEL CONSUMPTION

SOOT LAYER ON HEATING SURFACES

Fig. 6.32   Smoke testing oil fired boilers.

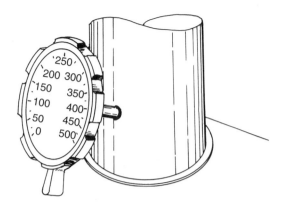

Fig. 6.33 Checking the flue gas temperature with a rotary thermometer. The aim should be between 200 and 250 °C with a CO_2 reading of 11 per cent.

It should be noted that the fluid contained in the gauge should be changed at regular intervals, as its ability to absorb CO_2 diminishes, depending on the incidence of use.

A reading of between 8 and 11 per cent, the highest number applying to modern boilers, should be the aim, a reading of less than 8 per cent will indicate an excessive air flow through the flue. Readings in excess of 11 per cent are likely to result in soot formation within the combustion chamber, and possible condensation in the flue. The actual efficiency of the boiler can be read directly from a standard scale which cross-references flue gas temperatures with the CO_2 percentage, see

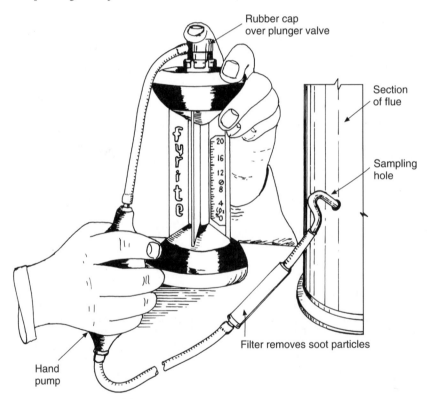

To ensure an accurate CO_2 test:

1. Zero the scale to the top level of the fluid.
2. Operate the burner for 5–10 minutes.
3. Insert sampling tube into test hole.
4. Place rubber cap over valve and depress.
5. The bulb of the aspirator pump is squeezed 18 times in succession.
6. The rubber cap is removed and the valve will automatically close.
7. The indicator is turned over twice to ensure thorough mixing of the fluid and CO_2.
8. Place the indicator on a level surface and read off percentage of the CO_2 on the scale.

Fig. 6.34 Sampling flue gases with a CO_2 indicator.

Table 6.3   Flue gas temperatures and percentage of CO_2.

CO_2 content (%)	Efficiency (%) at flue gas temperature of: (°C)				
	114	204	260	316	371
14	88	86	83	81	79
13	88	85	83	80	78
12	87	85	82	80	77
11	87	84	81	79	76
10	86	83	80	77	74
9	85	82	79	76	73
8	85	81	77	74	70
7	84	80	75	70	67
6	82	78	73	68	63
5	81	75	69	66	58
4	77	70	63	57	50

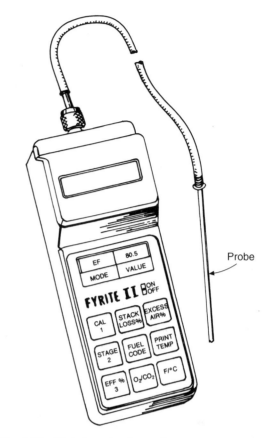

Fig. 6.35   Digital combustion analyser.

Table 6.3, or it can be calculated using the following formula:

Flue loss (per cent) =

$$\frac{0.477 + 0.072}{\text{per cent } CO_2} \times \left(\frac{\text{flue gas}}{\text{temp.}} - \frac{\text{ambient}}{\text{air temp.}}\right)°C + 6.2$$

(It should be noted that the figures 0.477, 0.072 and 6.2 are constants and relate directly to the type of fuel burned — in this case class C or D oil.)

Example: Calculate the efficiency of a boiler where the flue gas reading is 260 °C, a CO_2 percentage of 11 and an ambient air temperature of 15 °C.

$$0.477 + 0.072 = 0.484$$
$$0.484 - 11 = 0.044$$
$$0.044 \times (260 - 15) = 10.78$$
$$10.78 + 6.2 = 16.98$$

Approximately 17 per cent flue loss giving a combustion efficiency of 83 °C.

Digital combustion analysers

The method described for testing combustion efficiency using flue thermostats and CO_2 analysers, has been used for many years. If properly maintained, their accuracy is acceptable for most installations and they are widely used in the plumbing and heating industry. Where greater degrees of accuracy are required, sophisticated computerised analysers are obtainable, some directly showing the test results on a small screen — others actually print out the relevant details. When this equipment is used the batteries must be changed on a regular basis to ensure accurate readings. Figure 6.35 illustrates a typical example of this type of combustion testing equipment.

Flues for oil boilers

As with gas, the temperature of the products of combustion of oil are comparatively low and similar precautions must be taken to prevent condensation. This is especially so when burning class 'D' oil, as due to its higher sulphur content mixing with condensate, a dilute solution of sulphuric acid will form, seriously corroding the internal surfaces of the boiler. If a draught stabiliser is fitted, it must

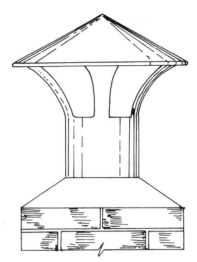

Fig. 6.36 Chimney terminal. Sometimes called a Chinaman's hat due to its shape. Its purpose is to prevent rainwater entering the flue. It is not suitable as a terminal for gas flues.

be carefully adjusted so that an excessive amount of cold air is not drawn into the flue, thus cooling the products of combustion below their dew point. Flues constructed to conform with Part J of the Building Regulations are satisfactory for oil burning, but in pre-1965 buildings, a suitable lining will be necessary. Flue terminals are not normally required, although it is recommended that some means of preventing rainwater entering the flue is employed. A terminal known as a 'Chinaman's hat' shown in Fig. 6.36, is suitable for this purpose.

In situations where downdraught is a serious problem, it can often be solved by fitting an OH cowl shown in Fig. 6.37. Generally they not only solve downdraught problems but also improve the flue updraught.

Servicing oil-fired boilers

Servicing schedules are always supplied by burner or boiler manufacturers, usually as an appendix to the installation instructions. They may differ slightly depending on the type of boiler and burner with which it is supplied. The following, however, lists the main servicing points relating to boilers fitted with atomising burners:

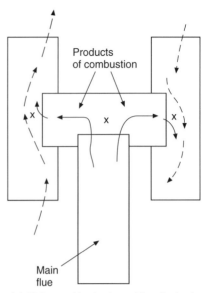

Products of combustion

Main flue

(a) This cowl is designed to eliminate downdraught and will increase the updraught in a flue. It will be seen that wind pressure shown by the arrows with dotted tails causes three points of negative pressure shown at x. This is due to the extensions of the main flue and horizontal section into the outlets which in effect form a type of Venturi effect
Suitable for all types of fuel, these cowls are made in galvanised and stainless steel, also terracotta. The latter are very heavy and require a proper scaffold for their installation. When used for gas flues the outlets must be provided with mesh guards to prevent the entry of birds.

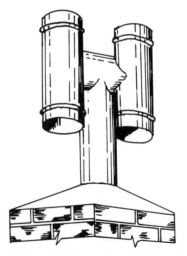

(b) Pictorial illustration of OH cowl

Fig. 6.37 OH cowl.

1. Inspect the general condition of the flue and its cleanliness. Any signs of excessive sooting will give some indication as to the efficiency of the installation. The owner of the premises should be notified if the flue requires repairing or sweeping, as this is not usually the responsibility of the maintenance engineer.

2. The boiler flueways and combustion chamber must be cleaned. Special wire brushes are necessary for this purpose, as some boilers, having small flueways, are difficult to clean. If the combustion chamber is lined with firebricks they must be replaced if badly cracked or damaged.

3. Isolate the electric supply, remove the burner and clean the nozzle assembly. Nozzles may be stripped down and cleaned, but great care is necessary to ensure the very small components are not scratched or damaged — some recommend changing the nozzle annually. The electrode insulation must also be inspected for damage such as cracks. When the assembly is replaced, its position in the blast tube and that of the electrodes in relation to the nozzle, must be carefully measured against the manufacturer's recommendations.

4. Clean the window of the photoelectric cell or resistor and ensure it is correctly located in its housing.

5. Apply some light oil to the motor bearings if necessary — some motors are lubricated for life.

6. All filters should be cleaned in paraffin.

7. Check all oil pipelines for leakage and rectify as necessary. This is important as any slight leak will vaporise, and apart from an unpleasant smell, could constitute a serious fire risk.

8. The electrical components and wiring of the installation should be checked for security, damage and correct operation.

9. The fire valve should be operated to ensure it functions correctly, and in the case of the drop-weight type, the cable and pulley fixings must be secure, free from corrosion and have freedom of movement.

10. The fuel tank should be visually examined for leakage and corrosion and any sludge should be drained off.

11. After ensuring the boiler is full of water (in unoccupied premises the system may have been drained) fire the boiler and after about 10 minutes when the flue is warm, combustion and efficiency tests should be conducted and recorded.

12. Any fail-safe devices on the burner must be operated to ensure they would function correctly in any emergency.

A service record should be retained showing the date of each service, the CO_2 reading and stack temperature.

Further reading

BS 5410:Part 1 Installation up to 44 kW for hot water and space heating installations.
BS 799. Oil-burning equipment.
BS 799:Part 2 Vaporising burners.
BS 799:Part 3 Automatic and semi-automatic atomising burners.
BS 799:Part 5 Oil storage tanks.
BS 4543:Part 3 Chimneys for oil-fired appliances.
BS 4876 Performance requirements and test procedures for domestic oil-burning appliances.

Boilers
Boulter Boilers Ltd, Magnet Works, Whitehouse Road, Ipswich, IP1 5JA, Tel. 01473 241555.

Burners and control equipment
NU-WAY Ltd, PO Box 1, Vines Lane, Droitwich, Worcester, WR9 8NA, Tel. 01905 794331.
Danfoss Randall Ltd, Perivale Industrial Estate, Horsenden Lane South, Greenford, Middlesex, UBS 7QE, Tel. 020 8991 7000.
Rielio Ltd, Unit 6, The Ermine Centre, Ermine Business Park, Huntingdon, Cambridgeshire, PE 18 6XX, Tel. 01480 432144.

Vaporising Burners Don Heating Products Ltd, The Trading Estate, Wellington, Somerset, TA21 8SS, Tel. 01823 663181.

Testing equipment
Shawcity Ltd, Pioneer Road, Faringdon, Oxon, SN7 7BU, Tel. 01367 241675.

Fire protection
Falcon Landon Kingsway, 1077 Kingsbury Road,
    Birmingham B35 6AD, Tel. 0121 327 1662.

Teddington Controls Ltd, Holmbush, St Austell,
    Cornwall, PL25 3HS, Tel. 01726 74400.

Oil storage vessels
Balmoral Mouldings, Balmoral Park, Larston,
    Aberdeen, Scotland, AB9 2BY,
    Tel. 01224 859100.

Technical information and training
Oil Firing Technical Association, Century Hse,
    100, High St, Banstead, Surrey, SM7 2NN,
    Tel. 01737 373311.

Self-testing questions

1. Identify the two main classifications and
   viscosity of oil used for domestic, commercial
   and light industrial use.

2. State the effects of low temperature on the
   viscosity of oil.
3. List two materials suitable for oil pipelines.
4. State why the oil storage vessel must fall
   towards the sludge cock.
5. Explain how a combustable oil mist is produced
   in an atomising burner.
6. List the procedure that must be followed to
   ensure the oil supply is purged of air from
   the storage vessel through to the nozzle.
7. Explain the term 'purge period' in connection
   with atomising boilers.
8. From Table 6.3 determine the combustion
   efficiency of a burner having a flue temperature
   of 204 °C and a CO_2 reading of 11 per cent.
9. State the effect on boiler efficiency if a smoke
   test gives a reading of nil on the smoke scale
   and the net flue temperature is found to be
   420 °C.
10. List the items that must be checked when
    carrying out an annual service on an atomising
    oil boiler installation.

7 Sanitary Appliances

After completing this chapter the reader should be able to:

1. Describe the installation and working principles of WC macerator units.
2. List and describe the functions of the appliances necessary in commercial and industrial buildings for sanitary and ablutionary purposes.
3. Explain the need for thermostatic mixing valves and describe their working principles.
4. Understand the methods of fitting shower units and their associated discharge pipework.
5. Identify the main types of urinal and their suitability for various situations.
6. State the basic principles of automatic flushing cisterns.
7. Understand the need for conservation of water and describe the methods used to avoid wastage with automatic flushing cisterns and ablutionary fittings.

Introduction

This chapter deals mainly with sanitary appliances more commonly found in industrial and commercial premises than in domestic use, the one exception being WC macerating units.

WC macerators

These units are designed for where it is not possible to make a normal full connection from the outlet of a WC to the main discharge stack. The function of the macerator is to break up the solids which may be discharged when a WC is flushed, simultaneously pumping them and the flushing water to the main stack or drain via a 20 mm or 32 mm pipe, depending on the type of unit used.

Pipe sizes and lengths depend upon the pumping capacity of the unit. Some manufacturers produce units for WCs only, others can be extended to pump the discharge from a washbasin and WC or a complete bathroom. Figure 7.1 shows the general arrangement of the unit in relation to the WC,

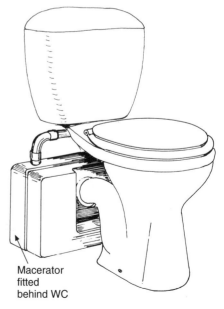

Macerator
fitted
behind WC

Fig. 7.1 Pictorial illustration of WC suite fitted with macerator unit.

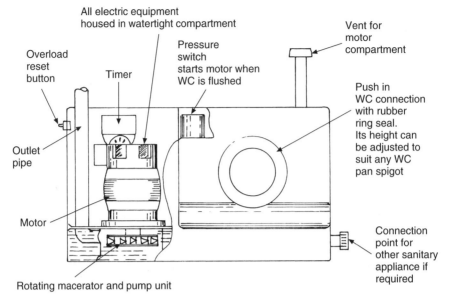

Fig. 7.2 Cutaway section of macerator unit.

which must have a P trap outlet. Figure 7.2 illustrates the main components of the macerator and pumping unit.

The following observations relate to all types of macerator irrespective of whether they are used with WCs, ablutionary appliances only or a combination of both. Prior to selecting a macerator, it is important to establish the type of appliance for which it will be used, and the lengths, both horizontal and vertical, of the main discharge pipes. It is important to refer to the manufacturer's specifications on these points.

Discharge pipes

As there are various types of macerator available the following should be used for general guidance only. With most types an increase in vertical height will reduce the length of horizontal run that is possible. One manufacturer states that for every 1 m of vertical pipework, 10 m of the maximum horizontal length must be deducted. This takes into account the extra work the macerator pump has to do. For example, if the maximum horizontal length pumped by a macerator is 50 m and the job requires a vertical lift of 2 m, the maximum horizontal length will be 30 m. Typical discharge

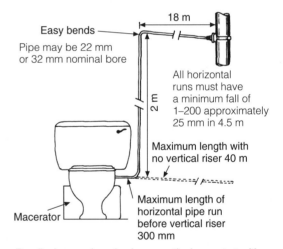

For discharge pipes having a vertical run, start with a base of 40 m maximum length and deduct from this 10 m for every 1 m vertical riser. In the example shown, the riser is 2 m high, 1 m must also be deducted for each bend.

Fig. 7.3

pipe arrangements are shown in Fig. 7.3. In most cases the diameter of the discharge pipe is 22 mm, and it must be copper or chlorinated PVC to BS 7291. The use of these materials will prevent sagging and the build-up of residual water in the pipe leading to blockages. Connections to the main discharge stack may be achieved by a suitable branch or a 100 mm × 32 mm clamp-on connector and reducing fitting to 22 mm. Horizontal runs should have a 1:200 fall and if they are longer than 12 m the diameter must be increased to 32 mm to avoid self-siphonage. Only one vertical lift is permitted and should be taken off not more than 300 mm from the macerator. If the macerator is at high level in the building, necessitating a vertical drop on the discharge pipe, self-siphonage could take place. In such cases an approved air admittance valve must be fitted at the highest point in the pipe run. The falls and pipe diameters from all appliances discharging into the macerator must comply with BS EN 12056 Pt 2, e.g. a basin discharge pipe will be 32 mm in diameter.

The operation of the motor is fully automatic, by means of a pressure switch and timing device. Units designed to deal with discharge of WCs and ablutionary appliances, are provided with the necessary connections, which also incorporate non-return valves to prevent the back-up of foul water into the appliance discharge pipes. To install the macerator unit it is placed in position below the WC and the pan spigot is pushed into the outgo socket on the macerator which is sealed by a synthetic rubber joint. Brass screws must be used to secure the WC to the floor, as although the motor can normally be serviced with the WC in position, should it be necessary to remove the unit as a whole, the WC must be capable of being withdrawn easily. The electrical connections must conform to the Institute of Electrical Engineers (IEE) standards, the usual arrangement being to provide a fused unswitched socket outlet having a 5 A fuse rating. The discharge pipe must be fitted in such a way that any vertical pipework is taken directly off the macerator and changes of direction made using large-radius bends. For those units with a 20 mm outlet it is recommended that 22 mm copper pipe with machine-made bends is used. It must be clearly understood that these macerators are entirely mechanical in operation and being liable to occasional failure they should not be installed in buildings where no alternative WC accommodation is available. The advice of the local authority building control officer should be sought in cases where any doubt exists concerning their installation.

Flushing troughs

These are used in buildings such as factories or schools where at peak periods the appliances may be in almost continuous use. A normal flushing cistern should be refilled in 2 minutes, but this may not be quick enough for peak usage. Flushing troughs enable this difficulty to be overcome because, due to the large volume of water held in the trough and the speed with which it is replaced, the WCs can be flushed almost continuously. The tanks themselves are made of galvanised steel or a suitable thermosetting plastic material with a base and height of approximately 255 mm. Lengths are determined by the number of WCs served and the width of the compartment. It is normal to restrict the number of WCs served from one trough to six. Due to the possible demand made on the water in the trough, it should be supplied via a float-operated valve with a minimum nominal diameter of 25 mm. It may be necessary to provide two valves if the supply is insufficient to meet the demand.

When filled with water these tanks are heavy and must therefore be well supported. If the WC partitions are constructed of brickwork, little support will be necessary, but if only lightweight partitions are provided then very strong additional supports will be needed. If there is any doubt about the strength of fixings obtained with brackets and screws, short lengths of galvanised angle iron should be built into the wall.

The actual flushing mechanism is very similar to that of an ordinary cistern, the only difference being that a measuring chamber is necessary to limit the volume of flush. Figure 7.4 shows a typical siphon for use with a flushing trough. When the cistern is flushed, water is drawn from the trough and the measuring chamber, in which is drilled a small hole. Water is siphoned out of the measuring chamber more quickly than it can enter, and when air is admitted to the siphon via the dip pipe, it breaks the siphon thus stopping the flush. If the siphon is made of metal, the small hole sometimes becomes corroded and fails to allow water to enter the measuring chamber. This will result in failure of the flushing arrangement and is the first thing to check if the siphon fails to operate.

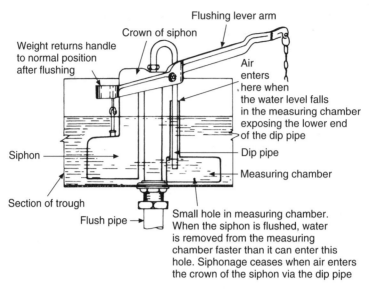

Fig. 7.4   Siphon for flushing trough. Do not attempt to reduce this volume of water to 6 litres by shortening the dip pipe, as this may be insufficient to clear the contents of the trap in existing WCs. Those produced now have been modified to function with a smaller flush volume and any new installations will have to be designed to comply with the Water Regulations.

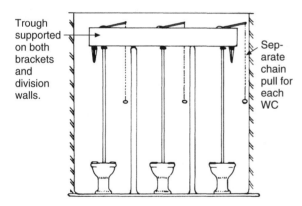

Fig. 7.5   Application of flushing trough.

Figure 7.5 illustrates a typical flushing trough serving three WCs.

Ablutionary fittings

Spray taps
The types of fittings used for ablutionary purposes in domestic dwellings have been dealt with in Book 1. In public and commercial buildings, however, needs may differ depending upon the nature of usage and the availability of space. In office blocks the use of washbasins is predominant, these being fitted singly or in ranges. As they are normally required only for hand-washing purposes, special basins provided with spray taps are satisfactory. Only one tap hole is necessary in the washbasin when these taps are used, and no plug is provided with the waste fitting as for hand washing it is considered unnecessary to fill the basin with water. As the flow rate for spray taps is much less than that for pillar taps, considerable savings can be effected in the use of both hot and cold supplies. These taps are also considered to be more hygienic, as the hands are washed in running water — a similar advantage to that of a shower.

Spray taps vary considerably in design, the simplest single supply type supplying water at a thermostatically controlled temperature to the basin. In cases where this arrangement is not possible spray *mixing* taps can be used. A typical tap of this type is shown in Fig. 7.6. They are provided with both a hot and a cold supply which *must* be at the same pressure for efficient working, prevention of pollution and possible wastage of water. As these

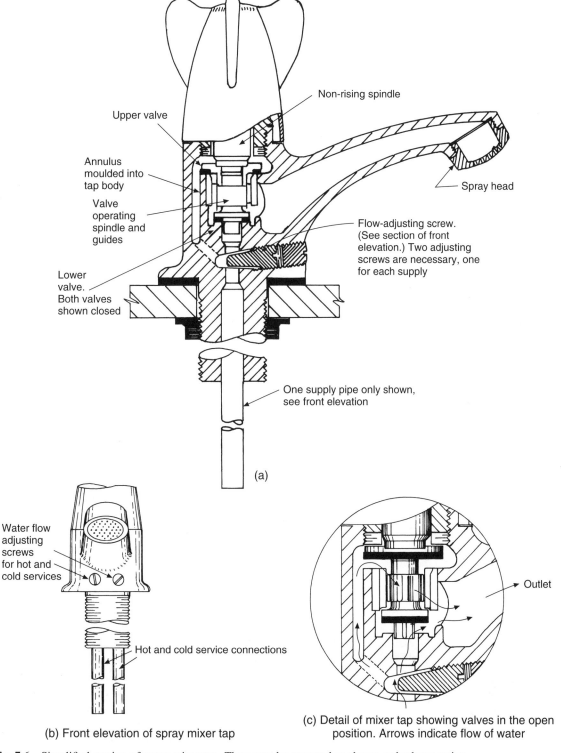

Non-rising spindle

Upper valve

Annulus moulded into tap body

Valve operating spindle and guides

Lower valve. Both valves shown closed

Spray head

Flow-adjusting screw. (See section of front elevation.) Two adjusting screws are necessary, one for each supply

One supply pipe only shown, see front elevation

(a)

Water flow adjusting screws for hot and cold services

Hot and cold service connections

(b) Front elevation of spray mixer tap

Outlet

(c) Detail of mixer tap showing valves in the open position. Arrows indicate flow of water

Fig. 7.6 Simplified section of spray mixer tap. They may be operated as shown or by lever action.

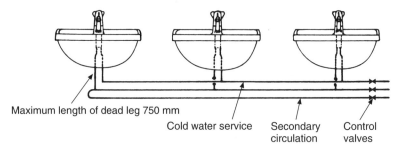

Maximum length of dead leg 750 mm

Cold water service   Secondary   Control
                      circulation   valves

Fig. 7.7 Use of mixer taps on a range of basins.

taps are seldom thermostatically controlled, they are made in such a way that the cold outlet is opened first, and hot water is only gradually admitted until the required temperature is achieved.

The hot supply to any spray tap must be taken from a secondary circulation, the maximum length of dead leg being 1 m. A dead leg of any greater length than this would unacceptably increase the time taken for hot water to reach the tap. A typical piping arrangement for these taps is shown in Fig. 7.7.

Basins used for hand washing only are smaller than those for domestic properties, and are usually supported on a purpose-made chromium-plated tubular frame or built in brackets. Fixings for any component in public buildings must be robust enough to resist carelessness and deliberate vandalism. The use of screws on brackets is seldom suitable in these conditions. Where it is necessary to seal the gap between adjoining basins, a glazed clayware cover strip is used as shown in Fig. 7.8. This prevents the ingress of filth to an area difficult or impossible to keep clean. These cover strips should be made good on the edges of the basins,

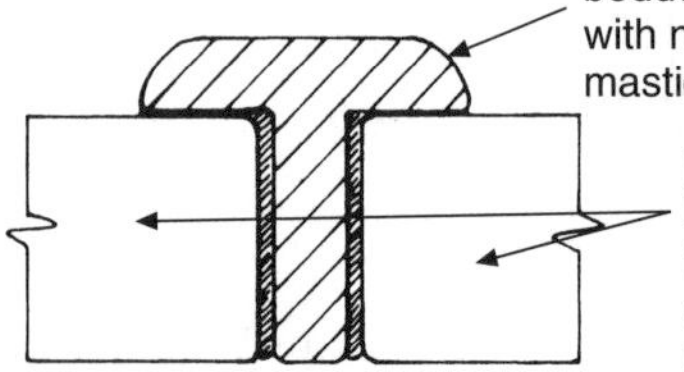

using a material that allows for their easy removal should a basin have to be replaced. A suitable non-hardening mastic should be used for this purpose as it is not easy to match these strips should they be broken.

Industrial hand-washing facilities
Washing facilities for industrial usage may have to withstand rough treatment, and for this reason ablution troughs made of heavy glazed clayware or stainless steel are commonly used. Figure 7.9 shows a typical installation. These troughs are suitable for wall fixing or, to save wall space, they may be fitted back to back and are then known as an island arrangement. An alternative to ablution troughs is the circular ablution fountain illustrated in Fig. 7.10. Due to its shape it allows the maximum number of people to wash at the same time. The water supply, which must be thermostatically controlled at a suitable temperature for hand washing, is discharged in the centre, forming an unbrella-like spray. The spray is operated by depressing the foot ring which is connected to a valve inside the column. Maintenance of the valve or discharge pipe is effected via the access panels in the column.

Non-concussive valves

In an effort to conserve water used with public washing facilities, non-concussive valves are sometimes used. They are designed in such a way that the head must be depressed to permit the valve to open. On releasing the head, the water will flow for a short period only prior to automatically closing thus avoiding the possibility of a tap being left running. The term *non-concussive* is derived from

Clayware cover strip
bedded on to basin edges
with non-hardening
mastic

Edges of wash-
basins. Note
that similar
arrangements are
made to seal the
edges of stall urinals

Fig. 7.8 Sealing the edges of basins in a range.

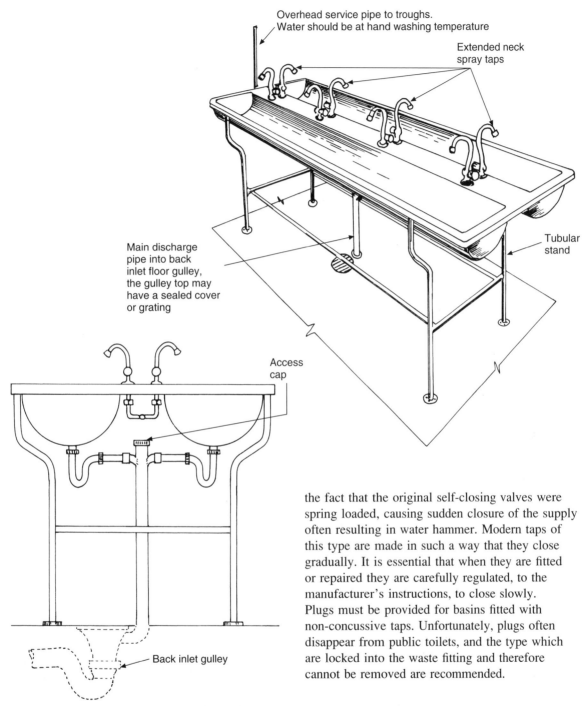

Overhead service pipe to troughs. Water should be at hand washing temperature

Extended neck spray taps

Tubular stand

Main discharge pipe into back inlet floor gulley, the gulley top may have a sealed cover or grating

Access cap

Back inlet gulley

End view of island showing discharge pipe and water supply arrangement. The back inlet gulley is not essential, but would be useful in cases where it is necessary to wash down the floor area.

Fig. 7.9 Typical industrial island-type washing trough.

the fact that the original self-closing valves were spring loaded, causing sudden closure of the supply often resulting in water hammer. Modern taps of this type are made in such a way that they close gradually. It is essential that when they are fitted or repaired they are carefully regulated, to the manufacturer's instructions, to close slowly. Plugs must be provided for basins fitted with non-concussive taps. Unfortunately, plugs often disappear from public toilets, and the type which are locked into the waste fitting and therefore cannot be removed are recommended.

Anti-scald valves

These valves are essential to prevent scalding in certain situations. They are becoming increasingly used in sheltered accommodation and public

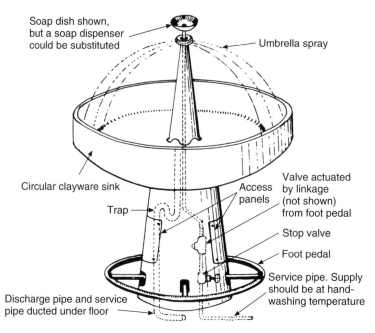

Fig. 7.10 Washing fountain.

buildings such as schools and hospitals. The usual form of temperature control used in these valves is a thermostatic bellows similar to that shown in Fig. 7.21. They can be used within certain limitations on supplies of differing pressures, usually up to 2 bar, although it is always preferable that the pressure on both hot and cold supplies is the same. The manufacturer's specifications should always be checked before fixing these valves. It should be noted that prior to the 1986 water by-laws it was mandatory that the hot and cold supplies in any mixer tap, with the exception of the 'biflo' type, should have equal pressure. The main reason for this was to (a) reduce the risk of pollution and (b) prevent water wastage arising from the possibility of a mains pressure cold water supply backing up into the low-pressure hot water system and causing the feed cistern to overflow. The reasons for prohibiting mixed pressures no longer apply due to the installation of check valves. It is, however, much easier to balance the mixer supply if both pressures are equal.

The valve shown in Fig. 7.11 is designed for controlling the supply of hot water to wash basins, baths and bidets. One valve can be used to control

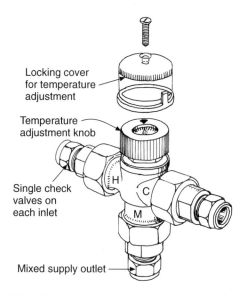

Fig. 7.11 Thermostatic mixer valve for wash basins and baths.

more than one appliance (see Fig. 7.12) providing that (a) an adequate supply is available and (b) the outlet pipe is no longer than 2 m. This is a Ministry of Health recommendation to avoid the incubation

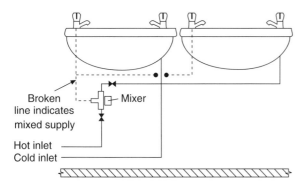

Fig. 7.12 Application of the thermostatic mixer valve. Note the maximum length of the mixed supply is 2 m.

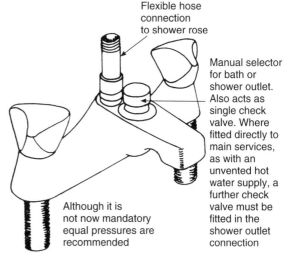

Fig. 7.13 Bath mixer and shower combination.

of Legionnaires' disease bacteria. The valves can be fitted in any position provided access is available for adjustment and maintenance. New pipework must be thoroughly flushed out before fitting, and if it is connected to an existing installation it is recommended that strainers are provided to both inlets. Service valves should also be fitted in all cases to avoid major shutdowns when maintenance is necessary. Always comply with the manufacturer's instructions relating to installation, commissioning and testing. The recommended temperature settings are 40–41 °C for wash basins and bidets and approximately 43–44 °C for baths. When the recommended temperature has been set the locking cap provided must be fitted to prevent any unauthorised tampering with the setting. After installation a check should be made within the period of time recommended by the manufacturer to ensure the mixed water temperature is within the prescribed limits.

Before dismantling a valve suspected of malfunctioning, first check the strainers are clean, the check valves are in working order and any isolating valves are fully open.

Showers

Showers have many advantages over baths, e.g. they occupy less floor space, and washing the body with running water is said to be more hygienic. In small domestic properties where there is insufficient space

for a shower cubicle, a shower and mixer tap combination as shown in Fig. 7.13 can be fitted to the bath. These are not thermostatically controlled however, and unless carefully regulated can cause scalding.

Showers operating under the pressure of a storage cistern in domestic housing are very economical. A shower lasting 4–5 minutes will use approximately 28–36 litres, whereas an average bath will use approximately 110–140 litres. However, some power showers are capable of producing flow rates of 20 litres per minute. Many of these operate at a minimum pressure of 1 bar and are often used with multiple shower heads. Such showers use more water than a bath and are not water conservation appliances. While this does not preclude the use of power showers, they should be selected very carefully in consultation with the customer.

Shower trays

Due to their heavy construction the use of clayware shower trays is limited mainly to public and commercial buildings. They have a rounded edge, sometimes greater in radius than the thickness of the wall covering. Figure 7.14 illustrates the possible problem. It is suggested that the method of fitting is the same as for clayware sinks illustrated

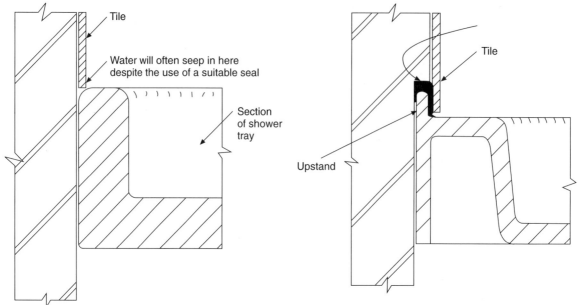

Fig. 7.14 Water seepage from a shower tray.

Fig. 7.16 This shower tray is manufactured with an upstand, which, properly installed, solves seepage problems.

in Book One Chapter 9, or that shown in Fig. 7.15. Some types of shower tray are produced with an upstand as shown in Fig. 7.16. Properly installed they also solve seepage problems. Acrylic and cast iron shower traps are normally provided with adjustable feet which allows the top edges to be carefully levelled; this is very important if purpose-made enclosures are to be fitted. When showers are installed on suspended wooden floors, shrinkage of

the timbers, especially when new, can cause many problems and for this reason one-piece shower units are recommended. These incorporate the tray and sides of the shower in one complete prefabricated unit, which has no joints so cannot possibly leak. An alternative to this is to fit the shower tray in a prefabricated lead safe as shown in Fig. 7.17, which properly installed is very resistant to leakage. After fitting the shower tray make sure it is protected

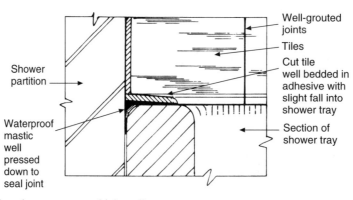

Fig. 7.15 Waterproofing shower tray to cubicle walls.

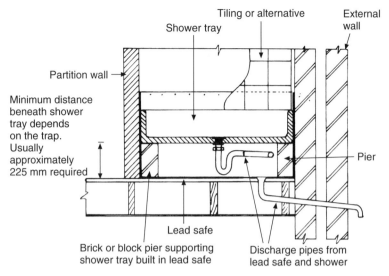

Partition wall →

Minimum distance
beneath shower
tray depends
on the trap.
Usually
approximately
225 mm required

Shower tray

Tiling or alternative

External
wall

Pier

Lead safe

Brick or block pier supporting
shower tray built in lead safe

Discharge pipes from
lead safe and shower

Fig. 7.17 Shower with lead safe on suspended floor.

from damage by other trades. Cardboard packing
secured with masking tape and covered with a dust
sheet is usually satisfactory.

Pollution risks
Showers are classified as a category 3 risk for back-
siphonage and require a backflow device complying
with type AU2; this can be an air gap or check
valves. Those with fixed shower heads have an air
gap well in excess of the minimum requirements.
Those with flexible hoses require double check
valves. Provided the shower mixer inlets and the
mixed outlet to the head are fitted with approved
single check valves this will meet the requirements
of the Water Regulations. An additional safeguard
is the use of a hose restraining ring shown in
Fig. 7.18. Fittings with flexible hoses that can
fall into a WC or bidet pose a category 5 risk of
pollution, and reference should be made to
Fig. 3.34.

Shower discharge pipes
The connection of shower traps to the discharge
pipe has always been a problem due to the shower
tray being fitted as low as possible. Because of
this very little access was provided to make the
necessary connections and to carry out any
maintenance required. Shower traps are very prone

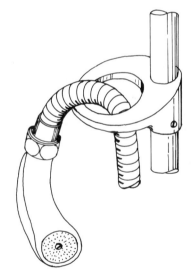

Fig. 7.18 Hose restraining ring. This must be fitted in
such a way as to prevent the shower rose falling below the
flood level of the appliance or into an adjacent WC or bidet.

to blockage with hair, and the very low flow rate
in the discharge pipe allows the build-up of soap
deposits, especially in the trap. The use of bottle
traps with a 50 mm seal, which can be dismantled
for cleaning by removing the grating and dip pipe,
makes maintenance a lot easier providing the trap
has been correctly installed and the discharge

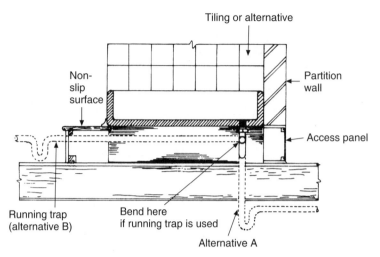

Fig. 7.19 Alternative shower discharge pipe arrangements. The use of a running trap above or below floor level renders it accessible for cleansing when necessary. Access to a trap may be gained under ceiling if the shower is over an outbuilding or garage.

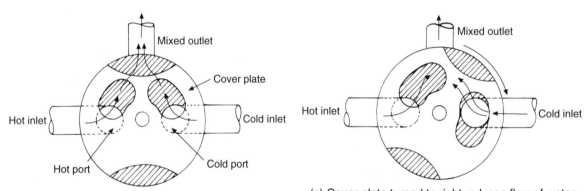

(a) Waterways of both ports equally exposed giving proportional mixing

(c) Cover plate turned to right reduces flow of water through hot port but exposes more of the cold port, reducing the temperature of the mixed flow

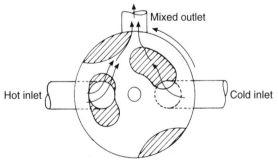

(b) Cover plate turned to left reduces flow of water through cold port but exposes more of the hot port, increasing the temperature of the mixed flow

Fig. 7.20 Mixing arrangements for manually operated shower mixing valves.

pipework checked for leakage prior to tiling or panelling the shower cubicle. Bottle traps, by virtue of their design, are not self-cleansing. In situations where traditional traps are fitted, Fig. 7.19 illustrates some methods of fitting which allow access to them.

Shower mixing valves
Although non-thermostatic mixers may be used, for the reasons given previously in this chapter, those with thermostatic control are preferable. Figure 7.20 illustrates the working principles of those operated manually. Thermostatic mixer showers work on the

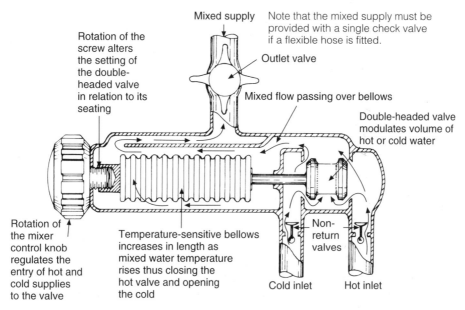

Mixed supply Note that the mixed supply must be provided with a single check valve if a flexible hose is fitted.

Rotation of the screw alters the setting of the double-headed valve in relation to its seating

Outlet valve

Mixed flow passing over bellows

Double-headed valve modulates volume of hot or cold water

Rotation of the mixer control knob regulates the entry of hot and cold supplies to the valve

Temperature-sensitive bellows increases in length as mixed water temperature rises thus closing the hot valve and opening the cold

Non-return valves

Cold inlet Hot inlet

Fig. 7.21 Working principles of thermostatic mixing valves.

movement of a heat-sensitive bellows or bimetallic coil, both of which automatically open or close the inlets to maintain a predetermined temperature. The temperature of water for showering is normally between 38 and 40 °C and thermostatic valves will maintain this temperature range plus or minus one or two degrees, thus ensuring safety. A typical bellows-operated thermostat mixer is shown in Fig. 7.21.

The position of the mixer unit in the shower cubicle is important. Its height from the base of the shower tray is usually 1.45 m. It should be positioned in the shower cubicle on the right or the left in such a way that the user's hand does not have to pass through the shower spray to adjust the temperature. The mixer valve and its associated pipework may be built into the wall providing the pipes are suitably protected against the corrosive effects of cement mortar or plaster. This avoids having the rather bulky mixer protruding into the cubicle. It is essential, of course, to check first if the wall is thick enough to accommodate the mixer unit. Most manufacturers supply a suitable casing with units designed for built-in fixing, which ensures that any servicing can be carried out with a minimum of inconvenience. Where it is not possible

or advisable to fit concealed units, face-mounting types are available. While such an installation does not present such an attractive appearance, both the pipework and the mixer unit are available for easy maintenance and this is the recommended practice in public or industrial premises. Polished stainless steel or chromium-plated copper pipe should be used where pipework is exposed on the surface of the shower cubicle.

The arrangement of the shower spray head has many alternatives, the two main ones being illustrated in Fig. 7.22. The type with a flexible pipe (Fig. 7.22(a)) enables the height of the shower rose to be adjusted and also permits its removal for hair washing. The fixed type (Fig. 7.22(b)) should be fitted, unless otherwise specified, so that the rose is a minimum height of 2 m from the floor of the shower tray. The minimum recommended head of water for the shower to be effective is 1 m from the shower rose to the base of the cistern (see Fig. 7.23). When the installation of a shower is considered in an existing building, this is one of the points that must be ascertained before work is commenced. If the minimum head is not available, several alternatives are possible. One is to raise the position of the feed cistern so that sufficient

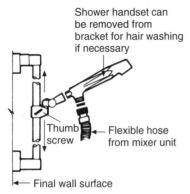

(a) Shower handset with adjustable bracket

This permits a variation of height between the base of the shower tray and the rose.

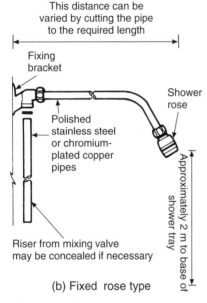

(b) Fixed rose type

Fig. 7.22 Shower head arrangements.

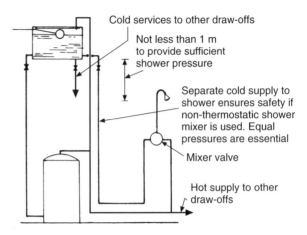

Fig. 7.23 Shower connections for domestic dwelling.

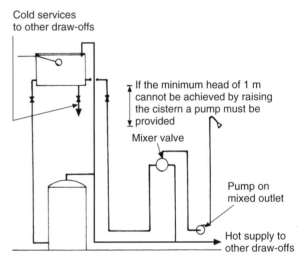

Fig. 7.24 Pumped shower outlets.

pressure is obtained; another is to fit a pump on the outlet of the mixer as shown in Fig. 7.24. The pump can be installed at floor level, in a false ceiling or even under the bath, the only prerequisite being that it is accessible for maintenance. Shower mixer may also be connected directly to a main water supply providing there are adequate safeguards against pollution risk.

When the hot storage vessel is a long distance away from the shower, it may be necessary to install a secondary circulation to avoid a long wait for hot water to reach the mixer valve. This will certainly be required in buildings where several showers are needed as, for instance, in changing rooms in a sports pavilion or in an industrial establishment where, due to the nature of the work, the employer provides bathing facilities. Such an installation is shown in Fig. 7.25. In most small domestic dwellings a secondary circulation should be avoided due to heat losses and wastage of water from long dead legs. In such cases the use of single-point instantaneous gas or electric heaters should be considered.

Figure 7.26 shows a typical gas-heated installation. Only certain types of heater are suitable

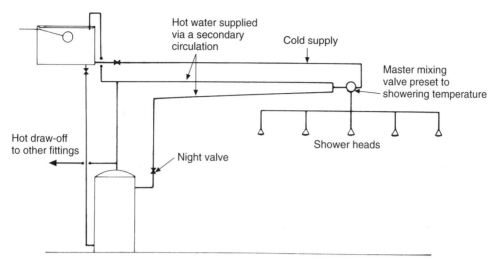

Fig. 7.25 Layout for a range of showers.

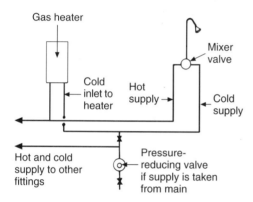

Fig. 7.26 Shower supplied by gas water heater. Note that some types of gas heater will work satisfactorily under lower presure than others and may be supplied by a feed cistern which is a better arrangement. If this type of scheme is contemplated the manufacturers of both the shower and the heater should be consulted.

for this purpose, and before such an installation is considered the advice of the local gas authority and the manufacturer of the shower should be sought. If the head of water from the cistern is insufficient to operate the gas valve on the heater, the supply may be taken from the main, but a pressure-reducing valve must be fitted and the installation must conform to the requirements of the Water Regulations regarding the possibility of back-flow and water wastage.

Electric showers

These showers have become very popular for domestic use in recent years, mainly due to their relatively low installation costs. They are also useful where an isolated shower unit is to be installed necessitating a long run of hot water pipe. Their rating in kW varies from 7.5 to 10.0 at normal mains voltage, the larger ratings giving quite a hot shower. Most electric showers will raise the temperature of the incoming water through 32–35 °C. The temperature of the main water supply is variable and is usually between 10 °C and 15 °C, but it can drop in very cold weather to 5 °C. This means that a shower with a low electrical rating might have difficulty in heating the water to showering temperature at high flow rates. A typical electric shower heater is shown in Fig. 7.27.

The electrical part of the installation must be carried out by a competent person and comply with the IEE Regulations. This is important because cable sizing, voltage drop and earthing have to be considered prior to installation. Electric showers are, like electric immersion heaters, high rating appliances and must be wired directly to the main consumer unit, with a separate fuse or MCB. An isolating double pole switch must be fitted to isolate the unit for repair or maintenance and a double check valve must be provided on the *outlet and not the inlet*. This allows the water in the heater

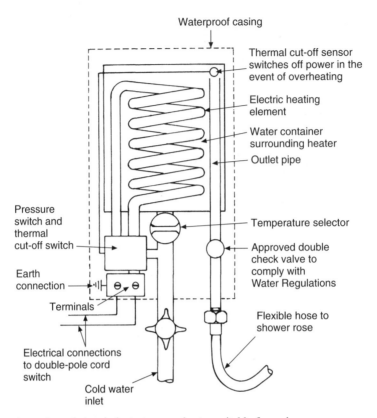

Waterproof casing

Thermal cut-off sensor switches off power in the event of overheating

Electric heating element

Water container surrounding heater

Outlet pipe

Pressure switch and thermal cut-off switch

Temperature selector

Approved double check valve to comply with Water Regulations

Earth connection

Terminals

Flexible hose to shower rose

Electrical connections to double-pole cord switch

Cold water inlet

Fig. 7.27   Diagrammatic section of electric instantaneous heater suitable for a shower.

to expand without causing a pressure build-up. Installation of a water treatment device or water softener is recommended in very hard water areas. Prior to installation the flow rate and static pressure of the water supply should be ascertained to ensure the compatibility of the unit. This is easily carried out using a pressure gauge and flow meter. Metered supplies do incorporate check valves and in such cases the manufacturers recommend the use of a mini expansion vessel on the water inlet to the shower unit. Most instant electric heaters incorporate a means of compensation for variations in supply pressures to maintain an even flow and stable temperature. This can be a simple device such as (a) a flow regulator (b) a pressure reducing valve or (c) a modulating solenoid valve. The last is the most effective in maintaining both a constant outlet supply and temperature. In situations where there is insufficient mains pressure to operate the shower, a pumped supply taken from a storage cistern will increase the pressure, but it is essential

that sufficient storage is available to supply any other existing drawoffs. A minimum storage of 120 litres is recommended by most electric shower manufacturers in such cases. If the appliance is to be used for the disabled, old or infirm, a temperature-limiting thermostat must be employed. Two other devices are also built into the appliance to prevent damage due to overheating:

(a) The thermal cut-off which automatically switches off the power supply if the water reaches a temperature of approximately 50 °C.
(b) The pressure switch ensures that an adequate supply of water is available before the heater element is energised.

Electrical switching
A fixed power supply with double pole cord-operated pull switch is essential. An ordinary pull switch simply breaks one conductor or wire, whereas the double-pole type breaks both the

conductors, isolating the heater completely, and provides the high degree of safety needed where electrical appliances are installed in surroundings subject to moisture and dampness.

Maintenance of shower units

Reputable manufacturers of shower equipment provide adequate information on maintenance, either with the unit or on request. The part of the installation which usually requires maintenance is the mixer where the O-ring seals and washers become worn after a period of use. Replacements are available from the manufacturer or a good plumbers' merchant. It will, of course, be necessary to quote the type and model number of the mixer. Before any maintenance work is undertaken the floor of the shower tray must be protected from scuff marks from shoes or boots with a suitable covering. The outlet grating should be sealed to prevent any screws or small parts of the mixer falling into the discharge pipe while work is in progress as these will be difficult if not impossible to retrieve. Any scale formation on the working parts of the unit should be carefully removed using a suitable descalent. The type used for descaling kettles is suitable for this purpose. Abrasive materials or acid solutions may do irreparable damage and should not be used. Where loss of temperature control occurs in thermostatic showers it is usual to renew the temperature sensing cartridge, as it is usually this that is at fault. Temperature must always be verified using

a suitable thermometer on completion of any maintenance work on the mixer unit.

Urinals

Urinals are made of the following materials: glazed clayware, stainless steel or, in the case of some bowl types, vitrified china. Fibre-glass urinals are made, but are considered by some to be unsuitable as they lack the strength and durability of the materials previously mentioned.

There are four main groups of urinals, known as the stall, slab, bowl and trough types.

Stall urinals

These were the original form of earthenware urinal and are still in use in older public buildings. While they have many advantages they are very heavy and expensive. Few, if any manufacturers now list them.

Slab urinals

As the name implies, this type of urinal is built up of rectangular slabs of glazed clayware which are bedded on to a separate channel. The usual flushing arrangement is a sparge pipe running the whole length of the urinal in which is drilled a series of holes. When the cistern is flushed, water passes through these holes to cleanse the slab. It is important that the sparge pipe is fitted in such a way that the water impinges on the slab to ensure that the whole surface is cleansed. Slab urinals are illustrated in Fig. 7.28; they may be supplied

Automatic flushing cistern

Sparge pipe, in which is drilled a series of holes, flushes the slabs when the cistern operates

Clayware slabs

Channel made up in sections

(a) Arrangement of slab urinals

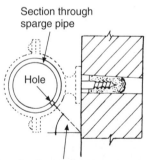

Section through sparge pipe

Hole

Water discharging from holes in sprage pipe at approximately 45°. A check should be made to ensure the slabs are thoroughly cleansed after each flush

(b) Fixings for sparge pipes

Fig. 7.28 Slab urinals.

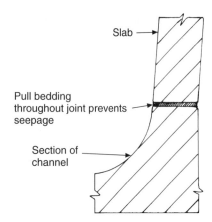

Fig. 7.29 Correct method of bedding urinal slabs.

even good bricklayers do not always realise the problems resulting from defective jointing. For this reason a plumber should be aware of the processes involved. Cement mortar should be of one part fine washed sand to one part cement to ensure that it is waterproof. Sometimes white cement is used to point the joints to improve their appearance. It is essential that all joints are properly bedded as in Fig. 7.29. Simply pointing the joints when the slabs are in position is not satisfactory as the pointing media is soon washed away leaving an unsealed area open to the entry of foul water. Some channels are made with a stepped rebate which reduces seepage problems. Channels for slab urinals are made having an integral fall and are numbered to indicate the correct fixing sequence.

Bowl urinals
No matter how carefully the joints on slab urinals are made, any movement of the structure in which they are installed may cause them to crack. While this is serious in any case, if urinals are fitted on the upper floors of buildings the result of leakage will be especially unpleasant. For this reason they are sometimes built into a lead safe in a similar way to shower trays. This is expensive and for this reason bowl or trough urinals are often specified because they are made as a single unit (see Fig. 7.30). This

with or without division pieces. From a sanitary standpoint, they are better without, as division pieces are not cleansed by the sparge pipe, but they do provide some privacy for the user. Although slab urinals are much cheaper than stall urinals, their main disadvantage is the number of joints required in their construction.

Joints for slab and stall urinals
It is usually the job of a bricklayer to fix and bed these urinals, and while they are concerned with giving a neat appearance to the finished joint,

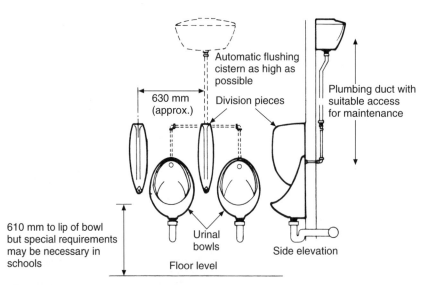

Fig. 7.30 Bowl urinals. Those parts of this illustration shown as broken lines may be concealed in a plumbing duct. Use of such a duct provides a neat installation, cleaning is easier and there is less opportunity for vandalism.

overcomes the jointing problem presented by slab urinals. The main objection to bowl urinals is the possibility of fouling the floor, but this can be overcome by fitting a floor gulley which enables the whole area to be periodically washed down. As with wall-hung urinals the branch discharge pipes can be fitted above the floor level, and if mechanical joints are used the discharge pipe system can easily be dismantled for cleansing. The discharge pipes should be large enough to accommodate the deposits of scale built up by uric acid and, in hard water districts, calcium carbonate. The number of urinals in the range also has an influence on the discharge pipe diameter.

Stainless steel urinals

This material has much to commend it for the construction of urinals as it is less prone to damage by vandalism. It is easily cleaned and complete units can be fabricated by welding under factory conditions making site jointing of sections unnecessary — except for fitting the outlet. Urinals made of stainless steel may be of the wall-hung trough type shown in Fig. 7.31, which gives access above floor level to the discharge pipework making maintenance easier. They are also constructed with an integral channel for floor fixing, and in this mode are not unlike slab urinals in appearance.

Bowl urinals can also be fabricated of stainless steel and, although expensive, they are less prone to vandalism than those made of clayware.

Waterless urinals

This is a comparatively new type of sanitary appliance which obviates the need for water for flushing urinals, thus saving water and reducing installation costs. A special renewable cartridge is necessary which contains an integral trap, a cross-section of which is shown in Fig. 7.32 together with its application in a bowl urinal. They are also suitable for installation in specially adapted trough urinals. Extensive tests have been successfully carried out in Germany and the USA to ensure they

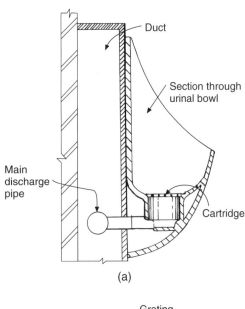

(a)

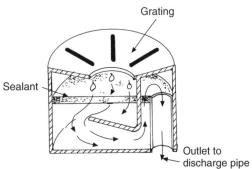

(b) Section through cartridge showing integral trap

Fig. 7.32 Waterless urinals.

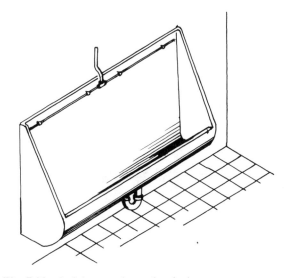

Fig. 7.31 Stainless steel trough urinal.

comply to the high standard of sanitation required. The absence of water prevents chemical reactions with urine, which avoids the objectionable odour associated with urinals. The cartridge contains a special biodegradable scalent which allows the urine to percolate into the cartridge and then floats on top of it. This provides a barrier preventing odours from the drain and the contents of the cartridge entering the washroom. A decrease in the speed of draining the urinal indicates the cartridge should be changed; frequency of changing will depend on the frequency of use. Heavy usage, such as in public toilets, may necessitate a change every 3–6 weeks. In areas of medium use such as offices and industrial premises, the cartridge may only need changing every 17–16 weeks.

Urinal waste fittings
The sizes of waste outlets recommended in BS 5572 are indicated in Table 8.1 (p. 246). Ranges of more than six urinals must be provided with two outlets.

The joint between the channel and the discharge pipe is made using a fitting similar to those shown in Fig. 7.33. They are designed to prevent debris that collects in the channel gaining access to the discharge pipe causing a blockage. If the outlet was flat it could become covered and cause flooding. The outlets shown may be made of good-quality brass, stainless steel or high-density polythene. The latter is not recommended as its service life, due to its inherent weakness, is likely to be shorter than metals. Domical gratings are hinged in such a way that they can be opened to expose the trap for rodding should this become necessary.

Due to the general lack of accessibility to discharge pipes from slab urinals, and the stresses to which they are subjected when cleaning is carried out, they should be made of a strong material such as copper or cast iron. With bowl or wall-hung urinals, the discharge pipes are usually exposed and sometimes PVC is used to reduce costs. Urinals can be a source of trouble, both

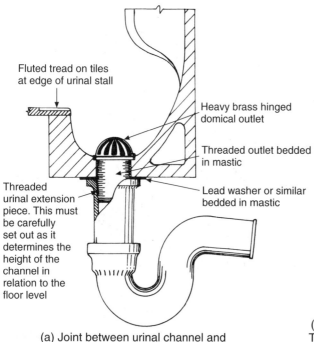

Fluted tread on tiles at edge of urinal stall

Heavy brass hinged domical outlet

Threaded outlet bedded in mastic

Threaded urinal extension piece. This must be carefully set out as it determines the height of the channel in relation to the floor level

Lead washer or similar bedded in mastic

(a) Joint between urinal channel and cast iron drain

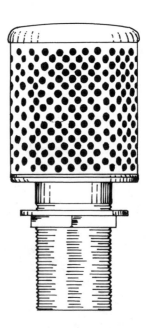

(b) Perforated stainless steel urinal waste fitting. These are more effective than traditional domical gratings in preventing blocked discharge pipes

Fig. 7.33 Urinal wastes.

from blockage in the discharge system and objectionable smells. Every care must be taken to ensure that the system is correctly installed with plenty of access to the discharge pipework and an efficient flushing cistern. A lockshield hose union tap should also be provided in urinal closets to enable the floor to be periodically washed down. The floor should have a slight fall towards the urinal channel, unless bowl urinals are installed, in which case, as stated previously, a suitable floor gulley must be provided.

Cleaning urinal discharge pipes

Due to the nature of urine, which can cause a build-up of lime deposit, and the ingress of debris, urinal traps and discharge pipes frequently become blocked. Reference should be made to Chapter 8 pages 262–3 for details on maintenance requirements.

Urinal flushing cisterns

Until the 1986 Water By-laws took effect, urinal flushing cisterns were regulated to flush three times per hour whether or not the building was occupied.

This obviously led to a great deal of wasted water, and since 1986 full automatic control has been mandatory to ensure urinals are flushed only when the building is occupied. The Water Regulations specify the volume of water permissible for urinal flushing. For a single urinal bowl automatically flushed, up to 10 litres of water per hour is permissible. With single appliances, however, it is more usual to use manual flushing apparatus such as a flushing cistern or valve. Where slab urinals or ranges of bowls are used 7.5 litres of water per hour per bowl is allowed, and for each 700 mm of urinal slab width. Where urinals are flushed by means of a valve it should not deliver more than 1.5 litres of water per bowl or slab position each time it is flushed.

In most cases where urinals are ordered, the supplier provides a suitable flushing cistern and all the associated pipework. The capacity of the cistern is normally 4.5 litres per bowl or 700 mm width of stall. The action of a typical automatic flushing cistern is illustrated in Fig. 7.34. The associated caption explains its working principles.

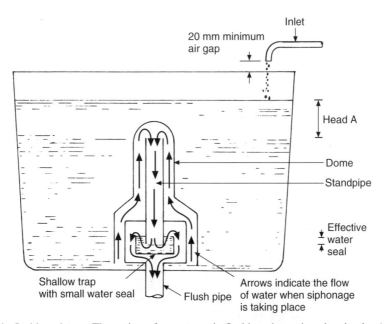

Fig. 7.34 Automatic flushing cistern. The action of an automatic flushing cistern is quite simple. As the water level rises pressure shown as head A increases until it overcomes the water seal in the shallow trap, causing it to overflow into the flush pipe. This causes a lowering of the air pressure in the standpipe and the dome and the water contained in the cistern is forced under the dome and into the flush pipe by atmospheric pressure. The action continues until the cistern is emptied and air enters the base of the dome. The shallow trap is resealed during the flush.

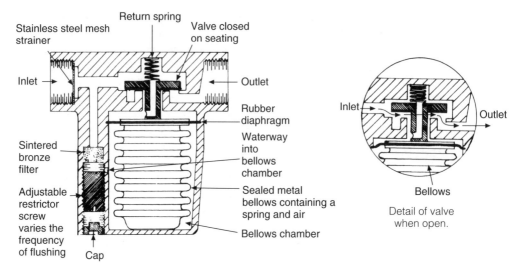

Fig. 7.35 Hydraulically operated valve. When draw-offs are closed the pressure of the water holds the valve down on its seating. Simultaneously water passes through the sintered bronze filter and into the bellows chamber where its pressure compresses the bellows. The water pressures on both sides of the diaphragm are now equal. The valve is held in the closed position by the return spring and the water pressure. If a tap is opened and the water pressure is reduced above the diaphgram the bellows can now exert an upward pressure, lifting the valve off its seating as shown in the inset. Water can then flow into the flushing cistern.

Automatic flushing cistern control

These cisterns will flush only when the water level reaches a point where its pressure 'blows' the shallow trap. By limiting the water inlet into the cistern to periods when the building is occupied, the flushing requirements can be controlled.

Pressure control valves (Hydraulic valve)

An effective device for saving water in urinal flushing cisterns is a valve patented under the name Cistermiser (see Fig. 7.35). This valve allows water to enter a urinal flushing cistern only when the building is occupied. It is automatic, passing water only when a pressure drop occurs in the pipeline to which it is fitted. A pressure drop normally only occurs when the building is in use and other taps and valves are opened. The valve operates in the following manner. When pressure is applied at the inlet port it will immediately press against the valve seat side of the diaphragm, thus holding the valve firmly shut. Water will then flow slowly through the adjustable restrictor into the bellows chamber, compressing the spring and air contained in the collapsible bellows until the pressure on both sides

of the diaphragm is equal. The valve is now held shut only by the return spring and water pressure on the valve seat. If the pressure drops on the inlet side of the valve this change will immediately be transmitted to the valve seat side of the diaphragm, and the valve will open causing a further pressure drop on the valve seat of the diaphragm. The valve will stay open until water flows back through the restrictor and pressure on both sides of the diaphragm again equalise and the return spring closes the valve. The length of time that the valve remains open, and consequently the quantity of water that flows into the cistern at each operation, is controlled by the setting of the adjustable restrictor which is accessible from the outside of the valve. If a valve of this type is fitted to an existing installation having a disc valve or pet cock, these should be removed as they interfere with its operation.

Electrical control

These work on the infra-red ray principle shown in Fig. 7.36 which, in very simple terms, detects movement, causing a solenoid valve to open

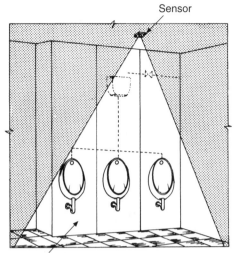

The valve will open only when movement is detected in the unshaded area by the sensor or once during a predetermined period to ensure the maintenance of sanitary conditions. The pipework and cistern are concealed in this illustration

(a)

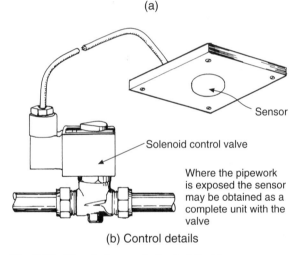

Sensor

Solenoid control valve

Where the pipework is exposed the sensor may be obtained as a complete unit with the valve

(b) Control details

Fig. 7.36 Electrical control of urinal flushing.

allowing water to flow into the flushing cistern for a predetermined period. Most systems using this type of control can detect, by means of an inbuilt microprocessor, whether or not the cistern has flushed during a period of 12 or 24 hours. This ensures that premises not normally used, for example at weekends, will flush at least once during the prescribed period to avoid unpleasant odours. This form of control is electrically operated, either by long-life batteries or mains supply using a step-down transformer. Those operated by batteries should incorporate a visual warning system to indicate when replacement is necessary. To ensure their correct functioning, it is essential that these components are correctly installed. Flushing intervals and cistern fill can be manually set by switches incorporated in the control unit.

Flushing valves
These are a type of equilibrium valve, and unlike a flushing cistern which requires time to fill, they can be used continuously. Until the 1999 Water Regulations took effect these valves were not permitted in the UK, mainly due to possible wastage of water and category 5 pollution risks. As the regulations now permit overflowing water from flushing cisterns to discharge into the WC, any malfunction of a flushing valve will only do the same. These valves are not permissible in domestic properties or any other building where a minimum flow rate of 1.2 litres per second cannot be achieved; this is the minimum required for these valves to function effectively. Any valve supplied by a pressure-flushing cistern which is connected directly to the main supply must be made or provided with a pipe interrupter permanently vented to the atmosphere. There are slight variations in the design of these valves, but the one shown in Fig. 7.37 is fairly typical and embodies their main characteristics. They may be supplied by a separate cistern designated only for use with these valves and such cisterns must be supplied through a type AG air gap. Figure 7.38 shows a suitable layout. They may also be used with an individual pressure-flushing cistern connected directly to the main supply. To avoid serious pollution risks only flushing valves provided with a pipe interrupter are permissible where this system of supply is employed. Although the use of these cisterns is a new concept in the UK, it is claimed they are very effective using the smaller volume of flushing water now mandatory.

Pressure-flushing cisterns
These are designed for use with flushing valves where the supply is taken from the mains. They

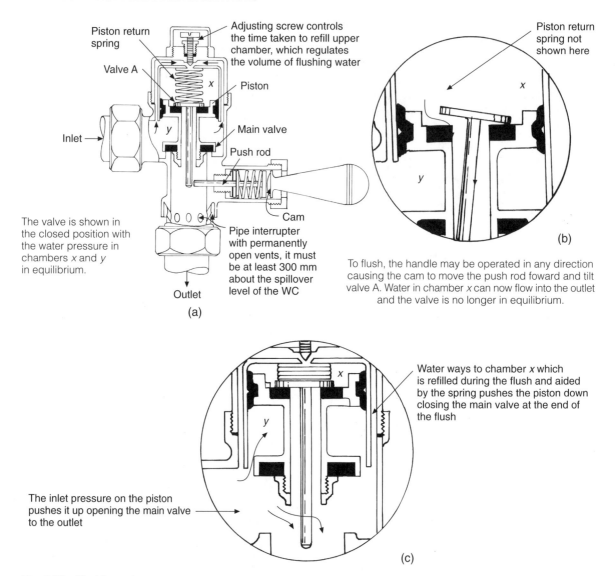

Fig. 7.37 Flushing valve.

are constructed and operate in a similar way to an expansion vessel in an unvented hot water system. When empty the vessel contains air at atmospheric pressure, but as it fills the air is compressed until it is at the same pressure as the water supply. At this point the automatic inlet valve will close. When the flushing valve is operated, water is released under the pressure of the compressed air into the WC.

Disposal of chemical wastes

Industrial wastes containing chemicals need special attention and the plumber is quite often involved with the discharge from laboratory sinks in schools and hospitals and should be aware of the methods used. Due to the action of concentrated acids and alkalis, it is important that these are diluted as much as possible before being discharged into the drain.

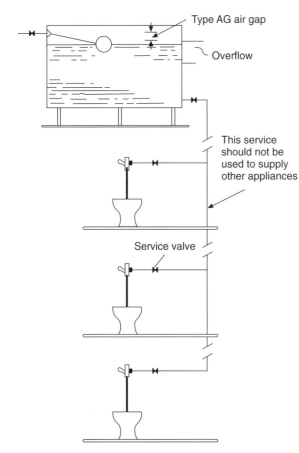

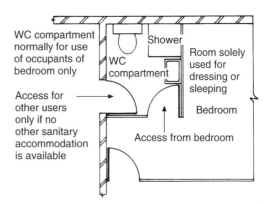

Fig. 7.39 Regulations relating to ventilation and access to WC compartments.

Fig. 7.38 Installation of flushing valves fed from a storage cistern.

Bottle traps having a large water capacity are sometimes used, which effectively dilute any corrosive chemicals. Another method is to use an acid receiver which is simply a large water container made of high-density polypropylene, glass or glazed earthenware. The discharge from the container is turned into an open channel leading to a suitable gulley. Pipes used for the discharge of chemicals are commonly made of high-density polythene with fusion-welded joints or chemical-resisting glass pipes and fittings are used with push-fit joints which permit the removal of the pipe periodically for cleansing. The use of chemical lead for this purpose is limited due to the fact that plastic materials having a high resistance to acids, are just as suitable and are much cheaper.

Sanitary accommodation

This subject is dealt with in the Building Regulations, 2000 in Part G of Schedule I which relates to hygiene. Sanitary accommodation may be defined as a room or space in a building which contains a water closet or urinal, whether or not it also contains ablutionary fittings such as a washbasin. No sanitary accommodation may open directly into a habitable room unless it is used solely for sleeping or dressing purposes. If the situation relates to a private dwelling and the WC is the only one available to its occupants, the accommodation must be constructed so that it is possible to enter it without passing through the bedroom or dressing room (see Fig. 7.39). The foregoing does not apply if there is other sanitary accommodation on the premises that can be used by the occupants. In simple terms, this means that, should there be another WC on the premises, the one in the bedroom or dressing room can only be used by the persons using this room. No sanitary accommodation may open directly into a kitchen or room in which food is prepared, or a room used for trade or business. The usual arrangements made to meet this requirement are:

(a) To site the accommodation so that it can be entered from the open air — not a very convenient method in poor weather conditions.

(b) To construct it so that it can only be entered through a ventilated lobby as shown in Fig. 7.40.

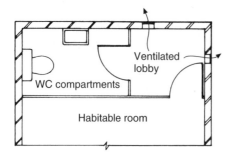

Fig. 7.40 Access to WC compartment only via a ventilated lobby.

Ventilation of sanitary accommodation
All sanitary accommodation must have a window, skylight or similar means of ventilation opening directly to the open air, which an equivalent of one-twentieth of the floor area must be capable of being opened. In buildings where the sanitary accommodation cannot be ventilated by natural means, mechanical ventilation will be necessary, giving at least three air changes per hour and discharging into the open air. In large buildings containing sanitary accommodation requiring ventilation, a ducted system is employed. It is essential with such an installation that two extractor units are provided so that if one breaks down the other can be put into service, providing continuity of ventilation. Whether or not dual extractors are required for one area of sanitary accommodation in private dwellings depends on the local authority. The foregoing relates mainly to public and multistorey housing, but it is not uncommon to encounter similar situations where mechanical ventilation is necessary in smaller properties, a typical example being where a WC apartment is constructed under a flight of stairs. There are many small extractor fans suitable for wall or ceiling fittings which are available for this purpose. The electrical supply can be arranged in such a way that when the light is switched on the extractor operates. These extractors are provided with an adjustable timing device which enables them to continue running after the light has been switched off. Adjustment of the timing device depends on the length of time the extractor should continue to operate to give the necessary air change. This can be calculated by dividing the volume of air removed per second into the volume of air contained in the apartment. When these units are situated in a ceiling the extracted air must be ducted into the atmosphere. It is not permissible to allow it to discharge into the roof space.

Further reading

Showers
Aqualisa Products Ltd, The Flyers Way, Westerham, Kent TN16 1DE, Tel. 01959 560008.
Mira Showers, Caradon Mira Ltd, Cromwell Road, Cheltenham, GL52 5EP, Tel. 01242 221221.

Sanitary fittings and appliances
British Standards BS 6465:1984:Part 1. Sanitary Fittings.
Armitage Shanks Ltd, Rugeley, Staffordshire, WS15 4BT, Tel. 01543 490253.
Ideal Standard, PO Box 60, National Avenue, Kingston-on-Hull, HU5 4HS, Tel. 01482 346461.

WC macerators
Saniflow Ltd, Howard House, The Runway, South Ruislip, Middlesex HA4 6SE, Tel. 020 8842 0033.
Technical Advice on Sanitary Appliances from the Building Centre, Store Street, London, WC1E 7BT, Tel. 020 7692 4000.
Thermostatic Mixing Valve Manufacturers Association (TMVA), Westminster Tower, 3 Albert Embankment, London SE1 7SL, Tel. 020 7793 3008.

Self-testing questions

1. (a) State the recommended fall for horizontal discharge pipes for a WC macerator.
   (b) Assuming a macerator to be capable of pumping a horizontal distance of 50 m state the actual length of discharge pipe if it includes a vertical run of 3 m and three bends.
2. Specify the type of flushing arrangements recommended for ranges of WCs with a high incidence of use.

3. List the advantages of spray taps for hand washing.

4. State the maximum length of a dead leg for spray tap installations and give the reason for imposing this limit.

5. (a) State the recommended temperature of water for showering.

   (b) List the advantages of a shower bath compared with a bath.

6. Specify the type of shower mixer you would recommend for use in an old people's home and state the reasons for your choice.

7. Identify the reasons for providing an independent cold water supply to a shower.

8. (a) State the type of urinal with which a sparge pipe is usually fitted.

   (b) Outline two ways of effecting savings on the consumption of water used in urinal flushing cisterns.

9. Explain the working principles of an automatic flushing cistern.

10. In what circumstances in WC compartments is mechanical ventilation mandatory?

11. Describe the requirements of the Water Regulations in relation to appliances with outlets fitted with flexible hoses.

12. Describe how backflow is prevented with flushing valves operated by a pressure cistern connected directly to the main supply pipe.

13. (a) Describe the maintenance checks necessary in the event of failure of a mixing valve.

    (b) State the final operation that must be carried out on completion of any maintenance.

14. Explain why double check valves are fitted on the outlet, not the inlet, side of electric showers.

8 Sanitary Pipework

After completing this chapter the reader should be able to:

1. State the basic requirements of the building regulations relating to sanitary pipework.
2. Explain the basic principles of discharge pipe systems.
3. Identify the causes of trap seal loss due to defective design.
4. Recognise the main features of above-ground sanitary pipework systems.
5. Sketch and describe simple details showing discharge pipe arrangements for single and ranges of sanitary appliances.
6. Recognise the importance of ventilating sanitary pipework where necessary.
7. Describe the correct procedures for testing and commissioning sanitary pipework installations.

Building Regulations relevant to sanitary pipework

Regulations are necessary in the construction industry to ensure that a building and its components are safe and suitable for the purpose for which they are designed and do not cause offence or nuisance in the environment. Part H of Schedule I to the Building Regulations 1991 and the associated Approved Document H covers the main legislation governing building drainage, and it is important for plumbers to be aware of the principal regulations related to their work. The following is a summary of the above-mentioned documents relating to discharge and ventilating pipes.

(a) Provision must be made in the drainage system (this includes above- and below-ground drainage) to prevent the destruction of trap seals, which would result in the admittance of foul air to the building.

(b) All discharge pipes must be of adequate size for their purpose and *must not* be smaller in diameter than the outlet of the fitting discharging into it.

(c) The internal diameter of a pipe carrying the discharge from a urinal must not be less than 50 mm excluding bowl types. Further information on urinal discharge pipes is included in Table 8.1. In all other cases of pipes carrying excremental matter the minimum internal diameter is 75 mm.

(d) All pipes and fittings used for the discharge of soil or waste and the ventilation of above-ground systems must be made of suitable materials having the required strength and durability for this purpose.

Table 8.1 Internal diameter and depth of seal of traps serving sanitary appliances, as recommended in BS EN 12056.PT2.

Type of appliance	Minimum nominal diameter (mm)	Seal depth (mm)
Washbasin	32	75
Bidet	32	75
Sink	40	75
Bath	40	50
Shower tray	40	50
Urinal bowl	32	75
Urinal stalls, 1 or 2 in range	50	75
Urinal stalls, 3 or 4 in range	65	50
Urinal stalls, 5 or 6 in range	75	50
WC	—	50

(e) All joints must be made in such a way as to avoid obstructions, leaks and corrosion.

(f) Bends must have an easy radius and should not have any change of cross-sectional area throughout their length.

(g) Pipes must be adequately secured to the building fabric without restricting their movement due to thermal expansion.

(h) The system must be capable of withstanding an air test when subjected to a minimum pressure equivalent to 38 mm head of water. (Details of testing procedures are illustrated later in this chapter.)

(i) Pipework and fittings must be accessible for repair and maintenance, and means of access must be provided for clearing blockages in the system.

(j) Every sanitary appliance must be fitted to its outlet, having an adequate water seal and access for cleaning. Where the appliance is made with an integral trap, e.g. a WC, it must be capable of being removed to provide access to the discharge pipework, unless other suitable cleansing provision is available. This precludes the use of cement mortar for bedding WCs on solid floors which was never good practice due to possible damage to the pan caused by expansion of the cement while hardening.

(k) No discharge pipes from sanitary appliances may be fitted on the exterior of the building except:
  (i) Where the building was erected before the building regulations came into force and is altered or extended.
  (ii) On low-rise buildings of up to three floors.

(l) No discharge pipes on the exterior of the building may discharge into a hopper head or above the grating of an open gulley. The regulations are in effect insisting that:
  (a) Only gulleys of the back inlet type are acceptable.
  (b) Discharge pipes must be connected directly to the underground drain or main discharge stack.

Standards relating to sanitary pipework

Apart from the Building Regulations, to which it is obligatory to comply, the other main document relating to this subject was BS 5572. It became the basis for the design of modern sanitary pipework systems, but due to the harmonisation of European standards, it has been superseded by BS EN 12056 Part 2. All reference in this chapter relates to Part 2. Those who are familiar with BS 5572 will notice few changes in the illustration of the basic pipework systems, although there have been changes in name. To avoid confusion, where relevant both names are shown, except in the fully ventilated system, where due to what appears to be an error in BS EN 12056, only the BS 5572 name is shown.

Requirements of sanitary pipework systems

The basic principles of overground sanitation and its associated pipework for lower-rise dwellings have been covered in Book 1. It is proposed in this volume to extend these principles to enable them to be applied, not only to multistorey dwellings, i.e. blocks of flats, but also to industrial, commercial and public buildings of up to five floors. In some cases the text and illustrations go a little beyond this requirement so that the subject can be treated in a comprehensive manner.

The object of any waste disposal system is to remove waste water from the building as quickly and quietly as possible and in such a way that no nuisance or danger to health is caused. Objectionable bacteria multiply very quickly in waste matter and sewage, hence the need for quick and efficient disposal. One of the most important factors relating to the removal of waste water is the discharge pipe diameter, and this has a direct effect, coupled with the fall of the pipe, on the retention of the trap seals. Any loss of the seals will admit foul air into the building, and the underlying principle of all sanitary pipework systems is to prevent this from happening. Pipes having a small diameter are the most likely to run full bore and cause self-siphonage.

Another factor that is often overlooked is the noise caused by the operation of sanitary appliances and their discharge pipes. It has been discovered that this can give rise to and aggravate nervous tension in some people, therefore systems should be designed having regard to this problem. Inefficient ventilation and anti-siphon traps are the principal causes of excessive noise.

Sanitary pipe sizes

The minimum diameter of traps for sanitary appliances is also shown in Table 8.1. In most cases the diameter of the waste is the same, although there are some exceptions, depending on circumstances, which are dealt with later.

It will be noticed that no mention has been made of WC outlet diameters. The reason is that, depending on their design and type, the outlet diameter may vary. A standard wash-down WC has an outlet of approximately 90 mm, and although good practice requires a discharge pipe to have the same diameter as the trap to which it is connected, it is not essential in the case of WCs. This is because of the relatively short period for which a WC discharges, and the fact that the pipe is unlikely to run full bore.

It is often considered good practice to oversize urinal discharge pipes so that they are of a larger diameter than the actual trap, as ammonia in urine, especially in hard water areas, causes scale to build up very quickly inside the pipe, and by increasing its diameter the frequency of descaling is reduced.

Trap seal loss

Pressure fluctuations causing seal loss are much more likely to occur in multistorey developments and public and industrial buildings than in small dwellings as the discharge pipes are longer, larger numbers of sanitary appliances are employed and the incidence of usage is greater. It is therefore necessary to design sanitary pipework to suit the type and purpose of the building in which it is to be installed.

Unstable air conditions in the discharge pipe system is the principal cause of trap seal loss, which causes self-siphonage, induced siphonage and compression. These three causes of seal loss are described and illustrated in detail in Book One, but as they are necessary for a full understanding of the following text, they are repeated here very briefly. Self-siphonage is said to happen when the discharge from an appliance causes loss of seal in its own trap, while induced siphonage relates to the loss of seal in a trap caused by the discharge of another appliance connected to the same discharge pipe.

Compression takes place when a falling body of water in a vertical pipe compresses the air below it, forcing out the trap seals of the lower appliances. Factors other than unstable air conditions in the pipework which affect trap performance can be summarised as follows:

(a) The shape of the appliance has a considerable bearing on its trap performance; the U shape of a washbasin makes it prone to seal loss as there is no 'rill' or tail off to reseal the trap.

(b) Where an appliance is fitted with an integral overflow, air can be drawn into the waste via the overflow and in some cases helps to maintain equilibrium of the air in the discharge pipework.

(c) The type of waste fitting used can influence the trap performance: if its grating offers resistance to the flow of water, it will reduce the volume of discharge and there will be less likelihood of the pipe running full (see Fig. 8.1). A special

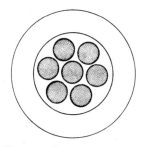

(a) Grating offers more resistance

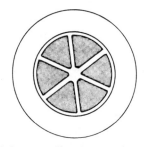

(b) Grating offers less resistance

Note that the total area of the outlet holes in the waste fitting (a) is less than for this waste fitting. This has the effect of restricting the volume of water discharged and reduces the possibility of self-siphonage.

Fig. 8.1   Basin waste grids.

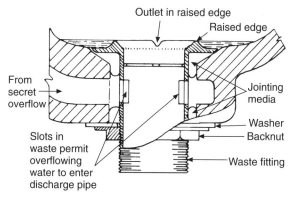

Fig. 8.2 Patent basin waste. The raised edge causes a small quantity of water to be retained momentarily when the basin is discharged. This water enters the waste after the discharge and reseals the trap should some seal have been lost due to self siphonage.

waste fitting with a raised edge is shown in Fig. 8.2; this holds back a small quantity of water from the final discharge which serves to top up the trap seal. Even if an appliance has an integral overflow, this should not be used as a substitute for a ventilating pipe.

(d) Bends should be reduced to a minimum as these slow up the flow, causing a solid plug of water to form more quickly than in a straight length of pipe.

Ventilation of sanitary pipework

A ventilating pipe may be defined as a pipe connected to a discharge pipe at one end, the other being open to the atmosphere which prevents positive or negative pressures being set up in the system, thus ensuring the retention of the trap seals. Providing branch wastes, especially those serving washbasins, are kept within the limits of length, number of bends, diameter and falls recommended later in this chapter, no ventilation of individual appliances will be necessary. Ventilation may be necessary, however, due to the number of appliances and their incidence of use, their distance from the main stack and the number of bends employed on the branch discharge pipe. The designation of modern sanitary pipework is in fact determined by the type of ventilation employed.

Compression and its prevention

Compression has been defined earlier in this chapter. Special precautions are essential to avoid a build-up of positive air pressure in the pipes, especially those serving high-rise developments. The ground floor appliances in tall buildings will be especially exposed to compression, and for this reason ground floor ablutionary appliances should be discharged into back inlet gulleys. The use of either a long-radius bend, or two 135° bends where the discharge stack joins the underground drain, is essential, permitting the air to flow easily into the drain in front of the incoming water. For the same reason, offsets should not be used in the wet part of the stack unless they are ventilated. Reference should be made to Book 1, Figs x.xx and x.xx, for further details. Generally, the effects of compression in very tall buildings can be avoided in one of two ways. One is to increase the diameter of the main discharge pipe to 150 mm nominal bore, which has the effect of reducing the possibility of the formation of a solid plug of water. There is, however, a disadvantage in using this method, as if the discharge pipe diameter is increased, the diameter of the underground drain must also be increased to the same size, which might have the effect of reducing the velocity of flow in the drain to such a degree that it would not be self-cleansing. The other and more common alternative is to use a nominal 100 mm pipe with selective venting. By siting these vents at strategic points in the system, the use of a 100 mm discharge stack can be extended to most applications in modern buildings. The single stack system and its vented modifications are described in the following text. Note that where the term 'group of appliances' is used, it should be interpreted as comprising one bath, one sink, one or two washbasins and one or two WCs with a flushing capacity of 7.5 or 6 litres.

Modern pipework systems

Prior to the adoption of the 1965 Building Regulations, one was permitted to fix discharge pipes to the exterior face of buildings. This was not only unsightly but it increased maintenance costs if the pipework required painting to protect it from corrosion. It was also often found that a dripping

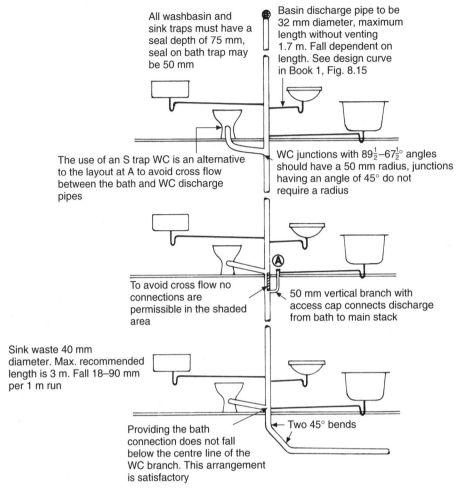

All washbasin and sink traps must have a seal depth of 75 mm, seal on bath trap may be 50 mm

Basin discharge pipe to be 32 mm diameter, maximum length without venting 1.7 m. Fall dependent on length. See design curve in Book 1, Fig. 8.15

The use of an S trap WC is an alternative to the layout at A to avoid cross flow between the bath and WC discharge pipes

WC junctions with $89\frac{1}{2}$–$67\frac{1}{2}°$ angles should have a 50 mm radius, junctions having an angle of 45° do not require a radius

To avoid cross flow no connections are permissible in the shaded area

50 mm vertical branch with access cap connects discharge from bath to main stack

Sink waste 40 mm diameter. Max. recommended length is 3 m. Fall 18–90 mm per 1 m run

Two 45° bends

Providing the bath connection does not fall below the centre line of the WC branch. This arrangement is satisfactory

Fig. 8.3 Primary ventilated stack system (Single stack system) in a three-storey building. The recommendations for sink discharge pipes also apply to bath wastes.

tap could, on a frosty night, cause an ice plug to form in the main discharge pipe causing it to split. The 1965 Building Regulations insisted that all discharge pipes should be fitted inside the building, with the exception of work on existing buildings or extensions where it might be difficult or costly to comply with this requirement. Subsequent regulations permit the exterior fixing of discharge pipes on buildings of up to three floors. Although this simplifies installations and releases floor space which would otherwise be occupied by a duct, it may involve the householder in a considerable amount of inconvenience, especially if the system is subjected to frost damage.

Primary ventilated stack system
(Previously known as the single stack system.) This system is used wherever possible, as the complete absence of trap ventilating pipes makes it less costly to install and enables it to be accommodated easily into internal pipe ducts. Due to the low flow rates in domestic dwellings, it can be used successfully in such buildings of up to five storeys, using a standard 100 mm main discharge stack, having one or two groups of appliances on each floor. Figure 8.3 illustrates this system in low-rise dwellings. Figure 8.4 shows a much larger scheme suitable for a block of flats. In buildings having more than five floors, the basic principles are the

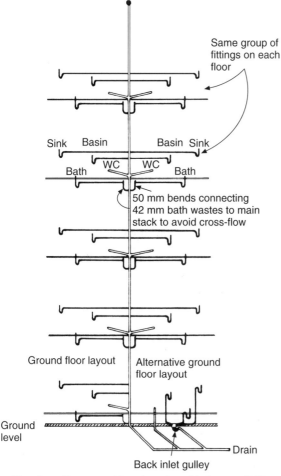

Same group of fittings on each floor

Sink Basin   Basin Sink
    Bath   WC   WC   Bath

50 mm bends connecting 42 mm bath wastes to main stack to avoid cross-flow

Ground floor layout | Alternative ground floor layout

Ground level

Drain

Back inlet gulley

The alternative ground floor layout avoids the possibility of compression occurring and detergent foam entering the lower fittings, and is essential in buildings of more than five storeys.

Fig. 8.4 Primary ventilated system in a five-storey block of dwellings. Note that two groups of fittings are permissible on each floor without stack ventilation. If the building has more than five floors or there are more than two groups of fitting per floor, stack ventilation will be necessary see Fig. 8.6 if a 100 mm diameter main discharge pipe is used. An alternative is to increase its size to 125 or 150 mm.

same, but when siphonage to an appliance or a group of appliances is likely, ventilation of the branch discharge pipe may be necessary. If compression is possible, the main discharge stack must be ventilated.

Stub stack system

The use of this system is confined to single-storey buildings or to a group of appliances on the ground floor. Traditionally, in such cases, the WC is connected directly to the drain, while the ablutionary fittings discharge into a back inlet gulley, and while this arrangement is acceptable, it does have some disadvantages. The possible exposure of discharge pipes are unsightly on the exterior of a building and can lead to blockage of the external pipes by formation of ice in frosty weather due to dripping taps. Such pipes are often run at a steep angle or vertically, resulting in noisy discharges. Where an appliance is situated some distance from the gulley, due to the length of the discharge pipe, self-siphonage of the trap seal may take place. The use of the stub stack system is illustrated in Fig. 8.5 and solves most of the foregoing problems, and in some cases can be more economical in cost. The only limitations imposed are:

(a) The centre line of the WC trap must not be higher than 1.5 m from the invert of the drain.
(b) The distance between the highest connection on the stack and the invert must not exceed more than 2.5 m.
(c) Arrangements must be made to ventilate the underground drain elsewhere in the system.

It is important to remember the underground drain is satisfactorily vented.

Secondary ventilated stack system

Shown in BS 5572 as the ventilated stack system and illustrated in Fig. 8.6. The main feature of this system is the cross-venting of groups of appliances, either from the WC branch or from the main discharge pipe adjacent to the appliances. This type of system is normally employed for high-rise housing of more than five floors. In buildings of this type, compression is the main cause of seal loss, especially when appliances on the upper floors are discharged. The use of the ventilating pipe permits the relief of any pressure build-up in the stack caused by a falling body of water. There is little danger of trap seal loss in each appliance if basic principles are embodied in the system, e.g. the careful grouping of appliances, limitation of the

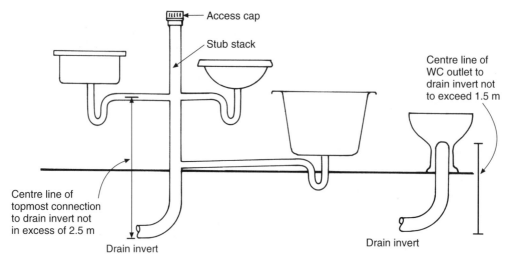

Fig. 8.5 Stub stack system. Note that an open vent must be provided at the highest point of the drain.

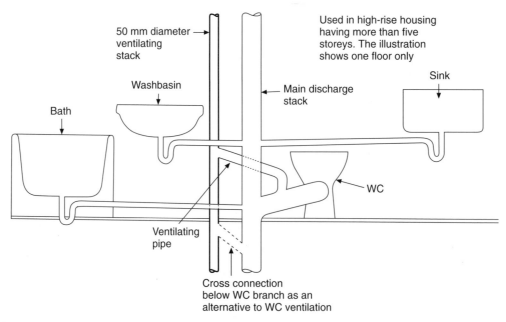

Fig. 8.6 Secondary ventilated stack system.

lengths and falls of branch discharge pipes and the use of the recommended branch waste diameters.

Modified primary ventilated stack system
Formerly known as the modified single stack system, it can assume many forms, depending on the number of floors it serves, the number of appliances connected to it, the length of the branch discharge pipes, the number of bends on each and the incidence of appliance usage. In general, however, it may be defined as a primary ventilated stack system with branch wastes of greater length

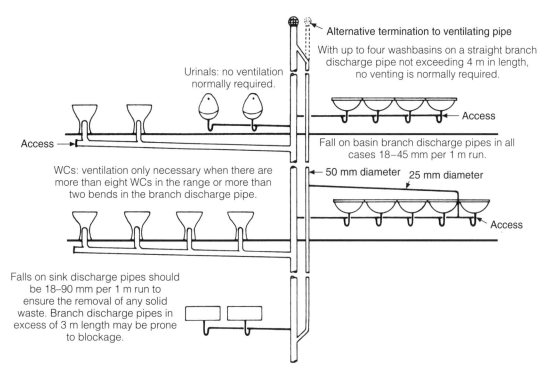

Fig. 8.7 Modified Primary ventilated stack system.

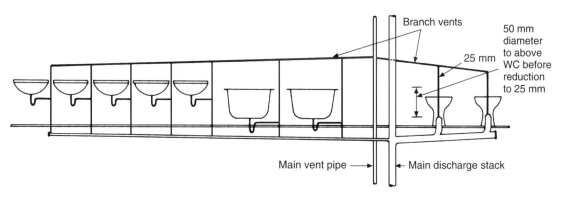

Fig. 8.8 Ventilated system. The general layout of pipework on one floor only of a hospital ablutionary annexe. If the main stacks serve a similar arrangement of appliances on three or more floors such a system of pipework may become surcharged; necessitating the ventilation of every trap.

than those recommended, or with branch wastes serving ranges of fittings where some danger of loss of trap seal exists due to siphonage. To overcome this, ventilation of certain appliances or groups of appliances is necessary as illustrated in Fig. 8.7. Note that with this type of system, the ventilating pipe is quite independent of the main discharge stack unless its upper end is turned into the main

stack above the highest appliance. In many cases there is obviously no need to carry two pipes above the eaves.

Ventilated system (Ref BS 5572)
The ventilated system is only used in the form shown in Fig. 8.8 when the incidence of usage is said to be 'congested'. In simple terms, this means

that the appliances, being in almost continuous use may cause surcharging of the pipework and subsequent seal loss unless each trap is ventilated. Apart from some modifications the system is very similar to the pre-1965, one-pipe, fully ventilated system. At the time of writing, there appears to be no clear listing in BS EN 12056 for this system, and until verification should be referred to as shown.

Branch positions in discharge stacks

The branch discharge vent connection for a range of WCs is shown in Fig. 8.9. The possibility of cross flow is greater in large multistorey buildings than in low-rise single domestic dwellings. To avoid this, all opposing branches must be carefully positioned in the main discharge stack as illustrated in Fig. 8.10. Note that the distance between opposed branches is greater when one of the opposing branches is a WC. This is because, unlike other sanitary fittings, a WC discharges water very quickly and increases the area in which cross flow can take place.

Multi-discharge pipe adaptor
These are usually listed in manufacturers' cataloguers as 'collar bosses'. They can be used in

multistorey dwellings where branch discharge pipes from a group of fittings such as a bath, basin, sink and WC, all on the same floor, are connected to the main discharge stack. A typical fitting of this type is shown in Fig. 8.11 and it will be seen that its use avoids the necessity of making several separate branches into the stack. What is possibly more important is the fact that branch discharge pipes can enter the stack in close proximity to the WC branch without the possibility of cross flow. Being a cross-sectional illustration only two connections are shown, there are in fact four at 90° to each other which allows for considerable flexibility in use. As is common with PVC fittings for sanitary pipework, these connections are blanked off, the blanks on the connections selected for a specific installation, being cut out with a hole or pad saw. Possibly the only disadvantage with fittings of this type is the fact that they are not self-cleansing and it is possible for soap deposits to build up in the annular chamber. To enable an obstruction to be removed all discharge pipe connections to the adaptor should be capable of being disconnected easily. Push-fit rubber ring joints or mechanical joints of the compression type are most suitable. Some thought should also be given to the siting of these fittings, as in most modern bathrooms they are housed behind panels, in which case means of access

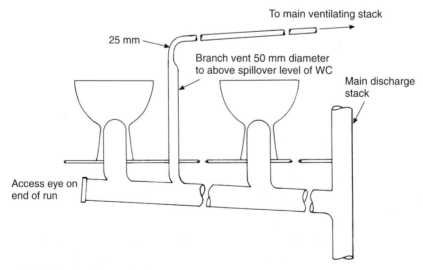

Fig. 8.9   Range of WCs showing branch discharge vent connection. Ventilation of WC branch discharge pipes is not normally necessary unless there are more than eight WCs or the pipe contains more than two bends.

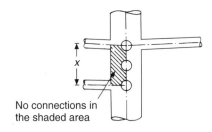

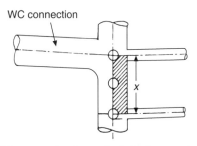

(a) Small diameter connections to vertical discharge stacks

Discharge stack diameter 75 mm 100 mm.
Minimum distance *x* 90 mm 100 mm.

(c) WC connection to vertical discharge pipe and its effect on other connections

In this case *x* must not be less than 200 mm, because of the greater volume of water discharged when a WC is flushed. Consequently the distance between opposed branches must be increased to 200 mm.

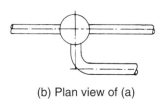

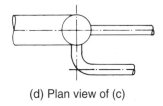

(b) Plan view of (a)

(d) Plan view of (c)

Fig. 8.10 Permissible connections of branch discharge pipes to main discharge stack to avoid cross flow.

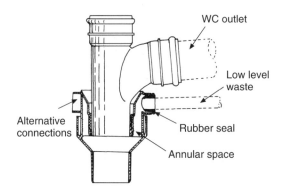

Fig. 8.11 Multi-discharge pipe adaptor.

should be provided. For details of other applications of these fittings reference should be made to the manufacturers listed at the end of this chapter.

Ventilation of branch discharge pipes to single appliances

It has already been stated that if the principles of the primary ventilated stack system are adhered to, no ventilation of branch discharge pipes is necessary. The following applies where single

appliances are situated some distance from the main stack, requiring longer branch pipes than recommended.

WCs and urinals

These appliances do not normally require ventilation when fitted singly as the discharge pipes are unlikely to become surcharged. The recommended limit on the length of a discharge pipe for a WC is 6 m, and there will be few instances in practice where it is longer than this. Urinal wastes should not exceed 3 m in length, due not so much to the possibility of trap seal loss, as to the build-up of scale on the inside of the pipe. Even pipes serving bowl urinals having such small diameters as 32 mm or 40 mm are unlikely to run full, but should siphonage take place the rill or tail-off of each flush is sufficient to reseal the trap.

Baths and showers

Siphonage does not normally cause trap seal loss with these fittings as the discharge pipes are adequately sized to ensure they do not run full. As baths and showers are flat-bottomed appliances, any seal loss, should it take place, would be replaced by the tail off of the flow of water which occurs at the end of each discharge. Because of this, and the difficulty of installing a trap having a 75 mm seal under these appliances, it is now permissible to use one having a 50 mm seal. The gradient or fall of the discharge pipe is not critical, being between 1 and 5° (18–90 mm per 1 m run). The main danger with these appliances is not so much loss of trap seal as build-up of soap deposits causing a blockage in long discharge pipes. For this reason they should not exceed 3 m in length.

Sinks

Because particles of solid waste are often discharged from a sink, common examples being tea leaves and vegetable peelings, a tubular trap (not a bottle type) should always be used as it is considered to be more self-cleansing. Normally, no special precautions are necessary to preserve the seal in a sink trap as sinks, like baths and showers, are flat-bottomed appliances, and the tail-off or rill reseals the trap should it be siphoned. The maximum recommended length of a sink waste is 3 m, this limitation being imposed not to avoid the formation of a solid plug of water but to minimise the possibility of blockage and reduce noise. The minimum fall of $7\frac{1}{2}°$ (135 mm per 1 m run) on the waste provides a greater velocity of flow ensuring that any solid particles of waste are removed. Where food waste macerators are fitted, the fall should be greater than $7\frac{1}{2}°$. These fittings reduce solid waste to a mass of particles creating ideal conditions for blockage. The reader is referred to Book 1, pages 198–9, where this topic is dealt with.

Washbasins

Detailed information relating to discharge pipes for washbasins is given in Book One, page xxx. Generally, the maximum length of waste permissible is 1.7 m, having a fall of not less than 1° (18 mm per 1 m run). If it is impossible to keep within these limits due to the distance of the basin from the main stack, or if an excessive number of bends are used, several alternatives are available to overcome the problem. A resealing trap can be used, although they tend to be noisy and for this reason are not recommended. Other alternatives are illustrated in Fig. 8.12(a), (b), (c). They all relate to

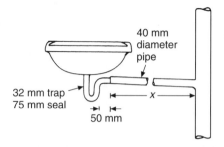

(a) Increase in discharge pipe diameter

All the arrangements shown here are suitable for discharge pipes exceeding 1.7 m in length, but less than 3.0 m (shown as *x* in the illustration).

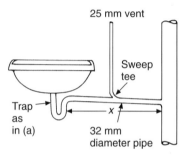

(b) Alternative to (a) using a 32 mm branch discharge pipe

Trap vent to be turned into main ventilating stack or vent portion of main discharge stack above the highest appliance.

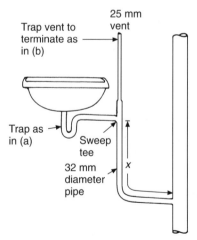

(c) As (b) but with a low-level connection to the main stack pipe

Fig. 8.12   Single basin connections.

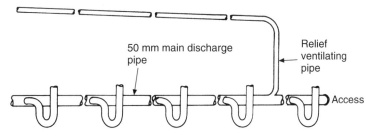

Fig. 8.13 Ventilation of a range of appliances using the modified single stack system.

single basins having a discharge pipe of not more than 3 m in length, diameter of 32–40 mm and a fall or not less than 18 mm per 1 m run. The one exception to this is Fig. 8.12(a) which may have a fall of up to $2\frac{1}{2}°$ or 45 mm per 1 m run. This is because it contains no bends. The introduction of bends in a discharge pipe having higher velocities than those recommended can cause the pipe to run full with subsequent siphonage of the trap. In all cases where the diameter of the discharge pipe exceeds that of the trap, as in Fig. 8.12(a), adequate access must be made to remove any sludge deposited due to the low-velocity flow of water.

Ranges of appliances

A series of similar appliances connected to a common waste is called a *range*, the most common examples being WCs, urinals or washbasins. Ranges occur only in commercial or public buildings where sanitary accommodation must be provided for large numbers of people. As the appliances have a higher incidence of usage than those in private dwellings there is a greater danger of trap seal loss, and some ventilation of sanitary pipework, especially that serving washbasins, will be required.

Ranges of basins
In factories and offices adequate provision for washing must be provided by law, and as many people are often employed in such premises washing facilities must be plentiful. In workplaces of an industrial nature wash fountains or washing troughs are quite common as they can be used by a greater number of people at the same time and

occupy less space than a range of washbasins. In general, washbasins are fitted in sanitary accommodation for offices because more space is usually available for ablutionary facilities. The best and most sanitary method of connecting a basin range to a main discharge stack is to use a common waste pipe like that shown in Fig. 8.13. Providing that not more than four basins are fitted to the common waste, which has a diameter of 50 mm and does not exeed 4 m in length, no vent is necessary. The low angle and the relatively large common waste ensure that it does not flow full bore. If a range of more than four basins is contemplated, or the main pipe exceeds 4 m in length, a vent must be fitted as shown in Fig. 8.13. The number of changes of direction in the branch discharge pipe must be kept to a minimum, and if any bends are necessary they must have a large radius. The reason for this is that they can reduce the flow rate to such an extent that a solid plug of water may form, leading to induced siphonage.

Spray taps are often used for hand washing and, as they have such a low flow rate, the discharge may not remove the grease and soap sediment in the discharge pipe. When these taps are specified BS EN 12056 recommends the use of a common discharge pipe having a diameter of not more than 32 mm maximum nominal bore as this has the effect of increasing the velocity of the discharge, washing out any soap deposits. The maximum length of the discharge pipe is unlimited, but for practical purposes it should be as short as possible to reduce problems with soap deposits. As the discharge pipes are unlikely to run full, traps of 50 mm seal depth are permissible. Basins fitted with spray taps should not be provided with plugs when small waste diameters are used as this could lead to surcharging if the basins were accidently filled. Although modern design has largely eliminated fully ventilated systems, such a system will be necessary where the incidence of usage is high and the branch discharge pipes run full. An experienced designer would anicipate these conditions and specify the type of ventilation required. Reference should be made to Fig. 8.8, part of which shows a fully ventilated range of basins.

Ranges of WCs

The normal size of a branch pipe for a WC range is 100 mm. Ventilation of these branches is seldom necessary except where more than eight WCs are fitted on one branch or where the branch has a significant number of bends. Even then, unless the usage of the appliances is 'congested' (in which case a ventilated system should be fitted), it is sufficient to ventilate the branch discharge pipe only, in a similar way to that of a range of washbasins where the main discharge pipe only is ventilated.

Ranges of urinals

As with WCs, it is seldom necessary to ventilate urinal wastes due to the relatively low flow rate. It is recommended that where bowl urinals are fitted, the individual pipe from each bowl to the main discharge pipe should be as short as possible with a fall of $1-2\frac{1}{2}°$ (18–45 mm per 1 m run). Falls on the main discharge pipe may be 1–5° (18–90 mm per 1 m run).

Special notes relating to ventilating pipes

Ventilating pipes do not normally carry waste water, the one exception to this being a wet vent (see Fig. 8.14). In systems constructed of cast iron, condensation in a long vertical section of the pipe may cause corrosion, resulting in particles of rust falling and blocking the lower bend in the stack. By connecting a branch discharge as shown, any debris will periodically be washed away. The position of the connection between a branch ventilating pipe and a main ventilating pipe is important and must always be made above the flood level of the

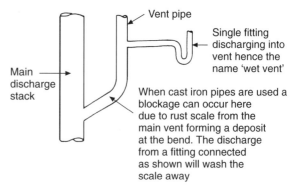

Fig. 8.14   Wet vents.

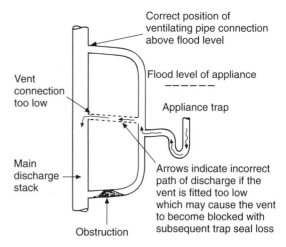

Fig. 8.15   The effect of incorrect vent pipe connections.

appliance, or appliances if a range is under consideration. Figure 8.15 shows the correct method of connection so as to avoid the discharge of waste water through the branch ventilating pipe should an obstruction form in the appliance discharge pipe. The lower end of the main vent stack can be connected to the drain in a convenient inspection chamber or, alternatively, above ground level in the main discharge stack as shown in Fig. 8.16. In buildings of more than five storeys, ground floor appliances should discharge into the underground drain. The object of these limitations is to avoid the base of the vent becoming surcharged, thus limiting its effectiveness in combating compression and also avoiding the entry of foaming detergent into the

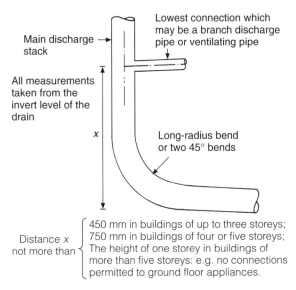

Fig. 8.16 Low connections on discharge stacks.

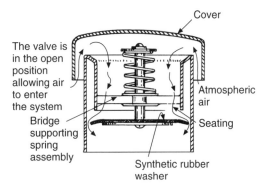

Fig. 8.17 Automatic air-admittance valve. These valves only admit air to a discharge pipe when it is subjected to a negative pressure. Atmospheric pressure acting on the top surface of the valve pushes it down, compressing the spring. When atmospheric pressure is restored the spring closes the valve to prevent the escape of drain air.

ground floor appliances. Any offsets or bends in the wet section of the main discharge stack should be avoided. If this is not possible they must be ventilated. Ventilating pipes should terminate in the open air in such a way that nuisance and health hazards are avoided. This is usually achieved by carrying them 900 mm above and not within 3 m of any opening, window or vent. For further details of ventilating offsets and ventilating pipe terminals, reference should be made to Book 1, Figs 8.20 and 8.24.

Air-admittance valves

These are designed to automatically maintain equilibrium in an unventilated part of a system and must comply with EN 12380. The Building Regulations permit only valves complying to this standard or those approved by the British Board of Agrément. It is essential that their installation complies with the terms of the certificate. It should be clearly understood they do not take the place of a vent pipe, which not only maintains stable atmospheric conditions in the drainage system, but also ventilates the drain. These valves are quite simple in operation and a typical example is shown in Fig. 8.17. The spring is adjusted during manufacture so that it just overcomes the weight of the valve and holds it in the closed position. The

slightest negative pressure in the discharge pipe causes the valve to open, thus maintaining atmospheric pressure in the discharge system. A typical situation where these valves may be an advantage is in a building having two main discharge stacks. It would only be necessary to terminate one externally of the building allowing for ventilation of the drain, the other could be accommodated in a suitable duct or the roof space using an air admittance valve. As they are mechanical devices it is important that they are fitted in an accessible position. A similar type of device is also available for branch discharge pipes where for various reasons the requirements of BS EN 12056 relating to trap ventilation and discharge pipe maximum lengths cannot be met. It should be noted that not all authorities accept such mechanical devices in connection with discharge pipe systems and their use should be treated with care.

Ventilating pipe sizing

This work is normally performed by a qualified plumbing designer or sanitary engineer and should not be carried out using rule-of-thumb methods. Circumstances alter cases and ventilating pipe sized for one job may not necessarily be satisfactory for another, even if the situation appears to be the same. Some of the factors that must be considered are:

Table 8.2 General guide to sizes of ventilating pipework.

Discharge stack or branch discharge pipe diameter	Size of main vent stack or branch ventilating pipe
Less than 75 mm	$\frac{2}{3}$ diameter (not less than 25 mm)
More than 75 mm	$\frac{1}{2}$ diameter

(a) The incidence of usage of the appliance.
(b) The number of appliances.
(c) The height of the building.
(d) Whether a group of single appliances or a range is being considered.
(e) Whether the building is domestic, public, or commercial.

Bearing in mind the foregoing, it must be stated that in many cases the plumber carries out work in smaller premises where no proper design of the sanitary pipework is available. For this reason the following information is given based on the recommendations of BS EN 12056. In dwellings of up to five storeys where the appliances are grouped together sufficient ventilation is provided via the main ventilation pipe. In buildings of greater height where the ventilated stack system is used, the main ventilating stack should be 50 mm. A 25 mm pipe is normally considered to be of sufficient size to ventilate single fittings and branch discharge pipes, but if it is 15 m or more in length, or contains more than five bends, then it should be increased to 32 mm in diameter. In the absence of specific information relating to the diameter of ventilation pipes, reference should be made to BS EN 12056. Table 8.2 gives a general guide which meet the requirements of the standard.

Access for cleaning and maintenance

It is essential that adequate provision is made to enable obstructions in the pipework to be cleared with the minimum of inconvenience to the client. One of the main problems with modern waste water disposal systems having low flow velocities in the branch discharge pipes is that they allow soap and other debris to build up on the invert of the pipe, which over a period of time frequently causes a

blockage. The incidence of blockage is higher in public or commercial buildings than in domestic premises and for this reason fewer access points are required in the latter type of building. It is not always necessary to provide access to every branch, and one cleaning eye in a suitable position may allow access to a group of appliances. Special consideration should be given to discharge pipes from appliances that are likely to cause trouble, i.e. urinals, public WCs, showers and sinks, especially those fitted with macerators, or those in canteen or restaurant kitchens into which large quantities of grease may be discharged. Good access is also necessary in multistorey buildings so that sections of the sanitary pipework system can be tested as they are completed.

Pipe ducts

In modern buildings having internally fitted pipework systems, some form of cover or ducting is often necessary to conceal the pipes. It is a fact that while many plumbers can appreciate a well-designed, neatly fitted pipe system, few members of the public can, and for this reason much plumbing pipework has to be concealed. Since ducts are used not only for plumbing systems but also for gas and heating pipework, they should be large enough to accommodate all the services, and provide sufficient space for maintenance work, especially the cleansing of discharge pipework, to be effected. Particular attention should be given to those items likely to require most maintenance, i.e. discharge pipes. Ducts should be constructed in such a way as to prevent the spread of fire and transmission of sound. This can be achieved by placing suitable packing material around the pipes where they pass through the floor of the duct. The material or method of sealing should not restrict thermal movement of the pipes. Water-mixed pastes and fire-resistant silicones are useful when irregular holes have to be filled. Special sleeves are also available containing flexible materials which swell, sealing any voids when heated.

In low-rise housing and conversions some builders give little thought to subsequent access to pipework, and plywood panels are often fixed over the pipes with nails and adhesives. However, if a plumber is involved at the planning stage, he should be sufficiently conscientious and competent to advise on the proper construction of small ducts, i.e. suitable panels *screwed* in position over access points. Otherwise, the householder may be faced with inconvenience and considerable expense in the future.

Inspection and testing

While work is progressing, care must be taken to prevent debris of any kind from entering the system. Any open ends of pipework should be fitted with temporary plugs such as a well-packed was of rag until the final connection is made. The installation must, on completion, be checked to make sure it has not been damaged in any way and has been securely fixed to the building fabric.

Testing for soundness

Sanitary pipework overground are tested with air. Testing should be carried out as one operation where possible, but on large installations it may have to be done in sections as the work proceeds. An air test is very searching and will detect any leak in the system. It will be found that it pays to take a little care when making the joints during installation in order to save a lot of trouble and expense searching for leaks on completion.

The procedure for testing is as follows. Plug any open ends with suitable plugs or stoppers, as shown in Fig. 8.18, and seal the plugs with water as air, being a gas, can pass through very small apertures. The water seals of any traps should be filled and a U gauge and test pump connected at a convenient point as shown, air being pumped into the system in a similar way to that used when testing a gas installation. The Building Regulations recommend a test pressure of 38 mm which must hold for a minimum period of 3 minutes. Should a leak be indicated by a drop in pressure, the joints should be tested with soapy water, again using the same technique as with gas installations. Leakage from cast iron stacks is usually due to defective joints. Blow holes in castings are not unknown and can be the cause of a drop in air pressure. If PVC pipe is used, any leakage is usually caused by the rubber seal having been displaced from its housing in the socket or damaged during the jointing process. Solvent-welded joints, if found to be leaking, would indicate very poor workmanship indeed.

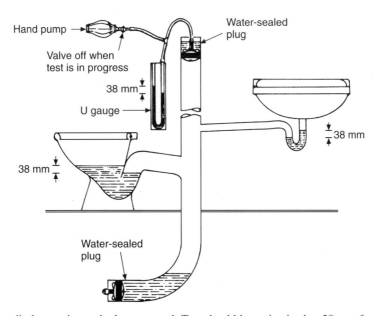

Fig. 8.18 Air testing discharge pipework above ground. Test should be maintained at 38 mm for a period of 3 minutes.

Water testing

Some authorities may request that a system be water tested up to the flood level of the lowest fitting. This simulates the effect of a blocked drain where, in practice, the lower part of the system could be filled with foul water. Should the local authority require such a test, the lower end of the pipe should be plugged prior to filling it to the flood level of the lowest appliance. The maximum head of water must not exceed 6 m as pressures in excess of this may cause the underground drains to burst.

Performance tests

The retention of trap seals should be tested under the worst possible working conditions. The appliances used for the test must be filled to their overflowing level and discharged simultaneously by pulling the plugs or flushing the cisterns. The seal losses due to positive or negative pressures in the stack should be noted. Each test must be repeated three times and a minimum seal depth of not less than 25 mm should be retained after each test. This can be checked by using a short length of clear plastic or glass tube and immersing it in the trap

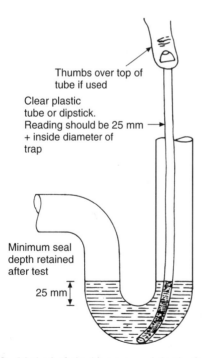

Thumbs over top of tube if used

Clear plastic tube or dipstick. Reading should be 25 mm + inside diameter of trap

Minimum seal depth retained after test

25 mm

Fig. 8.19 Method of checking trap seal depth after performance test.

after each test (see Fig. 8.19) or a suitable dipstick can be made by coating a small piece of wood or metal with a dark matt paint. Gloss paint may give a false reading. If plastic tube is used, make sure it is touching the bottom of the trap when the test is made, then place the thumb over the top end. This will retain the water in the tube, and when it is withdrawn a true reading will be obtained. Table 8.3 is based on the recommendations of BS EN 12056 and shows the number of appliances that should be discharged simultaneously during tests in different types of buildings under various conditions, i.e. domestic, public and congested. The selected appliances should normally be close to the top of the stack and on adjacent floors, as this gives the worst pressure conditions.

Maintenance

(a) Visually inspect all visible pipework, fixings and appliances for damage and security.

(b) Replace any access cover seals which may be damaged or decayed when the covers are removed for maintenance.

(c) Any hand-operated rods or cables used for removing blockages in discharge pipes should not damage the internal surfaces in any way. Powered equipment and kinetic rams should be used with extreme care. The rams are suitable for soft blockage only, as a stubborn blockage can cause a blows back and injure the operator. Before using powered equipment or kinetic rams, any automatic air valves should be removed as waste matter may be forced out of the openings causing damage to decorations. Their use can also blow out a rubber ring-sealed joint.

(d) Soft blockages such as soap and grease deposits can build up in a long run of pipe where the flow rate fails to maintain a self-cleansing gradient. Such obstructions can often be cleared using a plunger, but in such cases a more effective method would be to flush out the system using a solution of hot water and soda.

(e) Lime scale deposits form not only on the surface of appliances, but also on the inside of pipes. This is especially so in the case of urinals, and in extreme cases can completely

Table 8.3 Number of sanitary appliances to be discharged for performance testing.

Type of use	Number of appliances of each kind on the stack	Number of appliances to be discharged simultaneously		
		6/7.5/9 litres WC	Washbasin	Kitchen sink
Domestic	1 to 9	1	1	1
	10 to 24	1	1	2
	25 to 35	1	2	3
	36 to 50	2	2	3
	51 to 65	2	2	4
Commercial or public	1 to 9	1	1	
	10 to 18	1	2	
	19 to 26	2	2	
	27 to 52	2	3	
	53 to 78	3	4	
	79 to 100	3	5	
Congested	1 to 4	1	1	
	5 to 9	1	2	
	10 to 13	2	2	
	14 to 26	2	3	
	27 to 39	3	4	
	40 to 50	3	5	
	51 to 55	4	5	
	56 to 70	4	6	
	71 to 78	4	7	
	79 to 90	5	7	
	91 to 100	5	8	

Note: These figures are based on a criterion of satisfactory service of 99 per cent. In practice, for systems serving mixed appliances, this slightly overestimates the probable hydraulic loading. Flow load from urinals, spray tap basins and showers is usually small in most mixed systems, hence these appliances need not normally be discharged.

block the discharge pipe. The usual method of removal is by using chemical descaling agents. These are usually corrosive in nature and a check should be made to ensure they will not damage the material on which they are used. It is essential to carefully read any instructions shown on the container before use. Acid-based descaling agents in contact with cleansing agents, such as bleach, containing chlorine will produce chlorine gas. All pipework systems must therefore be flushed out prior to the application of the descaling fluid, and adequate ventilation must be maintained during its application. Operatives must always use protective clothing, e.g. gloves, overalls and eye protection when working with these materials.

Further reading

The Building Regulations 1985, Part H.
British Standards BS EN 12056.PT2 Sanitary pipework.
British Standards BS 3237:1981 Ducts for Building Services. Available from HMSO.
Building Research Digest No. 248 *Sanitary Pipework; design bases.*
Building Research Digest No. 249 *Sanitary Pipework; design of pipework.*
Building Research Digest No. 81 *Hospital Sanitary Services; some design and maintenance problems.*
Soil and Waste Pipe Systems for Large Buildings, Seminar Notes. Available from BRE Bookshop,

Building Research Establishment, Garston, Watford, WD2 7JR.

Information on fire-resistant sleeves: Palm Fire Products Ltd, 1 Station Road, Romsey, Hants, SO51 8DP.

Self-testing questions

1. State the two main causes of noise in sanitary pipework systems.
2. List four factors that influence trap seal loss in modern pipework systems.
3. State the recommended diameters of branch discharge pipes serving baths, washbasins and sinks in a multistorey building.
4. List and identify four pipework systems used for waste water disposal.
5. State which of these systems is the most economical and effective when the fittings are grouped round the main stack.
6. Describe and sketch three methods of preventing seal loss in a trap on a basin waste which is longer than the recommended design length.
7. Describe the type of trap recommended for use with a kitchen sink. Give your reasons.
8. Explain the term 'wet vent' and describe the circumstances in which it would be used.
9. Describe the procedure for air testing a system of discharge pipework.
10. (a) Describe the procedure for checking the depth of trap seal retained after a discharge has taken place.
    (b) State the minimum recommended depth of seal that must be retained.
11. Explain why a smaller discharge pipe is permissible for a range of wash basins fitted with spray taps.
12. Explain the term 'wet ventilating pipe'.

9 Underground Drainage Materials, Fittings and Systems

After completing this chapter the reader should be able to:

1. State the principal materials from which drainage pipework and fittings are made.
2. State the advantages and disadvantages of different drainage materials.
3. Describe and sketch the principal methods of jointing for each material.
4. Select suitable fittings for various applications.
5. Describe and sketch methods of support and protection for drainage pipelines.
6. State why ventilation of underground drains is necessary and how it is achieved.
7. Identify methods of providing access to drains including those under buildings.
8. Recognise special drainage fittings and understand their use.
9. Differentiate between types of drainage schemes.
10. Explain and illustrate the basic principles of setting out drains and the necessity of providing self-cleansing gradients.
11. Describe the methods and equipment used for testing underground drainage systems.

Regulations relating to building drainage

All work on drainage, whether new or alterations to existing schemes, is subject to the Building Regulations 1985 and the Approved Document H and comes under the jurisdiction of the local authority planning department. Advice should be sought from and notice in writing given to this department before any drainage work is undertaken. In most cases drawings of the proposed work will be required and, to avoid delay, they should be deposited well before the job is started. Although they are not at present mandatory, the relevant British Standards should be studied to ensure that the installation meets the minimum requirements. BS 8301 is the standard for underground drainage and although it has been superseded by BS EN 752 Parts 1–4, its content is still valid and is currently accepted by Building Control Officers.

Functions of drainage systems

The choice of materials now on the market for underground drainage is so vast that it is impossible to cover fully in this volume all the various joints, fittings and techniques used. Manufacturers of these materials, however, provide ample literature on the subject and the student who wants to obtain a greater depth of knowledge is recommended to contact them. Some of the companies prepared to send detailed information are listed at the end of the chapter.

The function of underground drainage systems is to convey waste water, which comes from the sanitary fittings via the overground discharge pipe system, to the sewer or, if applicable, other sources of disposal. It is essential that the pipes, fittings and methods of jointing should be suitable for the purpose for which they are used.

A good drainage system must be both self-cleansing and watertight. If it is not self-cleansing the system could become a breeding ground for harmful bacteria and would certainly give rise to unpleasant smells. If it is not watertight, the ground surrounding the drain would become polluted, again giving rise to unpleasant and unhealthy conditions.

The reader should understand that there are two main systems of drainage, the *conservancy system* and the *water carriage system*. The former is very rare in housing, the waste matter being retained in a suitable container and disposed of when necessary. As most permanent premises have some form of piped disposal system, the conservancy system is used only with temporary quarters such as building sites and touring caravans, etc., where no drainage is available. The water carriage system employs pipes to remove waste matter to the local authority's treatment works or, in an isolated area, a septic tank.

Although the topic of sewage treatment is beyond the scope of this volume it should be explained that a septic tank provides sewage treatment for single or small groups of premises. Its working principles are very similar to those of a large treatment plant and, properly maintained, it can be very effective in rural areas where no main drainage scheme exists.

Health and safety

The hazards related to working in trenches are dealt with fully in Book One of this series, but the main points are as follows:

(a) Shuttering of the trench sides will be necessary to prevent their collapse. Whether it is open or closed will depend on the composition of the earth strata and the depth of the trench.
(b) Suitable guard rails must be erected.
(c) Adequate warning notices must be displayed.
(d) Trenches should be inspected for safety on a daily basis by a responsible person. This is very important, especially where the soil is sandy or very wet.

Materials

Three main materials are used for underground drainage: clayware, cast iron and PVC. Drains made of clayware are seldom fitted by plumbers therefore this material will not be dealt with to any great depth. Cast iron drainage has always been, and still is, mainly installed by plumbers. They understand the basic principles of drainage and from a practical point of view they have the necessary skills for its installation. PVC is a comparatively new drainage material, sometimes installed by plumbers for both overground and undergound drainage. It has many advantages over traditional materials, being much lighter in weight and very easy to fit.

Cast iron drainage

One of the main characteristics of cast iron is its strength, and for this reason it is usually specified in areas such as car parks, shopping centres and public buildings, where drains are often exposed to damage by accident or an act of vandalism. It is fireproof, will not collapse under intense heat, does not give off toxic fumes and will not assist the spread of fire. While initially it may be more costly than other materials, it requires little or no maintenance and is likely to last the life expectancy of the structure in which it is installed.

Corrosion protection
The original method of protection against corrosion was by dipping pipes and fittings into a coal tar solution. However this had a limited life span and often resulted in encrustation on the inside surface of the pipe. Pipes and fittings are now treated as shown in Fig. 9.1, which complies with BS EN 877.

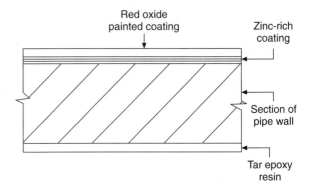

Fig. 9.1 Protection of cast iron pipes and fittings against corrosion.

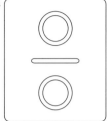

This symbol indicates the pipe is suitable for both above and below ground use

Fig. 9.2 Identification symbol.

Pipe cutting

Snap cutters are not recommended — the approved method is powered disc or wheel cutters. Cut ends should be treated with a zinc-based or other suitable coating to protect them from corrosion.

Identification as to use

Only those cast iron drainage systems identified by the symbol shown in Fig. 9.2 are suitable for both below- and above-ground use. This symbol is not shown for pipes and fittings suitable only for above-ground use, as they have thinner walls and are not as strong.

Jointing

The traditional method of jointing cast iron drains was the use of lead-caulked joints. Due to the requirements of the Building Regulations, which specify that underground drainage systems must be flexible, this method of jointing is no longer permissible for joints on long drainage runs. Most underground drainage systems use plain-ended pipe, the joints being made with clamping rings and a synthetic rubber seal. A typical joint of this type is shown in Fig. 9.3. Because electrical continuity may be specified, most producers of this type of jointing provide a bonding strip which bridges the rubber seal. The use of a silicon-based lubricant is recommended to enable the seals to be easily positioned.

One-off lead-caulked joints may be necessary when altering or extending an existing system. Figure 9.4 illustrates the method employed. A set of caulking tools will be necessary to make the joint.

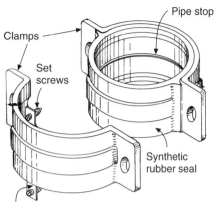

Pipe stop

Clamps

Set screws

Synthetic rubber seal

Earthing strip where specified bypasses rubber seal. The pointed set screws the protective coating of the pipe to ensure electrical bonding of the whole installation

Stainless steel bolfs (not shown) should be tightened with a torque wrench set at 20 mm on with a spanner until a suitable resistance is achieved.

Fig. 9.3 Typical clamp ring joint for plain-ended ductile cast iron pipes.

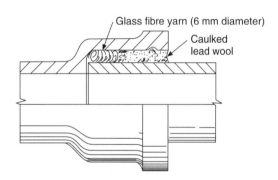

Glass fibre yarn (6 mm diameter)

Caulked lead wool

Fig. 9.4 Traditional lead caulked drain joint. This type of joint is normally used only when making connections to existing work.

PVC (BS 4660:1973)

The use of PVC underground drainage systems very quickly followed that of PVC discharge pipe systems. Both pipes and fittings are light in weight and easily handled, yet have sufficient strength to keep breakages to a minimum. Joints are quickly and easily made in all weathers and can be tested immediately in the same way as other materials using rubber ring joints. Pipes are available in 2, 3, 4 and 6 m lengths, so not a great many joints are necessary. PVC has excellent corrosion-resistent properties and a very smooth bore; reducing resistance due to drag or friction of the pipe walls to a minimum. This is a very important factor to consider with drains as they are laid with shallow falls and often carry very low flow rates.

PVC pipes should be given adequate support when stored on site. The ground should be level and free of stones or building rubble. Some authorities recommend that PVC pipes should not be stacked (especially in warm weather) more than three layers high, as the bottom pipes may distort and this would cause difficulty in aligning and jointing. Pipes having an integral socket should be stacked with the sockets in each layer at alternate ends of the stack, the sockets protruding. This will avoid imparting a permanent set in the pipes which may make their alignment difficult when laying. Do not forget to wedge the pipes to prevent them rolling outward.

The principle methods of jointing are the same as those used for overground systems embodying rubber rings or solvent welds. O-ring joints are made as follows. First, it is essential to cut the pipe square with a fine-tooth saw such as a hacksaw (see Fig. 9.5). Failure to do this and to file a chamfer on the spigot end of the pipe will cause difficulty when the pipe is pushed into the socket (see Fig. 9.6). An unchamfered edge may dislodge the O-ring. (It should be noted that a chamfered edge to the spigot is necessary on a joint on any material employing rubber rings.) Take care to remove with a sharp knife any whiskers that may have been left by the

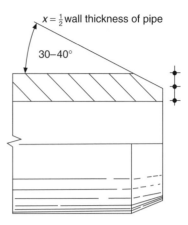

Fig. 9.6 Chamfers on pipe ends.

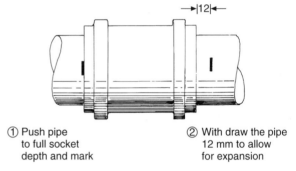

① Push pipe to full socket depth and mark ② With draw the pipe 12 mm to allow for expansion

Fig. 9.7 Making o-ring joints on PVC drains.

saw on the inside of the pipe. Prior to inserting the spigot into the socket, mark it so that its depth of entry can be seen so as to ensure that an adequate expansion gap is retained (see Fig. 9.7). The spigot should then be smeared with a lubricant supplied or recommended by the manufacturer, which will make its entry into the socket much easier. It is sometimes necessary to lever a long length of pipe into its socket. Figure 9.8 illustrates a suitable method of achieving this. Some manufacturers recommend that the open edges of O-ring joints are sealed with an adhesive tape such as Sealglass or Denso to prevent the entry of roots which may damage the socket or find their way into the drain, eventually causing a blockage. When solvent cement joints are used, the spigot and socket should be roughened with sandpaper or emery cloth, then cleaned with a recommended cleaning fluid to

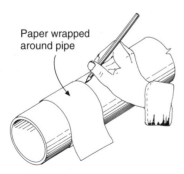

Paper wrapped around pipe

Fig. 9.5 Marking out a square cut.

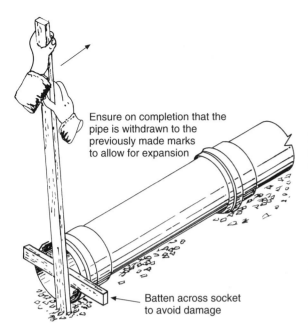

Fig. 9.8 Levering a PVC drain spigot into the socket.

Ensure on completion that the pipe is withdrawn to the previously made marks to allow for expansion

Batten across socket to avoid damage

Paint the area to be jointed with solvent cement and before it dries roll it in sharp (gritty) sand

Fig. 9.9 Preparation of 'key' to PVC pipes. This is necessary when joining PVC to other materials with a cement joint.

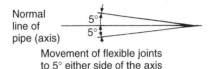

Normal line of pipe (axis)

5°
5°

Movement of flexible joints to 5° either side of the axis

Fig. 9.10 Minimum movement for flexible joints.

Flexibility

Due to the damage to rigid drainage systems caused by the settlement of building and subsidence of the subsoil, all underground systems of drainage must be flexible. Provision must be made for both angular and telescopic withdrawal movement: Figure 9.10 shows the standard to which all underground joints must comply. Prior to the 1965 Building Regulations, cast iron and clayware drains were laid on a concrete bed, but due to the success of the methods employed for bedding PVC drains, similar methods are now employed with all materials. Approved methods of bedding are covered later in this chapter.

Drainage fittings

General

The variety of drainage fittings produced is considerable, some only being encountered for special types of work. Those illustrated and described here are representative of the more common types the plumber is likely to come across in his normal work. Producers of clayware, cast iron and PVC systems of drainage manufacture a complete range of fittings in each material together with adaptors to enable connections to be made to other materials. Relevant catalogues should be referred to in order to ensure an effective connection.

remove any traces of grease. Only the solvent cement recommended by the manufacturer should be used as the period of time between its application and the point at which it sets is longer than that of similar materials used for smaller pipes. The surface area of a 100 mm pipe is comparatively large and a quick-setting cement would dry off before the joint could be made. The solvent cement is brushed evenly on to the mating surfaces of the joint. Sockets on drains are usually made with a slight internal taper to give what is called an interference fit, which means that the spigot tightens on to the socket as the joint is pushed home. Solvent-welded joints should be used on short lengths of pipe or connections to inspection chambers only, as they do not provide for expansion.

It is sometimes necessary to make joints between PVC drains and those of other materials using cement joints. As PVC is very smooth, a suitable key must be provided as shown in Fig. 9.9. All manufactured fittings e.g. channels, channel junctions and bends designed for use with traditional inspection chambers, are supplied with appropriate keying for use with masonry.

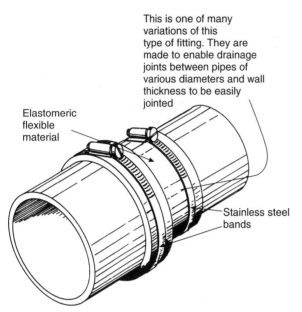

Fig. 9.11 Drainpipe adaptor.

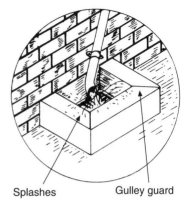

Fig. 9.12 Obsolescent method of waste discharge into gulley. Many older buildings have this arrangement. But splashing caused by the discharge pipe emptying over an open gulley grating causes fouling on the interior and top of the gulley guard leading to insanitary conditions.

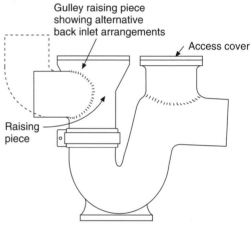

Fig. 9.13 Basic gully trap. May be obtained with or without access.

The adaptor shown in Fig. 9.11 is useful when adapting existing clayware drains having differing external diameters. These old drains are usually laid with cement mortar joints and it is near impossible to cut out the joint without breaking the socket. Adaptors such as this can be used for jointing socketless pipes of various materials and external diameters. Whenever connections have to be made between various materials, never attempt to 'make do' with non-standard adaptors, as this often leads to trouble in the future.

Gulleys
The purpose of a gulley is to provide a trap to prevent odours entering the atmosphere in cases where it is not possible, or desirable, to terminate a discharge pipe directly into the drain. A wide variety of types are available in all drainage materials for various purposes. In many older domestic dwellings, discharge pipes terminate over the gulley grating; this is provided with a guard made of brickwork or concrete (see Fig. 9.12) which directs the water into the grating. This method of discharge is no longer permissible and was always undesirable. Modern buildings employ the use of back inlet gulleys where the discharge

pipe terminates under the grating but over the water seal (see Fig. 9.13). A typical raising piece designed to fit a standard trap is also illustrated and shows how a gulley trap can be adapted to suit a variety of applications. Like other traps used on underground drainage systems, only a 50 mm depth of seal is necessary. It will be noted that this is less than that required for traps on small-diameter discharge pipes and is due to the fact that the risks of siphonage are almost negligible, as any discharge seldom permits the pipe to run full. Figure 9.14 shows two gulleys

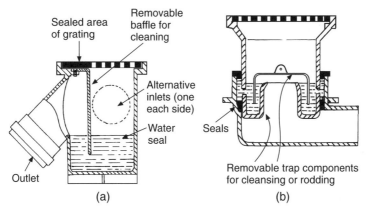

(a)   (b)

Fig. 9.14 PVC bottle-type floor gullies.

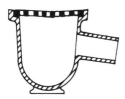

Fig. 9.15 Trapless gulley. Used for surface water only as an alternative to a rainwater shoe. Silt collecting in the base can easily be removed.

that are suitable for draining floors, such as commercial kitchens and sanitary annexes. Their short overall height makes them very suitable for this purpose. They are not self-cleansing but are provided with suitable access for routine maintenance.

Trapless gulleys (Fig. 9.15) and yard gulleys (Fig. 9.16) are often used for draining paved areas, as they provide for the easy removal of any silt that is washed off these surfaces. If silt is allowed to pass into the drain it will eventually build up and cause a blockage. Trapless gulleys must not be connected to foul water systems. They are usually used in conjunction with a trapped master gulley in the same way as rainwater shoes.

Rainwater shoes
These are trapless fittings used mainly to connect rainwater pipes to a surface water drain with access for cleansing both the drain and the rainwater pipe. Two or more such fittings are often used in conjunction with a master gulley (see Fig. 9.17).

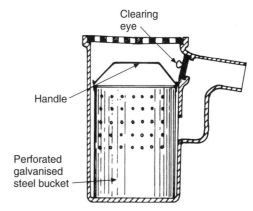

Fig. 9.16 Yard gulley. Used to collect the run-off from paved surfaces. The removable bucket prevents any silt being washed into the drain. It can be periodically removed for cleansing.

Intercepting traps
An intercepting trap is illustrated in Fig. 9.18(a), but they are seldom necessary in modern properties. They were originally intended to provide a trap to prevent sewer gas from entering the house drain. It should be explained that the house drain is that part of the system which is the responsibility of the householder. The sewer may be broadly defined as a main drain carrying the discharge from house drains to a sewage treatment plant; sewers are generally the responsibility of the local authority. In bygone days, the ventilation of sewers was not as thorough as it is today and, because sewage begins to decompose very quickly due to the action of

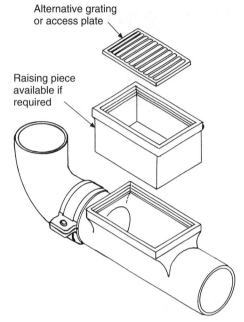

(a) Drain shoe. An alternative to a trapless gulley for surface water use

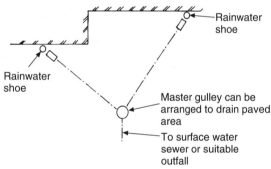

(b) Several of these shoes may be connected to a master gulley as shown

Fig. 9.17 Rainwater shoes.

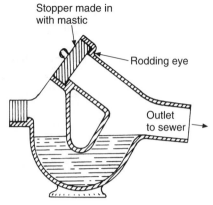

(a) Clayware intercepting trap. Made in clayware or cast iron. Some times called a disconnecting trap as its original function was to trap the house drain from the sewer

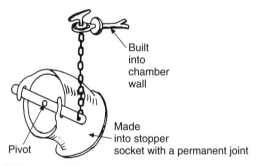

(b) Lever locking stopper for clayware interceptor

Fig. 9.18 Intercepting trap and lever locking stopper.

bacteria, sewer gas was produced which is both flammable and, in heavy concentrations, toxic. In these circumstances it was often necessary to prevent access of sewer gas to house drains. Due to better systems of drain ventilation, sewer gas nowadays seldom constitutes a danger. The main disadvantages of intercepting traps are that

(a) They are the most common cause of obstruction due to the collection of debris in the trap.

(b) They interrupt the smooth flow of water through the drain.

They are normally situated in an inspection chamber so that they are accessible when necessary. Figure 9.18(b) shows an improved interception stopper. As a result of back pressure in the sewer or incorrect fitting, the normal type provided often falls into the trap causing it to become blocked. This is impossible with the lever locking type.

Bends
A wide variety of bends is available in all drainage materials. They are specified by their radius and angles. They can also be obtained with access doors on either side or on the back. For full details of ranges of bends available,

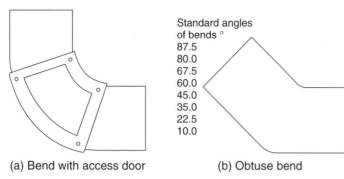

Standard angles
of bends °
87.5
80.0
67.5
60.0
45.0
35.0
22.5
10.0

(a) Bend with access door   (b) Obtuse bend

Fig. 9.19   Examples of typical cast iron bends.

reference should be made to manufacturers' catalogues. Some typical cast iron bends are shown in Fig. 9.19.

Duck foot or rest bends
These bends, made of clayware or cast iron, are provided with a foot or rest and are designed to give extra support to offset the weight of a cast iron discharge or vent pipe (see Fig. 9.20).

Bends and junctions used in chambers
Cast iron drainage systems were rarely used with open channels in inspection chambers and few, if any, manufacturers now make the necessary fittings. Sealed junctions of the type shown in Fig. 9.21,

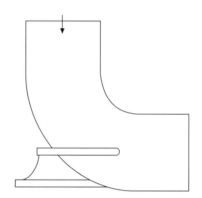

Fig. 9.20   Rest bend. Used where the bend takes the downward thrust of a stack.

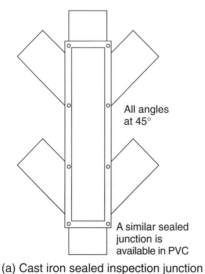

All angles
at 45°

A similar sealed junction is available in PVC

(a) Cast iron sealed inspection junction

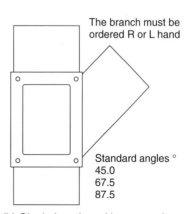

The branch must be ordered R or L hand

Standard angles °
45.0
67.5
87.5

(b) Single junction with access door

Fig. 9.21   Sealed junctions for access to cast iron drains.

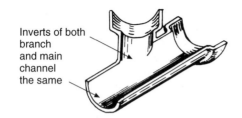

Fig. 9.22 Channel junction.

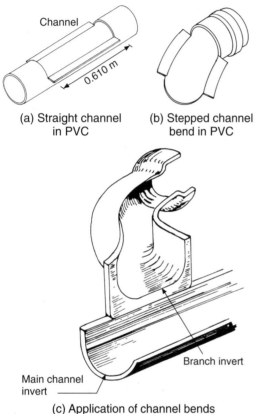

(a) Straight channel in PVC   (b) Stepped channel bend in PVC

(c) Application of channel bends

Used to connect a branch drain to the main channel as an alternative to a channel junction. Fitted in an inspection chamber and made in clayware or PVC.

Fig. 9.23 Straight channel, stepped channel and channel bends.

housed in brick or concrete chambers, are the normal method of providing access to underground and suspended cast iron drains. Producers of PVC systems also make sealed junctions and bends which are similar to those of cast iron and are mainly used for suspended drainage. For underground use, the open channel system, very similar to that of clayware, is generally used.

Level invent open channel junctions
A typical example of a clayware junction is shown in Fig. 9.22; those made of PVC are very similar. On all such junctions the branch joins the main channel at 45°. A wide selection of channel bends is available to enable a branch drain to enter the main channel at any angle at the same invert level.

Stepped channel bends
These are used in conjunction with the PVC channel shown in Fig. 9.23(a), or to make additional connections to an existing inspection chamber. The bend shown in Fig. 9.23(b) is used with this type of main channel and Fig. 9.23(c) shows their application.

Drain chutes
The purpose of these fittings (see Fig. 9.24) is to provide better access for rodding drains in deep manholes. They are normally only used with clayware drainage systems which employ open channels. Manufactures of PVC and cast iron drainage do not list these fittings.

Rust pockets
Rust pockets (see Fig. 9.25) are seldom necessary in modern drainage systems and few manufacturers list them. They are designed to retain the scale and rust deposits which collect at the base of cast iron vent pipes and which can cause the lower end of the pipe to become completely obstructed. The debris must be removed periodically. In cases where the vent

Designed to facilitate rodding in deep inspection chambers

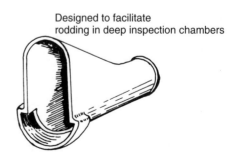

Fig. 9.24 Drain chute.

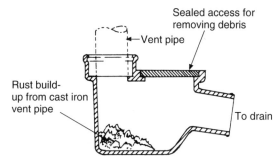

Fig. 9.25 Rust pocket.

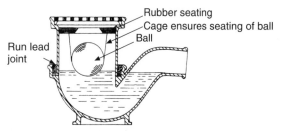

(a) Two-piece cast iron anti-flood gulley. Prevents flooding of surface-water drains when discharging into watercourses or tidal rivers

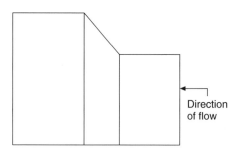

Fig. 9.26 Level invert taper pipe used to connect a smaller diameter drain to a larger one.

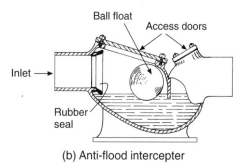

(b) Anti-flood intercepter

For use with surface water only. In the event of backflow the ball rises with the water until it seats on the rubber seal, thus preventing any further flow of water into the inlet.

Fig. 9.27 Anti-flood fittings.

has a branch discharge pipe fitted to it, a rust pocket is unnecessary as the flow of water will wash away any debris. These fittings are normally only found in older public and commercial buildings.

Level invert taper pipe
This is used when a change of diameter is made in a drain to avoid any disturbance of the flow in the drain (see Fig. 9.26).

Anti-flood fittings
These are used only on surface-water drains which discharge into a watercourse or tidal river. During periods of heavy rainfall or high tides, it is possible that water may back up the drain and cause flooding in or around the building. The gulley illustrated in Fig. 9.27(a) is made of cast iron with a rubber seating; they are also obtainable in clayware. If the water in the gulley rises the float engages on the seating making a watertight seal. The float seating and cage are usually made as a unit which

sits on lugs cast into the body of the gulley enabling it to be withdrawn for inspection. An anti-flood interceptor (see Fig. 9.27(b)) is a similar device which is housed in an inspection chamber and serves to prevent a complete system of drains from flooding.

Grease disposal
Grease and cooking oils have long been recognised as being a major cause of blocked drains, especially in catering establishments and commercial buildings such as hotels where large volumes of grease and cooking fats are discharged into the underground drainage system. As the waste water cools, the grease it contains solidifies and becomes deposited on the pipe walls, where it builds up until a solid plug of grease is formed, blocking the drain.

Grease traps Until comparatively recently the only method possible to separate the grease from the waste water was to install a grease trap, a

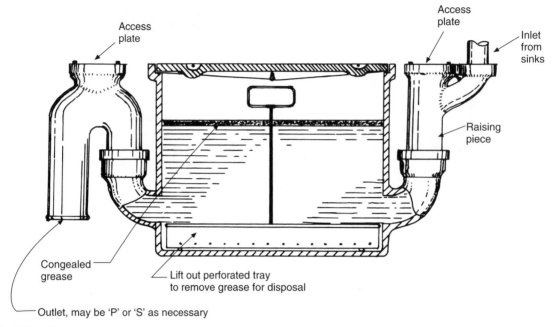

Fig. 9.28 Grease trap.

diagrammatic illustration of which is shown in Fig. 9.28. A grease trap is designed to hold a large volume of water so that the water discharged into it is cooled down, causing the grease to solidify and float to the surface, forming a solid cake. This is removed periodically by withdrawing the metal trap and disposing of the grease as refuse. This is an unpleasant task and is therefore often neglected causing the trap to become a breeding ground for bacteria. If the trap is undersized the water it contains will not cool sufficiently to allow the grease to congeal, the object of the trap being defeated and the grease still being carried into the drainage system. From the foregoing it will be seen that unless this type of trap is cleansed daily and is correctly sized, its use may make matters worse rather than better.

Grease converters These have been developed from research and investigation into the problems of effective grease disposal in large kitchens. The working principles are based on first separating the grease from the waste water effluent, then treating it chemically by a process called 'digestion'. A typical grease converter is shown in Fig. 9.29

where it will be seen that the effluent passes over the baffles which cause the formation of globules of grease. As the grease builds up it falls from the baffles and floats to the surface of the water in the container. Chemicals are added to the water producing active micro-organisms, which degrade and break down the grease causing it to become soluble in the water.

Maintenance The chemicals are poured into the converter directly or washed into it via a convenient sink. This operation should be carried out on a daily basis, but twice weekly may be sufficient, depending on usage. It is best done during a period when the kitchen is not in use, e.g. last thing at night, so that micro-organisms can grow and reproduce in a good environment. It will be necessary to periodically remove the top cover and clean out any solids deposited in the base of the container — ideally every 3 months. Both when commissioning and after cleaning, a culture of micro-organisms must be established before breakdown of the grease is effective, and the manufacturer's instructions must be complied with in this respect.

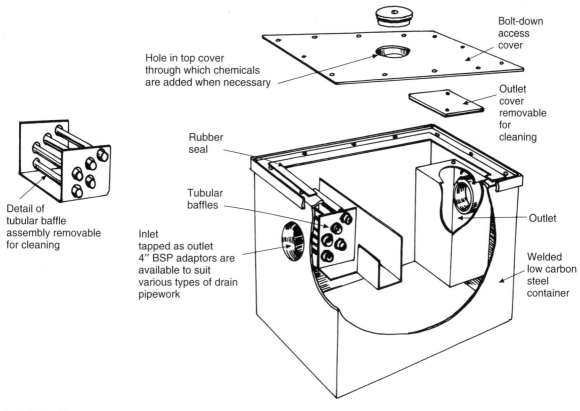

Fig. 9.29 Patent grease converter.

No type of grease disposal unit will function efficiently if it is incorrectly sized. The most reliable method of achieving this is to determine the requirements of the kitchen, e.g. number of sinks and type of usage, and to seek the manufacturer's instructions. If the grease converter is working effectively, there will be little or no odour, its content will have a consistency similar to a thick soup, with no surface grease or caking on the inside surfaces of the container.

Neither grease traps or converters should be situated in areas where food is prepared or served, and adequate access must be made for maintenance and cleaning purposes.

Support for above-ground drains

While drainpipes are usually laid underground, they sometimes pass through the basement of a structure;

they are then referred to as suspended drainage. Only cast iron or PVC is used in these circumstances as other materials would not be strong enough. Brackets for suspended drainage are usually purpose-made with provision for adjusting the falls on the pipeline to create a self-cleansing velocity. More will be said about this later. Figure 9.30 shows brackets, often made on site for supporting drainage in a building. They are made of black low-carbon steel bar or strip, angle iron and bright steel rod. Some method of heating will be required to forge the steel, oxyacetylene equipment being ideal for this purpose. Suitable dies will also be necessary to cut the threads on the steel rod. An alternative is to use *studding* which is simply threaded rod which can be cut to the length required. This material is available in various lengths, diameters and types of thread from good engineering stockholders or tool dealers.

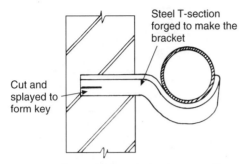

(a) Fixing a horizontal pipe into a wall

As there is no adjustment to these brackets
they would be set out with a line.

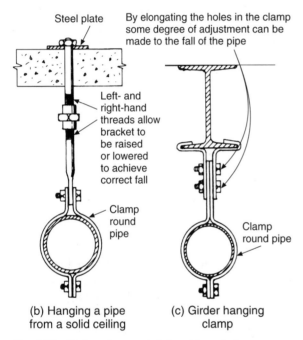

(b) Hanging a pipe from a solid ceiling

(c) Girder hanging clamp

Fig. 9.30 Fixings for suspended cast iron drainage systems.

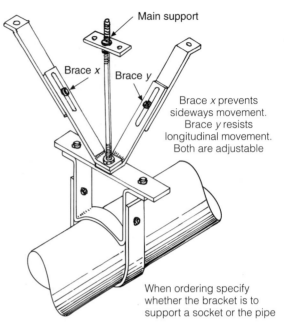

Fig. 9.31 Support bracket for PVC suspended drains.

Cast iron drains are very heavy and any brackets fabricated for their support should be stout enough to take their weight. If PVC is used, more brackets will be required due to its lack of rigidity, but they can be made of thinner material as PVC is not as heavy as cast iron. A typical bracket for PVC is shown in Fig. 9.31. The spacing of brackets for both cast iron and PVC suspended drainage is shown in Fig. 9.32. Apart from any anchor brackets, intermediate fixings should allow the pipeline to move when it expands or contracts. A method of supporting cast iron drainage just above ground is shown in Fig. 9.33. The pipe is simply bedded on brick piers which must be carefully set out so that the sockets are clear of the piers, otherwise it will be difficult to make the joints.

Bedding for underground drains

The word 'bedding' relates to the support of underground drainage and the material used for this purpose. It must be carefully laid in order to give a drain the necessary fall. For many years, when clayware and cast iron drains having rigid joints were mainly used, concrete was almost invariably employed as a bedding material. This 'traditional' method of bedding is shown in Fig. 9.34, but due to the types of joints and bedding materials currently in use and the emphasis on flexibility, rigid concrete beds are not normally recommended. The main disadvantage of a rigid bed is that any movement of the soil due to shrinkage or subsidence in made-up

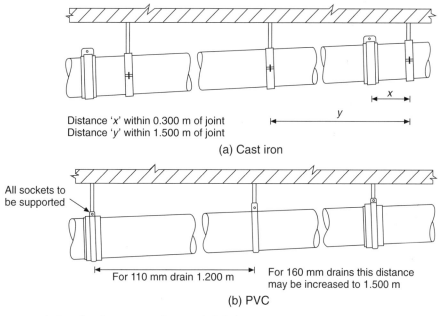

Distance 'x' within 0.300 m of joint
Distance 'y' within 1.500 m of joint

(a) Cast iron

All sockets to
be supported

For 110 mm drain 1.200 m

For 160 mm drains this distance
may be increased to 1.500 m

(b) PVC

Fig. 9.32 Recommendations for the support of suspended drains.

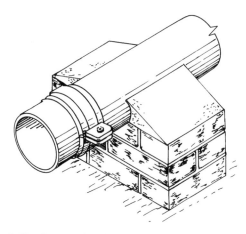

Fig. 9.33 Support for drains above ground level on brick piers.

ground may cause the concrete to crack. This would result in cracking of the drainpipe and subsequent leakage. Cast iron, due to its strength, is far more resistant to damage from this cause than clayware, and for this reason, despite its high cost, it is commonly used in public and industrial buildings. Although pitch fibre drains are now no longer made,

it was the use of this material — and a few years later PVC — that prompted the Building Research Centre to investigate the possibility of flexible bedding for drainage pipelines which were themselves flexible. After considerable research, bedding procedures similar to those shown were suggested by this body. It will be seen that in order to protect underground pipework from imposed loads and to ensure the movement of joints in the event of ground movement, great care must be taken to provide the correct type of bedding and methods of back fill. The first point that must be understood is that underground drainage pipework materials are either rigid, as in the case of clayware and cast iron, or flexible such as PVC. Due to the fact that it is softer than cast iron and clayware, incorrect methods of installation may cause it to become distorted or flattened, both faults leading to subsequent blockage in the pipe. Figures 9.34 and 9.35 show the bedding details for both rigid and flexible pipes which meet the requirements of the Building Regulations 1985. When socketed pipes are used a section of the bed must be removed to accommodate the socket so that the barrel of the pipe is fully supported on the bed. Any wooden

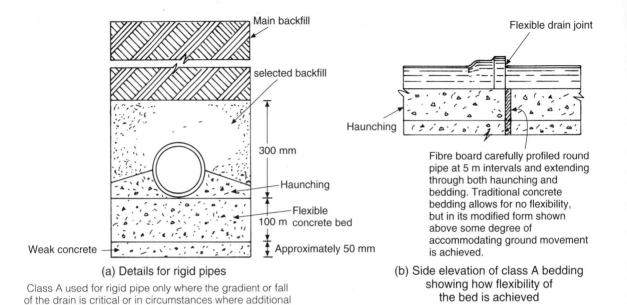

Main backfill

selected backfill

300 mm

Haunching

Flexible
100 m concrete bed

Weak concrete

Approximately 50 mm

(a) Details for rigid pipes

Class A used for rigid pipe only where the gradient or fall
of the drain is critical or in circumstances where additional
support is considered necessary such as under roads or
areas subject to vehicular traffic.

Flexible drain joint

Haunching

Fibre board carefully profiled round
pipe at 5 m intervals and extending
through both haunching and
bedding. Traditional concrete
bedding allows for no flexibility,
but in its modified form shown
above some degree of
accommodating ground movement
is achieved.

(b) Side elevation of class A bedding
showing how flexibility of
the bed is achieved

Fig. 9.34 Bedding for rigid drain.

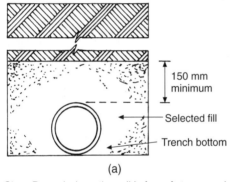

150 mm
minimum

Selected fill

Trench bottom

(a)

Class D used where the soil is free of stones and
reasonably dry and firm where the trench bottom can be
accurately trimmed by hand.

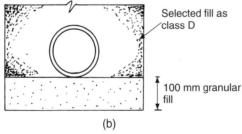

Selected fill as
class D

100 mm granular
fill

(b)

Class N, identical to class D except that the pipes are laid on
a granular bed. Used in similar conditions to those of class D
except where the trench bottom cannot be trimmed
accurately.

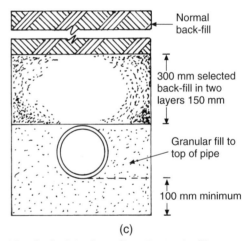

Normal
back-fill

300 mm selected
back-fill in two
layers 150 mm

Granular fill to
top of pipe

100 mm minimum

(c)

Bedding for flexible pipes. Class B granular fill to crown of
pipe gives sufficient support to walls of pipe to prevent
flattening and in the case of PVC pipes permits movement
due to expansion and contraction.

Note that in all cases selected fill must be free of large
stones, lumps of clay, pieces of timber, vegetable matter or
frozen material of any sort.

Fig. 9.35 Basic bedding systems for all types of drainage materials.

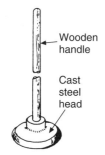

Punner used for light hand tamping
(consolidating) backfill.

Fig. 9.36

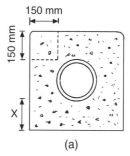

(a)

Recommended for drains above ground level, laid less than
1.2 m deep or under a building. The pipe is encased by a
150 mm surround of concrete.

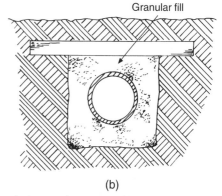

(b)

Shallow drains may be protected by means of a paving slab
laid across the trench. This method is not suitable where heavy
vehicular traffic is likely.

(c)

This illustration shows a flexible socketed clayware joint,
but the same techniques apply to all drainage
pipework materials.

Fig. 9.37 Concrete encasement of drains. This is only
necessary in the case of very shallow drains or those laid
above ground in locations where they may be subject to
damage.

pegs used for determining the gradient or fall of the
trench must be removed prior to laying the drain as
they would cause tilting of rigid pipes and distort
the circular profile of PVC pipes. The main feature
of current bedding systems is the use of granular
infill around the pipes such as sharp sand or pea
gravel. Sharp sand is easily recognisable as it is
much more gritty to the touch than that used for
bricklaying and contains a lot of very small sharp
stones. Do not, however, confuse it with aggregate
used for concreting, as this contains larger stones
which could damage the drain especially if it is
made of PVC. The object of the granular infill is
to distribute any loading more evenly around the
surface of the pipe thus avoiding its distortion,
or, in the case of clayware, cracked pipes due to
heavy loads. All drain trenches must be carefully
back-filled and consolidated with selected fill to
avoid damage to or distortion of the pipes. It is
generally recommended that this should be done
using a hand punner, similar to that shown in
Fig. 9.36, for a distance of at least 300 mm above
the top of the pipe.

Protection of drains near or above ground level or subject to imposed loads

It is no longer mandatory to encase pipes in
concrete if they are near the surface of the ground
unless it is considered necessary for their protection.
If, however, this method is used, flexibility of the
encasement must be maintained as illustrated in
Fig. 9.37(a). Where the soffit of the pipe is between
600 and 300 mm of the surface, and alternative to

complete encasement is to lay paving slabs over the granular fill (see Fig. 9.37(b)). This method has the advantage of allowing the pipe much more freedom of movement. Pipes laid under roads at depths of less than 900 mm from the soffit of the pipe to the surface may require a reinforced concrete slab extending beyond the sides of the trench to ensure the drain is not damaged by heavy traffic.

Drains laid below the bottom of a building foundation and within 1 m of it may be subjected to excessive stress. Reference should be made to the Building Regulations 1985, Approved Document H, which specifies the methods that must be employed to prevent damage under such circumstances. Figure 9.38(a) and (b) illustrate typical cases.

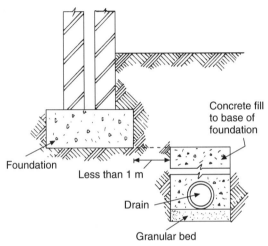

(a) Preventing instability of foundations less than 1 m from a drain trench

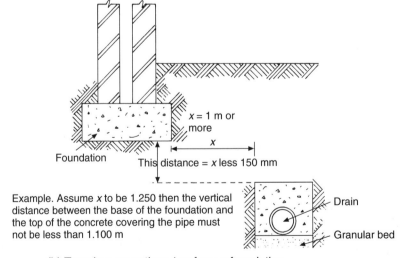

(b) Trenches more than 1 m from a foundation

Fig. 9.38   Drains adjacent to building foundations.

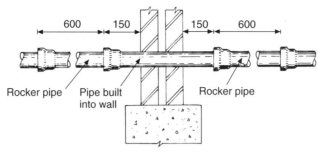

(a) In this case the pipe is built into the wall and
any downward movement of the structure likely to
cause damage on the drain is prevented by the
'rocker' pipes. All joints must be of the flexible type

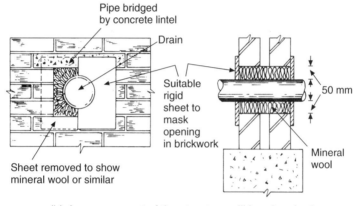

(b) Any movement of the structure will be absorbed
by the mineral wool without damage to the pipe

Fig. 9.39   Protection of drains passing through foundations.

Drains under buildings

In the past it was common practice to encase drain
pipes under buildings in concrete, but since all
drains must now be flexible, such encasement is
no longer recommended. Much better provision is
now made to prevent damage to the drain due to
settlement of the structure, as shown in Fig. 9.39(a)
and (b). In the case of Fig. 9.39(a) any movement
of the structure is accommodated by the rocker
pipes, while in Fig. 9.39(b) the 50 mm gap
around the pipe permits movement of the structure
independently of the pipe. This latter method has
the advantage that it is less likely to alter the
gradient of the drain. Special care is needed to
relieve the mass of the structure over the opening in

the brickwork through which the pipe passes. A
concrete lintel as shown, or a brick relieving arch
satisfies this requirement.

Types of drainage system

There are three types of drainage system in
common usage, known as combined, separate and
partially separate. Each has its merits and demerits,
but the local authority determines which type is
used, basing their choice on their overall knowledge
of local conditions. The reader should be aware that
not only has foul water drainage to be considered
but also that of surface water, e.g. water falling on
roofs and paved areas.

Key
RWG rainwater gulley
FWG foul water gulley
SVP soil vent pipe

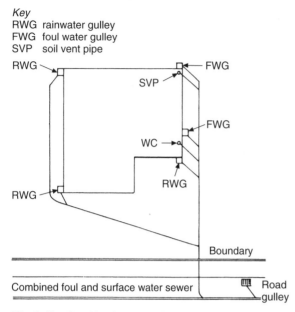

Fig. 9.40 Combined system of drainage.

Key
RWG rainwater gulley
FWG foul water gulley
SVP soil vent pipe

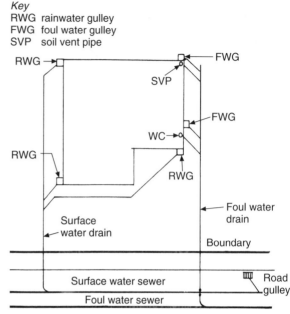

Fig. 9.41 Separate system of drainage.

Combined system

In this case, as its name implies, both foul and surface water are discharged into the same sewer (see Fig. 9.40). This system has the cheapest layout as it requires only one set of pipes, and during heavy rainfall both house drains and sewers are thoroughly flushed out. (Sewers are public authority drains to which private house drains are connected.) This system has often been used in the past where raw sewage was disposed of without treatment, e.g. discharged into the sea or a watercourse. Because of public alarm at this unhealthy state of affairs, it is now considered to be unacceptable and all foul water must be treated before the effluent is discharged in such a way. In many cases this, and the rapid growth of some urban areas, has put a serious strain on the capacity of existing sewage disposal plant, and most local authorities in urban areas now insist on the installation of separate systems. There is a further disadvantage: in storms and periods of very heavy rainfall, flooding and subsequent surcharging of the drains has been known to occur. Then again, in areas of undulating country it is often necessary to pump sewage from

one level to another so that it can be discharged into the sewage treatment plant. In the combined system it would be necessary to pump both foul and surface water, resulting in greater capital and maintenance costs for the equipment used.

Separate system

This system requires the use of two sewers (see Fig. 9.41), one carrying foul water to the treatment works, the other carrying surface water (which requires no treatment) to the nearest watercourse or river. It is expensive to install, but from the local authorities' point of view it is the most economical to operate because the volume of sewage to be treated is far smaller than the discharge from a combined system. The biggest danger is that cross-connections may be made, i.e. foul water may be connected to a surface water drain. There is little chance of flooding at times of heavy rainfall, but the foul water sewers are not flushed periodically with relatively pure water as in combined systems. It is the most commonly employed method of waste water disposal in new towns and urban areas, especially where large housing estates have been

Key
RWG   rainwater gulley
FWG   foul water gulley
SVP   soil vent pipe

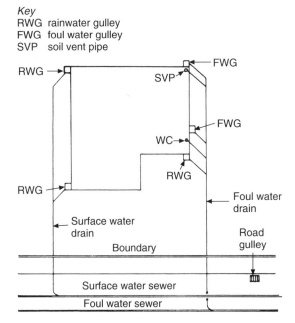

Fig. 9.42   Partially separate system of drainage.

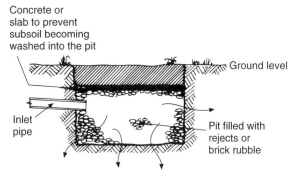

Fig. 9.43   Soakaway.

built, as pumping arrangements and sewage plants, which may already be overloaded, have only to cope with foul water.

Partially separate system
This system probably originated when towns began to grow in size and local authorities found it necessary to try to reduce the loading on the combined system, which in most cases had hitherto been employed. This is something of a compromise between the previously mentioned systems. It requires two sewers, one carrying water from paved areas and part of the roof, the other carrying foul water and water from the remainder of the roof as shown in Fig. 9.42. Some authorities permit the water falling on the front part of the premises to be discharged into the surface water sewer, water from other parts of the roof and paved yards at the rear of the premises being discharged into the foul water sewer. The disadvantages of this system are similar to those of the combined system, but to a lesser degree.

Soakaways   Soakaways are often used with partially separate systems to deal with water from a

roof not connected to the surface water drain (see Fig. 9.43). They usually consist of a pit dug well away from the building, filled with stones or brick rubble to prevent the sides caving in, into which surface water, usually from the roof, is discharged. Soakaways are only effective in ground that is sufficiently permeable to allow water to sink into the surrounding subsoil. They should be sited on sloping ground so that, in the event of flooding, water will flow away from the building. Soakaways can be used with advantage in country areas providing the soil is suitable, but they are seldom satisfactory in urban development or in heavy clay soil. Soakaways have to be approved by the local authority planning officer who often specifies their dimensions, the distance it must be from a dwelling and its construction.

Ventilation of drains

Underground systems of drainage must be well ventilated for two main reasons.

(a) To maintain an equilibrium of pressure between the air inside the drain and that of the atmosphere. If, for instance, the air pressure in the drain was lower than that of the atmosphere, the trap seals of gulleys and WCs would very quickly be destroyed and foul air from the sewer would be admitted to the building and its precincts.
(b) To prevent the build-up of foul air and possibly dangerous gases. Sewage is very quickly acted upon by bacteria which thrive and multiply in

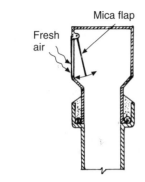

(a) Section of fresh air inlet

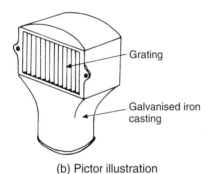

(b) Pictor illustration

These are normally used only in conjunction with intercepting traps in older buildings.

Fig. 9.44 Fresh air inlet.

dark airless environments. Their action on the sewage produces gases which not only have an objectionable smell but are also potentially dangerous due to their flammability. The action of the bacteria in breaking down the sewage is perfectly normal, but it must be confined to sewage disposal works and not allowed to happen in a house drainage system.

The system of ventilation used will depend upon whether or not an intercepting trap is used. Where it is, the inspection chamber in which it is housed must be ventilated to the atmosphere via either a fresh air inlet (see Fig. 9.44) or, in some cases where a building has no frontage, a second vent pipe. The upper end or highest point of the drain should also terminate as a vent pipe which in effect acts as a flue through which drain air can be discharged to the atmosphere. As hot water is often

discharged into a drain, the air it contains is warmer than that of the atmosphere, and air movement through the system occurs due to convection. Where no interceptor is fitted (and this is the case in most modern developments) both the house drains and the public sewers are ventilated in the following way. Air is admitted into the sewer via the sewer ventilating pipes, or more commonly through the holes that can be observed in sewer manhole covers. This flow of air passes through the drainage pipework system and re-enters the atmosphere via the main ventilating stack on each building. It ensures a thorough air flush of both sewers and house drains.

Figure 9.45 illustrates the two methods of drain ventilation. It must be emphasised that drain ventilation pipes should be fitted at the highest point of the drain, thus avoiding long unventilated branches. Both the vent pipe and fresh air inlet (if fitted) should terminate in such a way as not to allow foul air to enter the building via windows or ventilators.

The diameter of the main ventilating pipe is not specified in the Building Regulations, which simply state that it should be of adequate size for its purpose. It is normal in modern practice, however, to fit a pipe of the same nominal diameter as the main discharge stack, although older buildings may be found with pipe diameters of only 75 or 90 mm.

Access to drains

Although a good drainage system (whether carrying surface water or foul water) should be designed to avoid the possibility of a blockage, circumstances arise, often due to misuse, where this happens. It is therefore very important that adequate provision is made so that obstructions can be cleared with the minimum of trouble. Special attention should be paid to the areas where blockages are most likely, such as bends and junctions. The methods of obtaining access for rodding drains vary. Some drainage fittings are made to incorporate a rodding eye, or rodding eyes may be installed at suitable points in the drains. In some circumstances inspection chambers must be constructed. Where

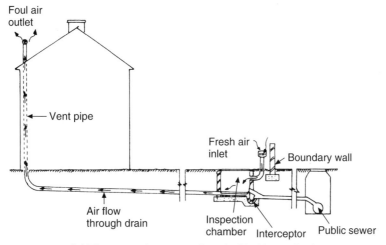

(a) The sewer is not ventilated with this method

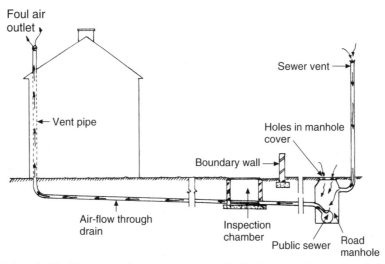

(b) Air admitted to sewer via sewer vents and holes in manhole covers allows
ventilation of the sewer and the house drain through the vent pipes on all premises

Fig. 9.45   Ventilation of drains.

possible, rodding eyes should be used in preference to inspection chambers, as they are cheaper and cause less interference to the drainage flow. The following list is a good general guide to the requirements of the Building Regulations for means of access, but whether a rodding eye or inspection chamber is used will be at the discretion of the local authority.

(a) Changes of direction of the drain requiring a 135° bend or less.

(b) Junctions where the branch joins the main pipe at an angle of less than 135° or bends occur with a similar angle or less.

(c) The point of connection between a drain and a sewer or within 12 m of such a connection.

(d) When a drain exceeds 90 m in length.

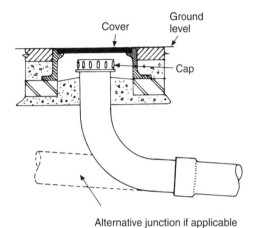

Fig. 9.46 Rodding eye bend.

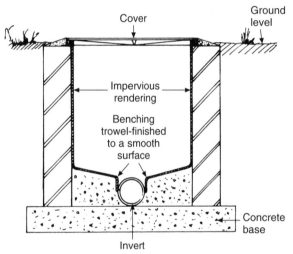

Fig. 9.47 Section through typical inspection chamber constructed of brick or precast concrete.

(e) Where a change of gradient occurs in the drain, for example with ramps or backdrops.

(f) The highest point or head of the drain.

Rodding eyes may be used with advantage where the drain is comparatively shallow as, for example, at its head. Figure 9.46 illustrates such an arrangement.

Inspection chambers are traditionally constructed of brick, either engineering bricks or good-quality stock bricks rendered with an impervious layer of cement mortar. It is essential that these chambers are watertight, and in cases where a drain is at great depth there must be sufficient room for a man to work in comfort. They are still often built using traditional methods, but for house drains and small schemes, due to the standard size of the chambers, precast concrete sections are used instead of bricks which saves time and often cost less. Figure 9.47 shows a section through a typical brick-built or precast chamber showing details of the benching. Benching must be high enough to prevent solid matter being washed up on it and putrifying. Where the depth of a drain is such that it is necessary for a man to enter a manhole for maintenance purposes, a steel ladder or step irons conforming to BS 1247 must be provided. When plastic systems of drainage are used, a chamber can be obtained as a complete unit having a number of blank sockets moulded into it which can be cut as required on site. The design of those illustrated in Fig. 9.48 is based on a traditional layout. As with PVC systems of discharge pipework all manufacturers of PVC drainage produce their own system, the parts of which are not always interchangeable with those of others. Before altering or extending an existing system, a check should be made that the appropriate components are available.

Connections to existing drains

It is sometimes necessary to make extra connections to existing drains; for example, a client may be having an extension built containing extra sanitary fittings. If an inspection chamber is already situated near by, the easiest method of connection is to use a channel bend, shown in Fig. 9.23. If no such chamber is available it will almost certainly be necessary to construct a new one. This is a comparatively simple matter in the case of clayware or PVC systems, as once the existing pipe has been exposed, the soffit can be carefully removed as shown in Fig. 9.49 using a disc cutter. Care must be taken not to let any spoil enter the drain, a channel bend being used to make the new connection.

There are times when a direct connection has to be made to an existing pipe, a typical example being where a house drain joins a public sewer. A fitting called a *saddle* is the usual means of making

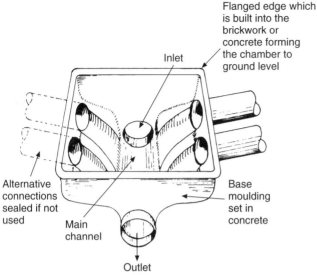

(a) Typical inspection chamber moulding made of PVC for plastic drainage systems

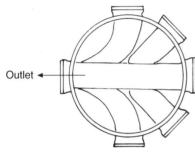

(b) Plan view of moulded base with five inlets and one outlet. Sutiable plugs are provided for inlets not used.

Fig. 9.48   Inspection chambers.

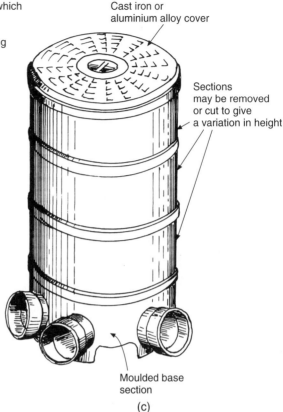

(c)

These inspection chambers are obtainable made of PVC, polypropylene or clayware. They are suitable for shallow drains only as if the invert is too deep difficulty will be experienced when it is necessary to clear blockages with drain rods. The initial cost of this type of chamber is relatively high, but their use reduces labour costs due to the comparative costs of traditional brick built chambers.

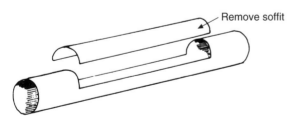

This leaves a channel around which an inspection chamber can be constructed, the new branch being connected by a channel bend. Note this method is not practicable for cast iron drains.

Fig. 9.49   Additional branches in existing drains.

such a connection if the main pipe is big enough (see Fig. 9.50). If it has a diameter of 225 mm or less this method should not be used, because a hole cut in a small pipe would weaken it to an unacceptable degree. In such cases a section of the pipe must be cut and a junction fitted using a slip socket. It should be noted that cutting a new junction into an existing pipe, especially if it is cast iron or clayware, is not an easy task and requires great care. However, the use of power cutting tools and the variety of fittings available have made it less formidable than previously. Remember that any connection using a saddle or junction, where there is no means of access, must enter the main drain or sewer at an angle of not less than 45°.

(a) Pictorial view of saddle

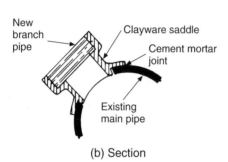

(b) Section

(c) Saddle for PVC drain connections

The area on the existing drain to which the saddle is fitted must be clean and dry after carefully cutting a hole in the main pipe (ensure it is large enough). The saddle is solvent welded in position.

Fig. 9.50 Saddles used for making a new connection to an existing pipe. Note that when cast iron or PVC is used an appropriate adapted will be necessary to connect to the saddle.

Local authorities usually insist that where a connection is made between a house drain and a public sewer, one of their own employees who specialises in this work carries it out.

Working principles of a good drainage scheme

Any drain or private sewer must be designed and constructed of a suitable size and gradient to ensure

Fig. 9.51 Explaining velocity. If the movement of a particle of water from A to B takes 2 seconds the flow rate may be expressed as 0.5 m per second.

that it is self-cleansing and efficiently carries away the maximum volume of matter which may be discharged into it. With this requirement in mind the plumber should be aware of two important facts which affect the efficiency of the drain. These are the gradient, inclination or fall of the drain, and the quantity of water flowing through it. (The latter has an important bearing on the fall as will be shown later.) If the water is flowing very slowly in the pipe, its velocity (speed of flow) is insufficient to carry with it any solid matter. If, on the other hand, the velocity is too great, the water leaves the solids behind. In both cases, solid matter can build up and subsequently cause a blockage. The quantity of water it carries and whether it is a constant or intermittent flow also influence the gradient of the drain. If, for example, the drain is laid to a very shallow fall, it may be self-cleansing if it is running half full, but if the rate of flow falls below this its velocity may be insufficient to keep the solids moving. The important point to understand from this is that the velocity is influenced by the quantity of water carried by the drain. To enlarge a little more on the term *velocity*, consider a length of drain 1 m long as shown in Fig. 9.51. Imagine a particle of water entering it at point A and the time it takes to reach point B. If it takes 2 seconds the water is said to have a velocity of 0.5 m per second; in other words it takes 1 second to travel a distance of half a metre. The minimum velocity for a self-cleansing gradient is approximately 0.75 m per second when the drain is flowing quarter full at a reasonably constant rate.

It is sometimes necessary to calculate the volume of water discharged by a drain at a given velocity, and the following formula is included for this purpose. It should be noted, however, that its use is restricted to pipes flowing full or half full only.

$$\text{Litres per second} = D^2 \times 0.0008 \times V,$$

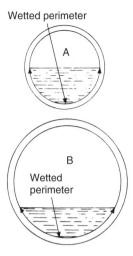

Fig. 9.52 Cross-sectional area of pipes for the purpose of comparison of flow. Although the volume of water shown in the cross-sectional area of drain B is slightly less than that of A the length of the wetted perimeter is greater in B resulting in more frictional resistance to the flow. For this reason it is clear that drain pipe diameters should not be oversized.

where
V = velocity in metres per second
D = diameter of pipe in millimetres
0.0008 = capacity in litres of a pipe 1 m long and 1 mm in diameter

The following example illustrates its use: calculate the volume of water in litres per second discharging from a pipe 150 mm in diameter, flowing full at a velocity of 2 m per second.

$$
\begin{aligned}
\text{Litres per second} &= D^2 \times 0.0008 \times V \\
&= 22,500 \times 0.0008 \times 2 \\
&= 36.
\end{aligned}
$$

Thirty-six litres per second will be discharged when the pipe is flowing full. The answer should be divided by two if the pipe is only flowing half full. If the velocity is quoted in minutes then the volume discharged will also be in minutes.

There are two practical points to bear in mind that relate to the sizing of drains. It is often asked, 'Why not have drain pipes of smaller diameter?' (100 mm nominal is the smallest permitted for foul water.) If this were so, the incidence of blockage would be increased, but, even more important, the drains might, on occasion, run full bore causing the

traps in gulleys and WCs to be siphoned out. The second point relates to oversizing drains. If a comparison is made between a 100 mm drain (A) and a 150 mm drain (B) carrying the same quantity of water (see Fig. 9.52 which shows a cross-sectional area of these two pipes), it will be seen that the frictional resistance offered by the portion of pipe wall with which the water comes in contact (called its *wetted perimeter*) in pipe A is much less than that of pipe B in relation to the volume of water carried because there is a smaller area of pipe in contact with the water. As frictional resistance slows up the flow of water, assuming that both pipes are laid to the same fall, the effect of oversizing might be that the smaller pipe has a self-cleansing velocity while the larger has not. From this one can also see how the inner surface of a drainpipe can influence its performance in terms of velocity, i.e. a pipe with a smooth internal bore will offer less resistance to the flow than one having a rougher internal texture.

Hydraulic mean depth (HMD)

While the calculation of the volume of flow in a drain in relation to its fall is the subject of advanced level work, the terms relating to it should be understood at this stage. The term *HMD* represents the relationship between the cross-sectional area of flow in a drain and the wetted perimeter. To find the HMD of a pipe, the following formula is used:

$$
\text{HMD} = \frac{\text{Cross-sectional area of flow in mm}^2}{\text{Wetted perimeter in mm}}
$$

The following example shows how to calculate the HMD of a 150 mm diameter pipe flowing half full. Taking the cross-sectional area first, it will be realised that what is required is half the area of a circle having a diameter of 150 mm. We use the formula area of circle = πr^2.

$$
\begin{aligned}
r &= 150 \div 2 \\
&= 75 \text{ mm} \\
r^2 &= 75 \times 75 \\
&= 5,625 \text{ mm}^2 \\
\pi r^2 &= 3.142 \times 5,625 \\
&= 17,673.75 \text{ mm}^2
\end{aligned}
$$

This is the full cross-sectional area of the pipe, but as the pipe is only running half full the area must be divided by two, thus:

$$17{,}673.75 \div 2 = 8{,}836.875 \text{ mm}^2.$$

The cross-sectional area of the flow is 8,836.875 mm^2. The wetted perimeter is half the circumference of the drain:

$$\text{Perimeter} = \frac{\text{Circumference}}{2}$$

Thus

$$\text{Wetting perimeter} = \frac{3.142 \times 150}{2}$$

$$= 235.65 \text{ mm.}$$

$$\text{HMD} = \frac{\text{Cross-sectional area}}{\text{Wetted perimeter}}$$

$$= \frac{8{,}836.875}{235.65}$$

$$= 37.5.$$

The value of HMD in this example is 37.5. HMD is used in conjunction with the fall or incline of a drain to determine drainage flow rates. While it is comparatively easy to calculate the HMD for pipes flowing full or half full, it is more time consuming to evaluate that of flows other than these. For this reason drainage engineers often use tables which give the HMD for any given cross-section of flow. All this may be a little advanced for the student plumber, but it is useful to have some knowledge of the essentials of basic drainage design. The plumber is normally only concerned with short branch drains, and as these are almost invariably only 100 mm in diameter the question of selecting a drain size is largely eliminated. What is important here is the fall of such drains. As they normally carry only intermittent flows, the incline must be greater than those of drains with a steady flow to ensure that they are self-cleansing. The recommended falls for such drains are shown in Table 9.1 which illustrates what is known as Maguire's rule. These figures indicate that a 100 mm drain must have a fall of 1 m in every 40 m, and so on.

Table 9.1 Maguire's rule.

Diameter of pipe (mm)	Recommended fall
100	1 in 40
150	1 in 60
225	1 in 90

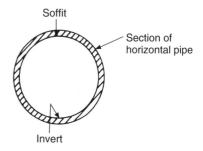

Fig. 9.53 Invert and soffit of drain section.

If the falls in Table 9.1 are adhered to, a self-cleansing gradient is assured with very low flow rates.

Invert and soffit

A clear understanding of these two terms is essential. Most drainage measurements are taken from the lowest point of the invert and such a measurement is referred to as the 'invert level'. If one can imagine the cross-section of a circular pipe with a horizontal line drawn across its centre, the curve or arch above the line is called the *soffit*. The half-section below the line will appear as an arch in the upside-down position and is the *invert*. Figure 9.53 illustrates these two terms.

Determining drain levels

All levels relating to a building are measured from the site 'datum'. It is sometimes called a temporary bench-mark, and consists of a peg firmly driven into the ground, usually protected with concrete and carefully levelled from the nearest Ordnance Survey benchmark. It is from the temporary bench-mark (TBM) that the invert levels of drains are taken. As an example, assume

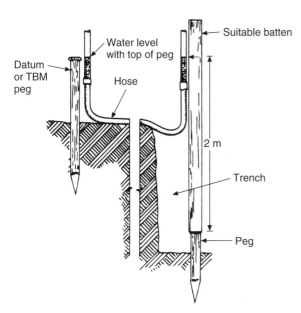

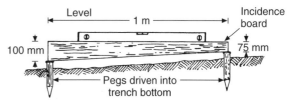

Fig. 9.55 Levelling in short branch drains using an incidence board. If the top of the board is kept level its bottom edge will give the fall found in the calculation.

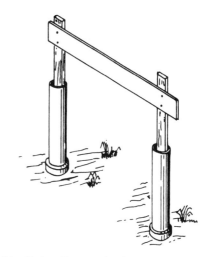

Fig. 9.54 Transferring invert level from temporary bench-mark (TBM). Note that TBM has a value of 100,000. Levels above this will therefore read 100,000+, levels below 100,000−, e.g. assume the drain invert level would be indicated in this case as 98,000−.

Fig. 9.56 Sight rail set up in pipes.

that the invert of a drain run is to be a given distance below TBM; a peg will be driven into the bottom of the drain trench and levelled by the foreman or site engineer using a surveyor's staff and theodolite. If, however, such equipment is not available, the job can be done quite as effectively with a water level. This is a simple device using the principle that fluids find their own levels and consists of two graduated glass tubes connected by a suitable hose. Its use in ascertaining levels is shown in Fig. 9.54.

Setting out the incline of a drain

For short runs the method involving the use of an incidence board is usually found satisfactory. To cut the board to the correct angle the required incline of the drain must first be found by the following simple formula:

$$\text{Incline} = \frac{\text{Fall (in m)}}{\text{Length (in m)}}$$

where the fall is the difference between the highest and lowest points of the drain.

For example, if a drain has a fall of 1 m in a 40 m length, the fall per metre run will be calculated like this

$$\text{Fall} = 1 \div 40$$
$$= 0.025 \text{ m (or 25 mm).}$$

The drain will fall 25 mm in every metre. An appropriate gradient incidence board with measurements is shown in Fig. 9.55.

For setting out longer runs, sight rails and boning rods are used. Figure 9.56 shows a typical sight rail set up in drainpipes. Two will be required, one at the highest point of the drain and one at the lowest, and the difference in height between the two will be the required fall of the drain. The boning rod looks rather like a large T-square and is illustrated in

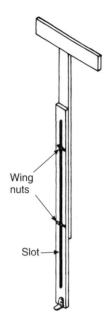

Fig. 9.57 Purpose-made adjustable boning rod.

Fig. 9.57. The procedure is as follows: a trench is excavated, the bottom of which follows the gradient as closely as possible; pegs are then driven into the trench bottom at intervals of about 1 m. The required gradient is achieved by placing the boning rod on top of each peg and checking that the top is lined up with the sight boards. If the top of the

boning rod shows above the boards then the peg must be driven down, if it is lower than the boards then the peg must be replaced by a longer one. Figure 9.58 illustrates the procedure. If necessary the trench itself can be excavated to the required gradient by boning in each section as it is dug. Where a granular bed is used the pegs must be removed as the drain is laid because they can damage pipes made of PVC if any movement of the ground takes place.

Ramps and backdrops

The gradient must be correct if a drain is to be self-cleansing. There are occasions where it is difficult to achieve a constant gradient, for instance where the drain passes through steeply sloping ground or when a drain discharges into a very deep sewer. Depending on the vertical distance between the inverts of the drains or sewers to be connected a ramp or a backdrop (sometimes referred to as a *tumbling bay*) can be considered. A ramp like that shown in Fig. 9.59 may be used if the difference in the invert levels is not in excess of 680 mm. If the difference is greater, a backdrop must be constructed (see Fig. 9.60).

In cases where the invert of the drain is lower than that of the sewer — for example, where the drainage of sanitary fittings situated in basements

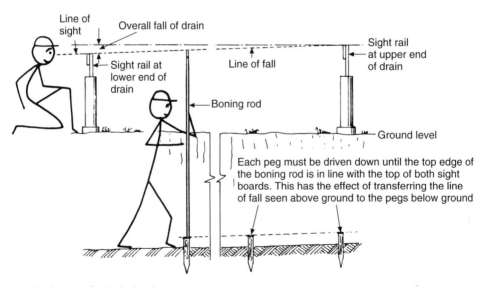

Fig. 9.58 Boning in pegs for drain levels.

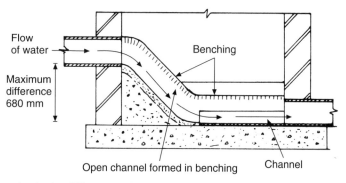

Fig. 9.59   Drain ramp, used only for differences of up to 680 mm in invert levels. Not all authorities will accept this arrangement due to possible fouling of the benching.

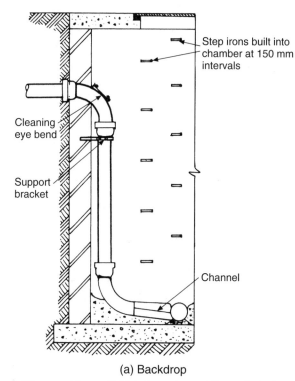

(a) Backdrop

Where cast iron or PVC drainage is used the vertical pipe may be fitted in the inspection chamber. In the case of clayware the pipe is fitted outside the chamber and is usually encased in concrete.

Fig. 9.60   Backdrop inspection chambers.

has to be considered — the use of a pump or sewage lift is essential. Due to both initial and possible maintenance costs of such equipment, drainage schemes should be designed in such a

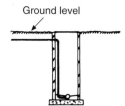

(b) High-level drain discharging into a low-level sewer

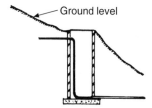

(c) Use of backdrop in steeply sloping ground

way that their use is unnecessary, certainly in all but very large buildings.

Mechanical sewage lifting equipment is usually installed by specialist contractors and is outside the scope of normal plumbing work.

Testing drains for soundness

For obvious reasons it is very important that, on completion of the work, a new drainage system is thoroughly tested. The two main methods of testing underground drains for soundness are water tests and air tests.

The building inspector will require the drain to be tested for soundness after back-filling of the trench

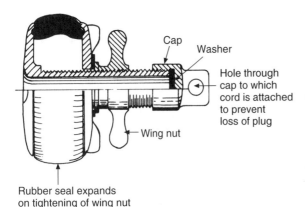

Fig. 9.61 Drain plug.

is completed. This is to ensure that any defects or damage sustained by careless back-filling will show up during the test. For this reason the contractor tests the drain to his own satisfaction both before and after back-filling. Back-filling should be done very carefully, avoiding large heavy objects such as pieces of brick or concrete, especially in the first layers covering the drain as it can easily be damaged at this stage.

Drain-testing equipment
Expanding plugs (see Fig. 9.61) are used to seal off the ends of drains under test. An alternative method of plugging, used when only one end of the drain is accessible, is to use the inflatable air bag or stopper shown in Fig. 9.62. This is floated down the drain and inflated with a pump similar to a bicycle pump when in the correct position. Great care should be taken to secure these plugs when water testing, as on completion of the test they can be swept away by the velocity of the water. When water testing lengths of drain sealed at the upper end of the trap, e.g. a gulley or WC, the air in the pipe must be removed as shown in Fig. 9.63. The only special items of equipment needed for air testing other than suitable plugs are a U-gauge (manometer) and hand bellows. This equipment has already been described in Chapter 8.

Testing procedures
The Building Regulations 1985, Schedule 1, Part H and BS 8301:1985 specify very clearly the requirements for testing underground drainage

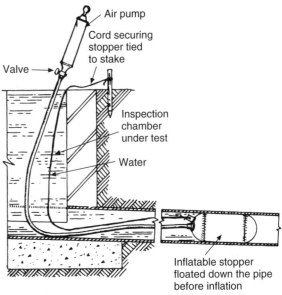

Fig. 9.62 Testing a length of drain with an inflatable stopper. This method is suitable when access can only be made to one end of the drain, e.g. last inspection chamber before sewer connection.

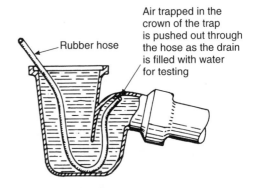

Fig 9.63 removing the air from the crown of a gulley trap when water testing drains

systems. Drains may be water or air tested, but water is usually preferred as it relates more closely to the actual working conditions. It is not so severe as an air test, since air, being a gas, can penetrate smaller apertures or cracks than a fluid. Should an air test reveal a leak, it is not easy to find, especially if it is at some point underneath the drain. It will be noted in the following text that some latitude is allowed when testing with either air

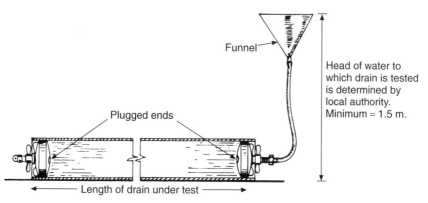

Funnel

Head of water to
which drain is tested
is determined by
local authority.
Minimum = 1.5 m.

Plugged ends

← Length of drain under test →

Fig. 9.64   Water testing of drains.

or water. If a small loss of water or air pressure is shown up during the test, it does not necessarily mean the drain is leaking. Several factors may account for small losses, such as temperature changes resulting in expansion or contraction of the water, air or pipes. This especially applies to PVC, which, having a high expansion rate, has the effect of lengthening the pipe run, thus providing more space for the testing media. In the case of water tests, trapped air and seepage into the jointing material can sometimes show a loss of water. Leaking plugs or drain stoppers should also be checked if a loss of air pressure, or water in the case of a water test, becomes apparent.

Water testing (see Fig. 9.64)   When testing drains with water a minimum head of 1.5 m is required, measured from the invert of the drain at its highest point. The drain is then filled with water. Any air in the crown of trapped gulleys should be removed after which a gulley stopper should be clamped in position. The head of water to which the drain is subjected should not exceed 4 m including the 1.5 m above the invert of the pipe, as pressure in excess of this may result in split pipes. In some cases, to avoid subjecting a long drain to excessive pressure, it may be tested in sections. A period of approximately 2 hours is allowed for the water to settle and for any absorption by the pipe or jointing materials to take place. The drain should then be topped up and left for 30 minutes during which time any loss of water will be observed. The permissible loss of water during this test period is shown in

Table 9.2   Permissible loss of water.

Nominal diameter of pipe (mm)	Permissible loss of water in litres per metre run
100	0.050
150	0.08
225	0.12
300	0.15

Table 9.2. To give an example, assume a drain has a nominal diameter of 100 mm and has a length of 20 m, the permissible loss will be

$$0.050 \times 20 = 1 \text{ litre}$$

If the water level has fallen during the 30-minute test period, and in the case of this example not more than 1 litre of water is required to bring it up to its original level, the test result is acceptable.

Air tests   Air tests are conducted in a similar way to those on discharge pipework with all open ends on the system suitably plugged and air blown or pumped in until a pressure slightly in excess of 100 mm is recorded. Allow 5 to 10 minutes for the air to stabilise, after which adjust the air pressure as necessary. During a period of 5 minutes the pressure should not fall more than 25 mm for the test to be acceptable. If any gulleys or WCs are fitted prior to testing, a test pressure of 50 mm is the maximum obtainable without blowing the water seal in the trap. The test period is the same, but in

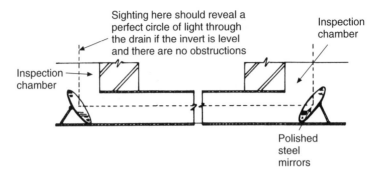

Fig. 9.65 Mirror test.

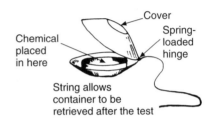

Fig. 9.66 Chemical test. The chemical in liquid or soluble tablet is placed in the container, the two halves of which are then sealed with gummed paper. The container is then flushed down the drain and will open due to the action of moisture on the gummed paper, allowing the chemical to enter the drain.

this case a permissible pressure loss of only 12 mm is allowed. The reader should note that the test pressure and period of time is slightly different from the tests conducted on discharge pipework.

Mirror and ball tests   Tests to check the alignment and general condition of the inside of the drain are sometimes conducted. They are seldom required by the local authority inspectorate but may be necessary to satisfy the client or his representative, the clerk of works, on the standard of workmanship. The mirror test illustrated in Fig. 9.65 permits the drain to be sighted through by eye to check the invert level. A clear circle of light should be seen if the drain is absolutely straight and true. This test will also show up any defect such as material protruding from a joint or deposited on the invert. A similar test may be conducted with a ball slightly smaller in diameter than the drain. When it is rolled down the drain, any obstruction will, of course, arrest its progress. Both of these tests can only be applied if access is available at both ends of the drain.

Testing existing drains   Chemical or scent tests are sometimes used to detect leaks in existing drains which should not be subjected to pressure testing. These smell or olfactory tests, as they are called, are often applied to determine whether a leaking drain is responsible for causing unpleasant odours inside premises. Chemical substances such as calcium carbide or oil of peppermint, when in contact with water, give off an unmistakable smell. Both these chemicals can be washed down

the vent pipe with water. Special containers are also available for introducing the chemical to the drain. When charged they are washed into the drain via a WC or gulley. Figure 9.66 illustrates this equipment and how it operates.

Tracing drains

It is sometimes necessary to determine whether a drain is carrying foul or surface water. Such an occasion could occur if a connection is to be made to an existing drain. Obviously, allowing foul water to discharge into a surface water drain could give rise to untreated sewage being deposited in local rivers or streams, thereby causing a nuisance and the possibility of danger to health. For this purpose a solution of fluorescein is poured into the drain via a gulley, WC or convenient inspection chamber and, when added to water, it imparts to it a bright green colour. Checks can subsequently be made at other points of inspection to determine the passage of water through the system. Any method of colouring the water may be used in the absence of fluorescein, a suitable substitute being whitening or a little coloured emulsion paint. Another chemical, potassium permanganate, which when added to water colours it purple enabling it to easily be identified.

Maintenance of underground drainage systems

In a well-installed self-cleansing system of drains there is little that can go seriously wrong — when it does it is usually due to misuse.

In the case of public buildings such as hospitals and large commercial and industrial buildings where the possibility of a breakdown could be serious, consideration should be given to a planned maintenance scheme to ensure regular inspection of the drainage components. Each building complex will require its own individual schedule which is usually drawn up by the person in charge of maintenance. Some of the more general items are listed as follows:

(a) All gullies should be checked to ensure the base of the trap is clear of debris prior to washing out with a suitable detergent. Where applicable, ensure all gratings are clean — remove and clean buckets in yard gulleys.

(b) Where an interceptor is fitted, check the trap is not blocked with debris and make sure the stopper in the rodding eye is sealed. Check that the flap in the fresh air inlet is unbroken and in working order.

(c) Rust pockets may be found in old public buildings and these must be cleaned if necessary.

(d) Brick and concrete inspection chambers are normally parged internally with cement and sand mortar. This should be checked for cracks and lack of adhesion, as broken pieces of parging could fall into the channel causing a blockage. In deep chambers inspect the step irons for deterioration and replace if necessary. Any double-sealed chamber covers inside a building should be cleaned and repacked with grease.

(e) To inspect the internal surfaces of drain pipes, the mirror test shown in Fig. 9.65 is a simple method but it can only be used on straight lengths of pipe with adequate access. A more effective method is the use of closed circuit television where a very small camera is drawn through the pipe recording pictorially its condition. Companies with the necessary expertise and equipment are normally employed to carry out work of this nature.

Decommissioning

Where possible, any drains no longer in use should be removed. This may not be feasible for those underground, in which case the ends of the pipes must be securely stopped and filled as far as possible with lime slurry, which provides some degree of disinfection.

Further reading

Building Regulations Part H
BS 8301 Building drainage
BS EN 752 Drain & Sewer Systems outside Buildings Part 1 generalities & definitions; Part 2 performance requirements; Part 3 planning; Part 4 hydraulic design environmental considerations.
BS EN 877 Ductile iron pipes and fittings
BS 4660 Specifications for UPVC. Drain pipes and fittings for gravity drains and sewers.
BS 6087 Specifications for flexible joints for cast iron pipes and fittings.

Drainage Systems and Materials

Cast iron systems
Glynwed Foundries, Sinclair Works, Ketley, Telford, Shropshire, TF1 4AD, Tel. 01952 641414.

Plastic systems
Wavin Building Products Ltd, Parsonage Way, Chippenham, Wiltshire, SN15 5PN, Tel. 01249 654121.
Marley Extrusions, Lenham, Maidstone, Kent, ME17 2DE, Tel. 01622 858888.

Grease disposal
Wade International UK Ltd, 20 Broton Road,
Halstead, Essex, CO9 1HE, Tel. 01787 475151.

Building Research Establishment, Publications CRC
Ltd, 151 Rosebury Avenue, London, EC1R
4GB, Tel. 020 7505 6606.
BRE Digest 292 Access to domestic underground
drainage systems.
BRE Digest 365 Soakaways.

Self-testing questions

1. State the procedures that must be complied with
   before any new drains are laid, and before
   extensions are made to those in existence.
2. Make a sketch showing the invert and soffit of
   an underground drain pipe.
3. List the principal materials from which
   underground drain pipes and fittings are made.
4. State the maximum angular movement either
   side of the horizontal position for flexible joints.
5. (a) State the maximum recommended stacking
       height for PVC drain pipes.
   (b) List the advantages of PVC drainage
       systems.
   (c) Describe in detail the procedure for making
       an O-ring joint on PVC drainage systems.
6. (a) State the minimum depth of seal in a
       gulley or intercepting trap and describe the
       purpose of each.
   (b) In what circumstances would trapless
       gulleys be used?
7. State the two important details that should be
   included on any order for cast iron drain bends.

8. State the purpose of flexible bedding and
   outline its main features.
9. (a) Identify the difference between combined
       and separate systems of drainage.
   (b) Explain why the separate system now more
       common in urban areas?
10. (a) State the circumstances in which a
        soakaway might be employed.
    (b) Describe a suitable site for a soakaway
        attached to a private house.
11. Give the reasons for ventilating drainage
    systems, and explain how this ventilation takes
    place.
12. List the building regulations' requirements for
    means of access to drains.
13. State the minimum permissible diameter for a
    foul water drain.
14. (a) Define the term 'self-cleansing velocity'
        and identify its relationship with the fall
        of a drain and the volume of water it
        carries.
    (b) State the recommended fall for a branch
        drain carrying intermittent flows.
    (c) What might be the effect of an excessive
        fall on the contents of a foul water
        drain?
15. (a) Make a sketch showing how air is removed
        from the crown of a gulley trap when
        testing drains.
    (b) Explain why it is sometimes necessary
        to trace drains, and state the methods
        used.
16. State the procedures for water testing an
    underground drain to the requirements of
    BS: 8301.

10 Sheet Lead Roof Weatherings

After completing this chapter the reader should be able to:

1. Describe and illustrate with sketches how lead sheet gutters are fabricated.
2. Describe the various methods of discharging water from lead gutters into rainwater pipes.
3. Describe the techniques of forming solid and hollow rolls and joints for flat and pitched roofs covered with sheet lead.
4. Select and illustrate with sketches the methods of weathering dormer windows.
5. Mark out sheet lead for forming all types of welded joint.
6. Select suitable welding techniques for fabricating sheet lead weatherings.

Introduction

Metal roof weatherings have been used throughout the ages both for covering whole roofs and for forming the components necessary to make a watertight joint between building materials such as brick, stone, timber, tiles or slates. This subject is very extensive due not only to the number of materials that can be used, but also to the varying techniques employed with each one. For many years the methods of weathering with sheet lead remained unchanged, but relatively recently many improvements have been made enabling thinner sheet to be used than hitherto. A greater emphasis is now given to lead welding, which when employed by skilled craftsmen can result in considerable saving of time and materials. The basic joints, however, remain the same and it is perhaps a tribute to the plumbers of long ago that the skills evolved by them are still in use. Much of this chapter deals with traditional methods that plumbers may encounter in their everyday work.

The skills and knowledge in working with lead sheet are still necessary, especially for those engaged in jobbing or renovation work. This is recognised by the lead body in plumbing, the British Plumbing Employers Confederation and an NVQ unit dealing with lead flashings is required for the completion of level 2. For those requiring a more comprehensive range of skills an NVQ level 3 module is available at some colleges or training centres. The recommendations given in this chapter are based on those quoted in the Sheet Lead Manual volumes 1–3 published by the Lead Sheet Association (LSA). This body is the leading authority on good practice using lead sheet in buildings.

The text here relates to larger areas of lead sheet than those used for flashings. It is therefore important that the following facts are understood relating to the free movement of lead when it expands and contracts due to temperature variation. Figure 10.1 shows the basic principles that apply

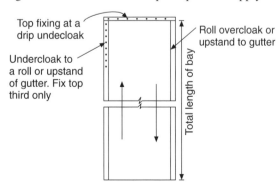

Fig. 10.1 Basic principles for fixing large areas of lead, showing how it can expand and contract freely.

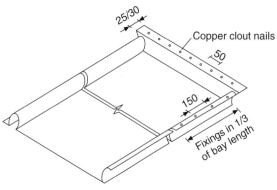

25/30

Copper clout nails

50

150

Fixings in 1/3 of bay length

(a) Roofs of up to 3° pitch

The undercloak to a drip is shown. Upstands at abutments should have a minimum height of 110 mm.

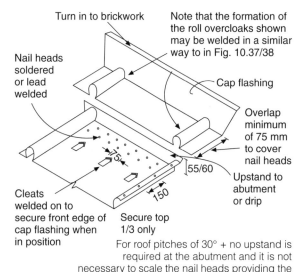

Turn in to brickwork

Note that the formation of the roll overcloaks shown may be welded in a similar way to in Fig. 10.37/38

Nail heads soldered or lead welded

Cap flashing

Overlap minimum of 75 mm to cover nail heads

Upstand to abutment or drip

55/60

75

150

Cleats welded on to secure front edge of cap flashing when in position

Secure top 1/3 only

For roof pitches of 30° + no upstand is required at the abutment and it is not necessary to scale the nail heads providing the overlap gives a cover of 75 mm vertical height.

(b) Method of fixing bays on roof pitches of +3 to 10°

An abutment detail is shown but the same fixing technique applies to drips on low pitched roofs.

Fig. 10.2   Fixing bays.

when fixing large areas of sheet such as flat or pitched roofs, gutters and dormer tops. Figure 10.2(a) and (b) illustrates some of the practical applications of this principle. Note also that the roof pitch has an influence on the methods of fixing employed and the thickness of lead used. This is especially important when weathering vertical or steeply pitched areas, and smaller pieces of lead than those used on low pitches should be considered to avoid creep. Tables 10.1 and 10.2 show that generally the thicker the lead used the greater will be the area

Table 10.1   Maximum superficial areas recommended for bays in lead gutters

BS EN 12588 code no	Length between drips (mm)	Overall widths (mm)	Drip heights (mm)
4	1500	750	55
5	2000	800	55
6	2250	850	55
7	2500	900	60
8	3000	1000	60

Note: 1. The minimum fall in all gutters is 1:80.
2. Increase in the drip height for code 7/8 accommodates the increased lead thickness.

Table 10.2   LSA recommendations for bay sizes on roofs pitch at 10° or less

BS EN 12588 code no	Spacing of joints with the fall (rolls) (mm)	Spacing of joints across the fall (drips) (mm)
4	500	1500
5	600	2000
6	675	2250
7	675	2500
8	750	3000

Note: the spacing for drips is the same as for gutters.

that can be covered in one piece. When costing a job the extra cost of thicker lead must be weighed against the greater number of joints required when using thinner sheet. This is especially important when considering valley gutters, as the number of drips used increases the width of the sole of the gutter at the upper end. This will require a larger area of lead and the possibility of centre roll, thus increasing labour costs. See Fig. 10.5.

Contact with other materials

Lead can be used with copper, stainless steel, aluminium and zinc or galvanised products without any significant electrolytic corrosion, with one exception. Where lead and aluminium are used together in a marine environment, the oxide on the surface of the lead reacts with the sodium chloride in a salt-laden atmosphere. This creates a caustic run-off which attacks the aluminium, especially where any water is trapped in crevices. For this reason aluminium or its alloys should not be used in a marine environment unless protected by a suitable paint.

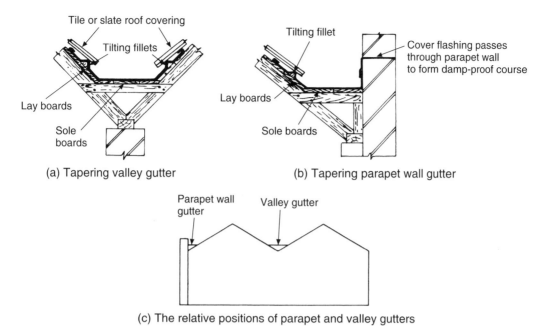

(a) Tapering valley gutter

(b) Tapering parapet wall gutter

(c) The relative positions of parapet and valley gutters

Fig. 10.3 Sections through tapering valley and parapet wall gutters.

Contact with timber

The dilute solutions of organic acids found in hardwood, especially oak, can cause very slow corrosion of the lead. The use of underlays such as inodorus felt or building paper largely overcomes this problem. Any preservative or fire retardant solution with which the timber is treated is unlikely to cause corrosion, providing it is dry before the lead is laid. In wet weather the run-off from new cedar or oak shingles forms a dilute acid solution which will cause corrosion. The lead should be painted with a bituminous solution for a few years until the free acid has leached out of the shingles.

Contact with mortars and concrete

Where lead is in contact with new cement or concrete it should be treated with bituminous paint, typical examples being where lead is used as a damp-proof course or for cladding concrete walls. Where lead is turned into brickwork joints, e.g. for flashings, no protection is necessary as the free lime in the cement, which causes the corrosion, is carbonised very quickly.

Metal lining of gutters constructed of timber

Gutters formed with timber and lined with metal are usually referred to as *tapering valley* or *parapet wall*

gutters, sections of which and the positions they occupy on a roof are shown in Fig. 10.3. The term *tapering* is used to denote the difference between these and the valley gutters dealt with in Book one, Figs 11.50, 11.52 and 11.53. Both tapering valley and parapet wall gutters are usually found in old buildings constructed before the introduction of purpose-made profiled gutters. A variation may occur where two flat roofs drain into what is called a *box gutter* (see Fig. 10.4(a)).

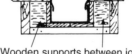

Wooden supports between joists
supporting sole boards and sides of gutter

(a) Section of a box gutter between two flat roofs

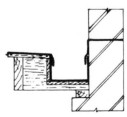

(b) Section of a box gutter between a flat roof and a parapet wall

Fig. 10.4 Box gutters.

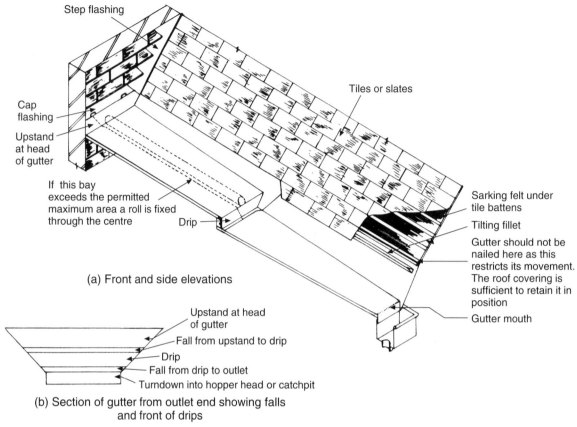

Step flashing

Cap flashing

Upstand at head of gutter

Tiles or slates

If this bay exceeds the permitted maximum area a roll is fixed through the centre

Drip

Sarking felt under tile battens

Tilting fillet

Gutter should not be nailed here as this restricts its movement. The roof covering is sufficient to retain it in position

Gutter mouth

(a) Front and side elevations

Upstand at head of gutter
Fall from upstand to drip
Drip
Fall from drip to outlet
Turndown into hopper head or catchpit

(b) Section of gutter from outlet end showing falls and front of drips

It becomes narrower towards the lower end hence the name 'tapering'. A box gutter retains the same width throughout its length.

Fig. 10.5   Section through a tapering valley gutter.

The plumber must always advise the carpenter responsible for the wooden structure about the positioning of any drips and falls required. Assuming the plumber is working with lead sheet which is 2.4 m wide, the most economic way to cut it will be across its width, so the distance between the drips must be less than this to allow for the upstand and turn down. If the drips are too far apart the cutting of the lead will be uneconomical and its movement may be restricted, possibly causing it to crack at a later date. It cannot be stressed enough that most of the failures experienced with sheet metal weathering of gutters or flat roofs are due to insufficient freedom of movement caused by fixing the sheets too rigidly or covering excessively large areas with one piece of material. The fall of the

gutter is also important as it is essential to remove the water as quickly as possible to avoid overflowing.

Falls in gutters and flat roofs should never be less than 1 in 80, although pitch a of $2\frac{1}{2}°$–3° is recommended for roofs to avoid ponding. The depth of drips in gutters with a capillary groove is 40 mm, but those are not recommended, as instances have been found where the groove has become choked with debris causing the ingress of water into the building. The minimum recommended depth of drips in gutters is 50 mm. Figure 10.5(a) illustrates a typical tapering valley lead-lined gutter laid between two sloping roofs. In the case of parapet wall gutters only one side will taper. A study of the front elevation in Fig. 10.5(a) and (b) will indicate that the taper of such gutters is due to its fall and any drips in its length. In long gutters it is sometimes necessary to split the upper bays by means of a roll to avoid exceeding the maximum recommended total area of the material. This does not apply to box gutters as they do not increase in width due to the fact that the upstands on both sides are at 90° to base. In all gutters most expansion will take place longitudinally and it is important that the upstands are not rigidly nailed. Any cleats used must allow freedom of movement throughout their length. When the tiles or slates are laid their weight is normally quite sufficient to keep the edge of the gutter in position.

Gutter outlets

Two types of outlet are used to allow water from the gutter to be discharged into the drainage system. The chute method is used when the gutter discharges through a parapet wall into the hopper head as shown in Fig. 10.6. The other method is known as a catchpit, shown in Fig. 10.7, and is generally used where the rainwater pipe is fitted inside the building. The exception to this is where the outlet from the catchpit is offset through the wall directly into the rainwater pipe. The advantage of using a catchpit is that as the water falls into it from the gutter its velocity is increased enabling it to be discharged more rapidly. An overflow is desirable with this type of outlet, as if the pipe

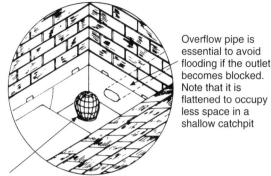

Overflow pipe is essential to avoid flooding if the outlet becomes blocked. Note that it is flattened to occupy less space in a shallow catchpit

Outlet and overflow pipes soldered or lead welded in catchpit

Fig. 10.7 Catchpit. Generally used to remove water from roofs via an internal rainwater pipe.

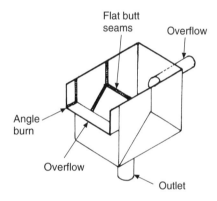

Flat butt seams

Overflow

Angle burn

Overflow

Outlet

Fig. 10.8 Lead-welded catchpit.

became blocked, the building could be subjected to flooding. A catchpit made of lead sheet is shown in Fig. 10.8 and lead welded joints have been used in its construction. They may be bossed, but this is time consuming and the application of lead welding techniques would be a more practical and economic proposition.

Covering large flat areas with lead sheet

There are two main systems used for weathering large areas. Wood-cored rolls and drips are used on pitches of up to 10°–15°. On steeper pitches lap joints and hollow rolls are usually employed as this method is less labour intensive. The lower limit of fall recommended for flat roofs is $2\frac{1}{2}$°. Although low-pitched roofs are often called flats, a fall is

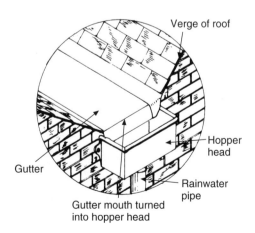

Verge of roof

Hopper head

Gutter

Rainwater pipe

Gutter mouth turned into hopper head

Fig. 10.6 Rainwater chute. Used to discharge rainwater from a gutter directly into a hopper head.

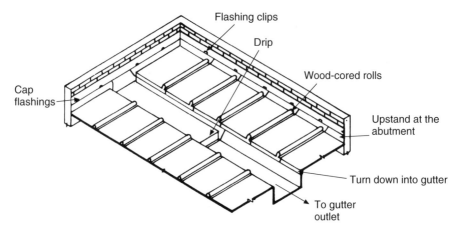

Fig. 10.9 Typical layout of a simple flat roof discharging into a box gutter.

essential to enable the water falling on them to run off as quickly as possible; insufficient fall will result in 'ponding'. The amount of fall on a roof also influences the type of joint that may be made laterally across the flow of water. For those with a very low pitch drips will be necessary, but as the roof angle increases a lap joint may be used. The amount of fall on a flat roof is determined by what are called *furring pieces* which are cut and fixed by the carpenters across the joists before the decking is laid. The specifications recommended in Book One should be adopted, and the preparation of the surface, including laying the felt, should be carried out as described there prior to fixing any metal coverings.

As with gutters, the thickness of the lead used for covering large flat areas will influence the maximum superficial area that should be laid in one piece. The type of joint used with the fall on roof pitches of up to 15° is almost invariably the wood-cored roll. Figure 10.9 illustrates a part of a lead-covered flat to enable the reader to visualise the various details to which reference is made and their relative positions on the finished job. Figure 10.10 shows how to set out one 'bay' of sheet lead with approximate allowances for the joints. Table 10.2 gives the recommended distances between the joints on the roof. Do not forget the allowances for the joints which have to be added to the measurements given in the table.

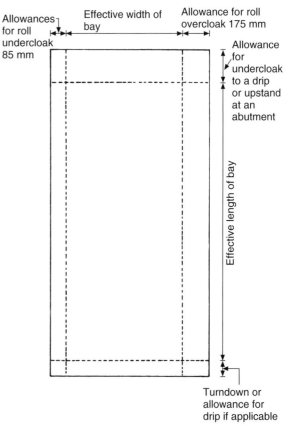

Fig. 10.10 Setting out a bay for a lead flat showing the necessary allowances. Note that the allowances for the undercloak and overcloak will vary slightly according to the size of the wood roll. Refer to Table 10.2 for overall width and length.

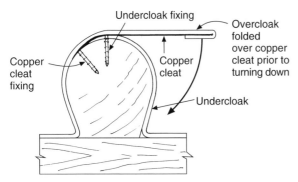

Copper
cleat
fixing

Copper
cleat

Undercloak fixing

Overcloak
folded
over copper
cleat prior to
turning down

Undercloak

Fig. 10.11   Alternative wood roll finish without splashlap.

Wood rolls

The traditional method of forming lead sheet over wood rolls with splash laps is described in Book 1 Chapter 11. The purpose is to give the edge of an overcloak a greater degree of rigidity, preventing high winds lifting the lead. Rolls and splash laps are still widely used but the roll shown in Fig. 10.11 is an alternative when the appearance of the splash lap is undesirable, e.g. on or near vertical cladding. When using lead codes 4,5,6 the lead is folded over copper cleats fixed at intervals of 450 mm which resist the tendency of the edge to lift in high winds. When using codes 7,8 in moderately sheltered areas the copper cleats are unnecessary, as the thickness of the lead is considered sufficiently stiff to resist lifting. It is important that the fixings for the copper cleats do not penetrate the undercloak, as this will restrict its free movement.

Forming wood-cored rolls at abutments

Rolls can be worked by the bossing process or by lead welding which is dealt with later in this chapter. Figure 10.12 shows the stages of bossing roll overcloaks at an abutment. This is a more difficult procedure than working undercloaks as the lead has to be worked all the way round the wooden roll. The formation of an undercloak is not a great deal more difficult than that of an external corner, and for this reason details are not included. During the final stages of working the overcloak the lead must be firmly held down with a wooden block as shown or it will lift and a hollow will form under the lead. This will cause the lead to be stretched,

sometimes resulting in the formation of a hole when it is finally dressed to shape around the roll. The plumber's mate usually stands on the timber block during the whole operation, holding down the top of the partly formed roll with a piece of batten. This 'holding down' is essential and must be done during the final stages of driving the lead into the corner. This is shown in stage 4 and it must be carefully carried out to avoid splitting the lead. A lot of practice is required to produce an acceptable job. Many plumbers have specially shaped chase wedges for 'driving in', as this is difficult to accomplish with the ones normally supplied due to their rounded edge. The use of a well greased drip plate is essential during this operation to enable the lead to 'slide' easily over the undercloak when it is finally driven home.

The finished job should appear as shown in stage 5 with the lead driven home into the corner and trimmed prior to dressing out all the tool marks.

Roll ends at the eaves

The stages of bossing the roll end are shown in Fig. 10.13. Some plumbers use a short-handled dummy over which the lead is bossed out. During this operation it is essential that the lead bay is not drawn forward leaving a gap at the abutment and this must be watched continually. As with the formation of the overcloak at the abutment, it is helpful to have a mate standing on a suitable timber in the bay to prevent this happening. Figure 10.14 illustrates the detail of a bossed roll intersection with a drip on a low-pitched roof. The rolls may be 'inline', or 'staggered' as shown in the inset. This is one of the most difficult bossing operations to do as lead has to be gained to successfully work in the corners shown at 'B'. For this reason the roll inline method is preferred as the lead bossed out of the roll end can be worked into areas 'B'.

Hollow roll work

Hollow rolls are used on roofs covered with sheet lead having pitches upward of approximately 20°. These joints are easily damaged and must not be used in situations where foot traffic is anticipated. Joints across the flow of water are lapped, drips or welts being unnecessary on pitched roofs of this

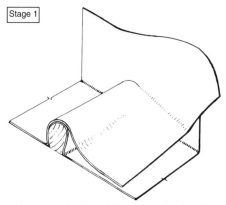

Stage 1

The lead is bossed to the height of the roll using the same technique as for an external corner. It must be worked to a little less than 90° to allow for the taper in the wood core.

Surplus lead from corner worked over the top of the roll. This will be needed when the roll is finally driven home.

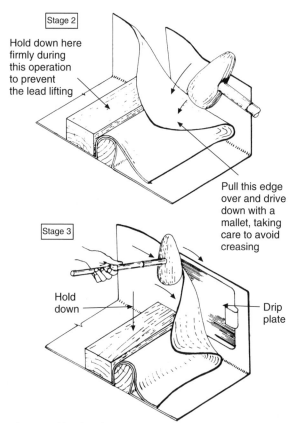

Stage 2

Hold down here firmly during this operation to prevent the lead lifting

Pull this edge over and drive down with a mallet, taking care to avoid creasing

Stage 3

Hold down

Drip plate

Insert a drip plate between the undercloak and overcloak and drive the lead along with a mallet as shown. It may be necessary to repeat this, and the operation shown in stage 2, two or three times prior to driving in as shown in stage 4.

Fig. 10.12 Bossed roll ends at abutments.

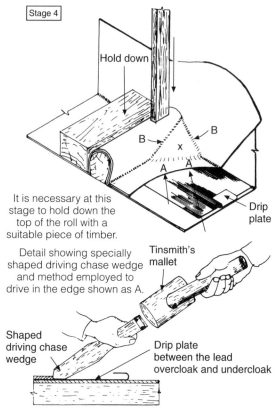

Stage 4

Hold down

B

B

x

A A

Drip plate

It is necessary at this stage to hold down the top of the roll with a suitable piece of timber.

Detail showing specially shaped driving chase wedge and method employed to drive in the edge shown as A.

Tinsmith's mallet

Shaped driving chase wedge

Drip plate between the lead overcloak and undercloak

The final operation requires the triangle x to be worked into the corner with great care, by driving in the edge shown as A, triangle x is gradually diminished, chasing in at arrows B simultaneously. It is essential that the drip plate is well oiled or greased to enable the lead to slide over it. The lead must be worked all the way and any attempt to stretch it in the final stages will cause it to split.

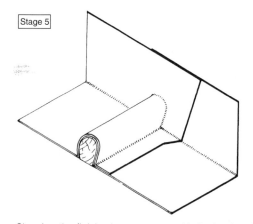

Stage 5

Showing the finished appearance with the lead worked neatly in position.

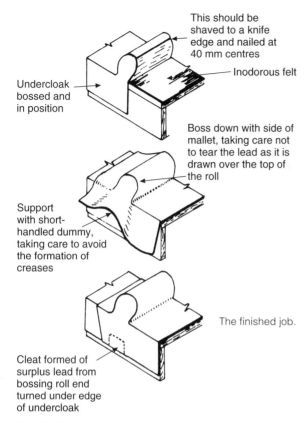

This should be shaved to a knife edge and nailed at 40 mm centres

Inodorous felt

Underclock bossed and in position

Boss down with side of mallet, taking care not to tear the lead as it is drawn over the top of the roll

Support with short-handled dummy, taking care to avoid the formation of creases

The finished job.

Cleat formed of surplus lead from bossing roll end turned under edge of undercloak

Fig. 10.13 Forming bossed roll ends.

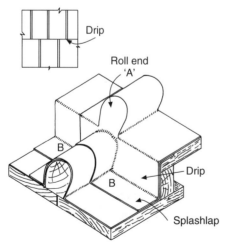

Drip

Roll end 'A'

B

B

Drip

Splashlap

The rolls are 'in line' as distinct from those shown inset as the lead gained from bossing down roll end 'A' can be worked into areas 'B' where it is necessary to gain lead.

Fig. 10.14 Detail of bossed roll intersection on a low-pitched roof.

angle. Figure 10.15 illustrates the general method of weathering using this technique. The top edges of each sheet must be well secured by a double row of nailing to prevent it slipping or pulling away from its fixings. As this method of weathering is almost invariably confined to double pitched sloping roofs, it is seldom necessary to form a hollow roll to an abutment. Where necessary, however, they can be bossed over a short length of wooden roll, using it as a former, or they can be lead welded using similar techniques to those used for wood-cored rolls. Details of these operations are not shown, but can be obtained from the Lead Sheet Association.

Hollow roll finish at eaves and ridge
One method employed to turn the roll over the eaves is to use a 32 mm bending spring which is

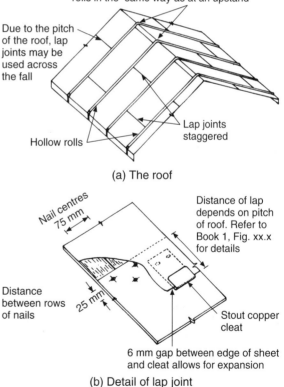

Hollow rolls over the ridge on low pitches are turned over a 32 mm bending spring. If the pitch is steep it may be necessary to fix a wood roll along the ridge and finish the hollow rolls in the same way as at an upstand

Due to the pitch of the roof, lap joints may be used across the fall

Lap joints staggered

Hollow rolls

(a) The roof

Nail centres 75 mm

Distance of lap depends on pitch of roof. Refer to Book 1, Fig. xx.x for details

Distance between rows of nails

25 mm

Stout copper cleat

6 mm gap between edge of sheet and cleat allows for expansion

(b) Detail of lap joint

Fig. 10.15 Double pitched lead covered roof using hollow roll jointing.

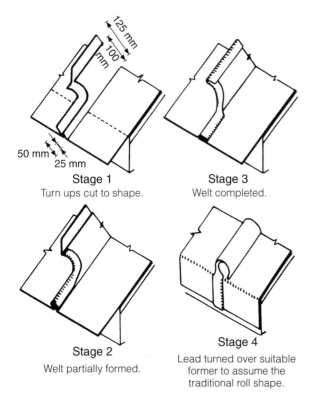

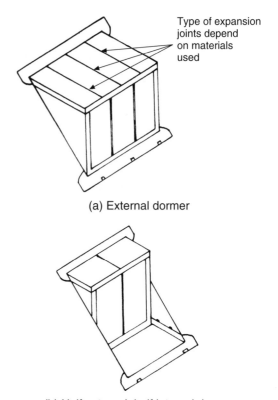

Fig. 10.16 Stages in forming hollow roll ends using the welted method.

inserted into the roll and supports it as it is turned. The same method is used to turn it over the ridge where applicable, but if an angle of more than 30° is required in either case it will be difficult to remove the spring. An alternative method of finishing the roll at the eaves, by means of a welt, is illustrated in Fig. 10.16 and is both simple and effective. Another alternative is to weld 'in' a cover to the end of the roll.

Dormer windows

These are used to give natural light to rooms built into a roofspace as in chalets or bunaglows. There are three main types of dormer: external, internal and partially internal. The external type is the most common as it generally affords more light and often allows for the enlargement of the room in which it is situated. The reasons for using the other types of dormers are usually aesthetic as they may enhance

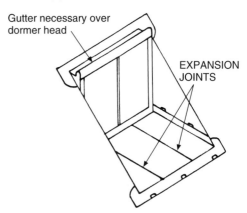

(c) Internal dormer. Not favoured in modern buildings except those with mansard roofs. Tends to restrict daylight in rooms

Fig. 10.17 Types of dormer window.

the appearance of the building. For example, partially internal dormers are often used with mansard roofs and can give a very pleasing

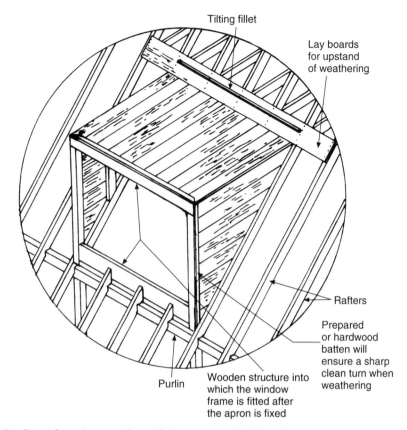

Tilting fillet

Lay boards
for upstand
of weathering

Rafters

Prepared
or hardwood
batten will
ensure a sharp
clean turn when
weathering

Wooden structure into
which the window
frame is fitted after
the apron is fixed

Purlin

Fig. 10.18 Wooden frame for a dormer prior to fixing apron and window frame.

effect. The three types of dormer are shown in Fig. 10.17.

As external dormers are the most commonly encountered, it is this type that will be discussed in detail although the basic principles of weathering all types of dormers are very similar. Tile hanging (roof tiles fixed vertically) or shiplap boarding is sometimes used to weather the cheeks (the triangular sides of the dormer) and in such cases the roof tiles are weathered to the dormer cheek by means of soakers or cover flashing.

Dormer weatherings
It is an unfortunate fact that some builders, to save expense, often use the window frame as a structural part of the dormer. This usually involves nailing the frame through the apron, allowing water to penetrate the roof. The implications of this are not

always immediately apparent, but in the long term it can result in serious timber defects which are expensive to correct. It is essential that a timber frame forming the dormer is first constructed by the carpenter, into which the window frame is fitted after the apron has been fixed. Figure 10.18 illustrates the wooden structure forming the dormer prior to fixing any weathering.

The bottom of the frame must be weathered first by means of the dormer apron. Figure 10.19 illustrates both the timber framework of a dormer and a section through the apron showing how the sill of the window frame should be fixed to avoid nailing through the apron. An alternative method that can be used with a sill having no groove is to use metal cleats screwed to the window frame and the dormer trimmer. They should be made of galvanised steel, but short lengths of 25 mm copper

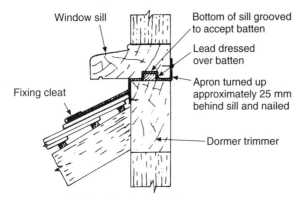

(a) Section through dormer apron

Shows how the window sill should be secured to the dormer frame without resorting to nailing through the weathering.

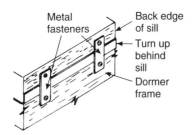

(b) An alternative method of securing the window frame to the dormer frame avoiding the necessity of grooving the underside of the sill

Fig. 10.19 Dormer sill details.

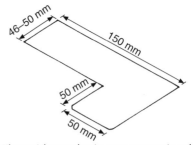

Setting out (approximate measurements only)

(a) Fixing the front edge of dormer apron: setting out

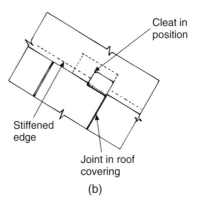

(b)

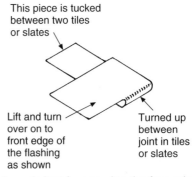

(c) L-shaped cleat for securing the free edge of aprons

Fig. 10.20 L-shaped cleat for securing the free edges of aprons.

tube flattened to form a cleat provide a useful substitute.

It is important to secure aprons for two reasons: (a) to prevent creep and (b) to prevent high winds lifting the free edge. The fixing cleat nailed to the wooden frame or turned over a tile batten will prevent creep, and the L-shaped type shown in Fig. 10.20 secures the front edge. One end of the dormer apron is shown in Fig. 10.21 and it will be seen that it not only weathers the underside of the wooden sill but also the corner post. This is important as any seepage of water into the building at this point, although it may not show, will cause rapid deterioration of the wooden structure.

Dormer cheeks

The cheeks on small dormers may be fixed in one piece, providing the maximum superficial area does not exceed approximately 0.5 m². Where larger dormers are required the cheeks should be fixed within the limitations of those shown in Fig. 10.22. For larger areas a greater number of welts and laps will obviously be necessary. The method of fixing

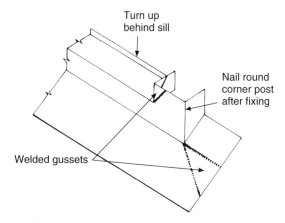

Turn up
behind sill

Nail round
corner post
after fixing

Welded gussets

Fig. 10.21 Apron detail round corner post.

the top edge of the cheek is shown in Fig. 10.23 where copper or stainless steel nails are used as prescribed. Three method of securing the lower edges of any form of lead cladding including dormer cheeks are shown in Fig. 10.24. However, the most common method employed for dormers is to form an 'S' cleat, tucking one end behind the soakers, the other turned to secure the free edge of the cheek. Figure 10.25 illustrates a method of securing large vertical (or near vertical) sheets, including dormer cheeks, by means of a concealed copper cleat worked into the welt. They should be spaced at 450 mm intervals. Rigid fixings in the centre of the sheet, often called intermediate fixings, are avoided where possible due to the restriction

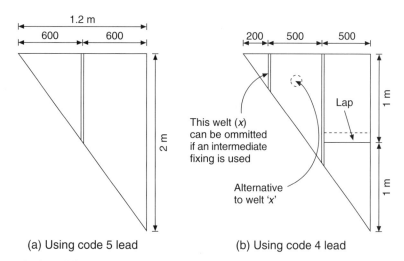

1.2 m

600   600

2 m

(a) Using code 5 lead

200   500   500

Lap

1 m

This welt (x) can be ommitted if an intermediate fixing is used

Alternative to welt 'x'

1 m

(b) Using code 4 lead

Fig. 10.22 Dormer cheek cladding: recommended sizes.

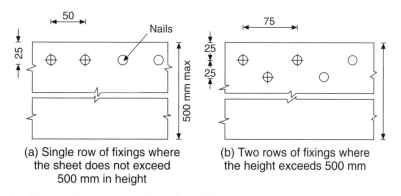

50

Nails

25

500 mm max

(a) Single row of fixings where the sheet does not exceed 500 mm in height

75

25

25

(b) Two rows of fixings where the height exceeds 500 mm

Fig. 10.23 Top or head fixings for dormer cheeks and cladding.

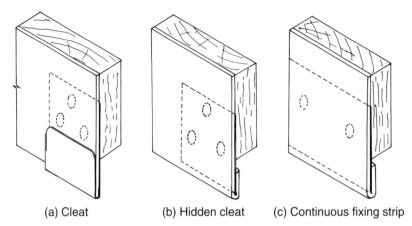

(a) Cleat (b) Hidden cleat (c) Continuous fixing strip

Fig. 10.24 Methods of clipping the bottom edge of lead weatherings.

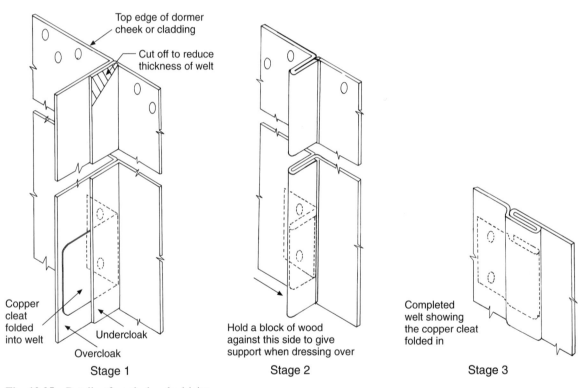

Fig. 10.25 Details of vertical welted joints.

on the movement of the lead. The traditional solder dot is typical of this undesirable method of fixing. The method shown in Fig. 10.26 is far more satisfactory and providing the screw is not too tight, the elongated hole in the lead under the washer permits sufficient freedom of movement.

The sides of the cheeks are turned round the dormer frame so that the joint between this and the window frame can be weathered. The lead can then be nailed to the frame and finished by covering with timber moulding or allowing enough material to turn back on to itself, concealing the nails as shown

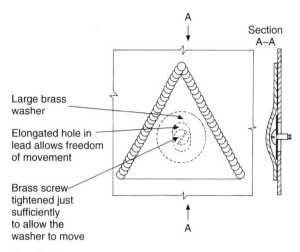

Large brass washer

Elongated hole in lead allows freedom of movement

Brass screw tightened just sufficiently to allow the washer to move

Section A–A

Fig. 10.26 Fixing a lead sheet using a brass screw.

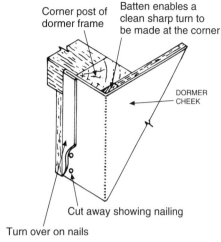

Corner post of dormer frame

Batten enables a clean sharp turn to be made at the corner

DORMER CHEEK

Cut away showing nailing

Turn over on nails

(a) Suitable for small dormers only

in Fig. 10.27. An alternative method using a welt to secure the facing lead to the cheek is preferable (Fig. 10.27(b)). Although the welt is shown on the front it is often more convenient to turn it back onto the cheek. The wooden moulding shown is not used where the lead is turned completely round the post.

The right-angled triangle The reader will see that the cheeks of a dormer have the form of a right-angled triangle, and it is useful to be able to calculate the length of the slope C (called the *hypotenuse*) between the two known lengths of the right angle A–B (see Fig. 10.28).

The following will show how measurement C can be determined. If a number is multiplied by itself it is said to be squared, e.g. $4 \times 4 = 16$. Another way of writing this is $4^2 = 16$. The number 4 is said to be the square root of 16. The square root of number 36 is 6, and in this case the square root of the number is said to have been extracted. It can be indicated thus:

$$6 = \sqrt{36}.$$

This symbol over a number indicates that its square root must be found. If the foregoing is understood it is not a big step to determine the length of C in the right-angled triangle using the following formula:

$$C = \sqrt{A^2 + B^2}.$$

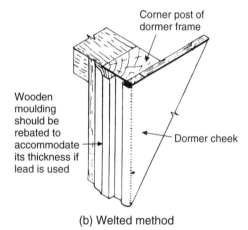

Corner post of dormer frame

Wooden moulding should be rebated to accommodate its thickness if lead is used

Dormer cheek

(b) Welted method

If copper or aluminium weatherings are used this corner should be turned using a welt as shown.

Fig. 10.27 Methods of finishing the cheek to the corner post.

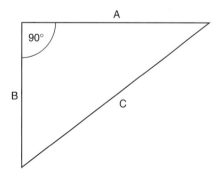

A

90°

B

C

Fig. 10.28 The right-angled triangle.

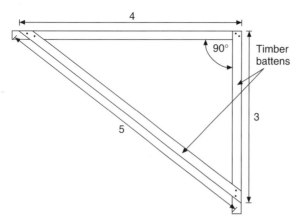

Fig. 10.29 Construction of a site set square.

If the alphabetical symbols are given a numerical value of 3 and 4 respectively, the equation will become:

$$C = \sqrt{3^2 + 4^2} \, .$$

The numerical value of C can now be ascertained by first squaring 3 and 4 thus:

$$C = \sqrt{9 + 16}$$
$$= \sqrt{25} \, .$$

All that remains is to extract the square root of 25, which can be seen at a glance to be 5; thus C, the unknown side of the original triangle, will be 5. It is a fact that when the ratio of the measurements of a right-angled triangle are 3, 4, 5 they form a 90° angle, often called a *perfect square*. This knowledge is very useful as it enables one to remember how to determine the hypotenuse of a right-angled triangle, and also how to set out 90° angles. If a large square is required on site it can be made with three pieces of batten as shown in Fig. 10.29. This principle is often called the *3, 4, 5 rule* in the building industry and is known to have been used by builders in ancient civilisations. It must be said that in practice the ratios of a right-angled triangle are seldom 3, 4, 5. These numbers have been used because their square roots can be seen at a glance, more difficult examples can easily be ascertained using a calculator.

Dormer tops

Figure 10.30 shows one side of a small dormer with all the components so far discussed in position. It will be seen that the top of the dormer is similar

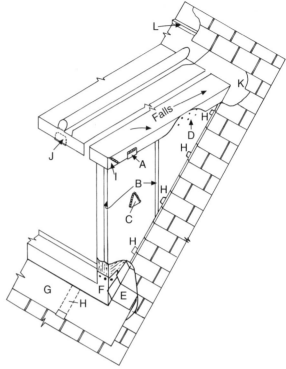

Key to details
A Cleat welded to turn
   down the methods
   shown in Fig. 10.24
   are an alternative
B Welts
C Intermediate fixing
D Head fixings
E Soakers
F Apron fixing detail
G Apron
H Cleats, cheeks and apron
I   Corner welded or bossed
J Surplus lead from bossed
   overcloak turned under
   the undercloak
K Roof coverings cut away
   allowing water to clear
L Tilting fillet

Fig. 10.30 Small dormer window showing the main details. The falls shown are suitable only if a single roll is provided.

to a small flat roof or canopy and the procedure for weathering is exactly the same. Whether the dormer top falls to the front or back depends on several circumstances. If the dormer cheeks are finished with tile hanging or shiplap boards, the usual procedure is to provide a fascia board and gutter as for the eaves of a building, as shown in Fig. 10.31. In such cases it is usual to arrange for the dormer top to fall towards the front so that water can be collected in a normal eaves gutter. A roll or batten is often fixed to the edge of the dormer top to avoid the necessity of fixing gutters all round. In some buildings the use of eaves gutters

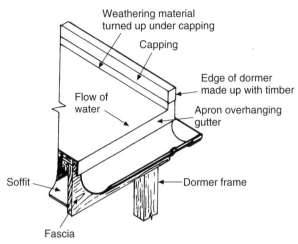

Fig. 10.31 Finishing the verge of a dormer with tile-hung or shiplap boarded cheeks.

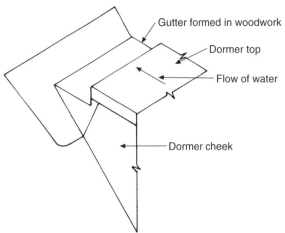

Fig. 10.32 Dormer gutters. Where it is undesirable to discharge the flow of water into an eaves gutter on the front of the dormer, a proper gutter must be constructed and weathered in a similar way to a chimney back gutter.

is avoided as they require a certain amount of maintenance, and in cases where the metal roofing and cladding are features, their appearance would be objectionable. In these circumstances it is necessary to arrange for the dormer top to fall to the back to avoid water form the roof discharging over the windows. If the dormer is a small one with not more than one longitudinal joint in the roof covering, the fall can be arranged so that the water runs off the edges, but if there is more than one longitudinal joint, a gutter will be necessary at the intersection of the dormer top and the roof (see Fig. 10.32).

Mansard roofs

The mansard roof is sometimes called a double pitched roof, and it is thought that this type of structure originated in Holland. Mansard roofs can also be constructed with a flat top (see Fig. 10.33(a)), although these are uncommon in the United Kingdom. The double pitched type is illustrated in Fig. 10.33(b) from which it will be seen that it provides more living space than one having a single pitch. When hand-made clay tiles were common it was sometimes possible to pick out those having more curve than others so that the angle between the two pitches could be weathered completely without flashing. With mass-produced tiles the curves are all identical and the point where

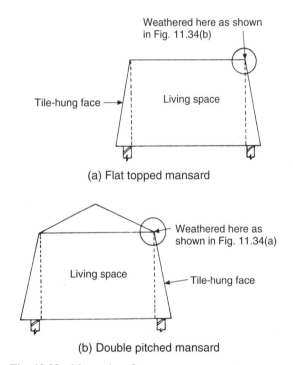

Fig. 10.33 Mansard roofs.

the two pitches intersect must be weathered with a flashing as shown in Fig. 10.34. The flat topped mansard roof (see Fig. 11.33(a)) must be weathered like a flat roof, using traditional materials or asphalt. If the latter is chosen, it will still be necessary to

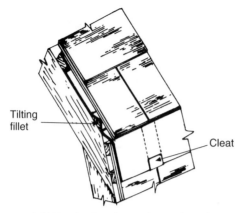

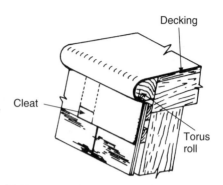

(a) Weathering the roof covering of a
double-pitched mansard roof

(b) Flat-topped mansard roof weatherings

The torus roll shown here is a common method
of finishing the edge when lead sheet is used.

Fig. 10.34

use a metal edging strip, similar to that shown and
having a doubled back edge left open to provide a
key for the asphalt.

Lead-welded details

Due to its very low melting point, lead can be
easily welded to fabricate roofing components,
both on and off the site, the latter enabling much
of the work to be prepared in favourable conditions.
A good example of this is where a number of
chimney or pipe flashings are to be made. Most
of the work can be done out of position in one
operation which saves a considerable amount of
time. Lead components do not need to be tailor-
made quite so precisely for a specific piece of work
and can be eased one way or the other to suit, for
example, a number of chimneys which may have
slight variations in size.

Chapter 1 refers to the two main techniques of
lead welding; here it is shown how they are applied.
These simplify some of the traditional methods,
and the increased use of welding techniques reduces
the necessity for many labour-intensive bossing
operations. It should be noted, however, that not all
bossed details take long and many experienced lead
workers prefer bossing to lead welding for such
details as undercloaks to rolls and drips and
working down a roll and into a gutter or onto a
fascia board. Conversely, most would agree that
some details can be achieved more economically

using welding. It must be stressed that the use of
welding requires absolute competence to avoid the
early failure of any joints.

Where possible welded joints should be made
off the job where the most convenient position can
be used and to avoid fire risks. If the joint is to
be made in position use a lap joint, as the welding
flame does not penetrate the lead. If this is not
possible soak any woodwork with water or a flame
retardant solution prior to welding. Always carry
out any positional welding early in the working
day so that plenty of time is left on completion
to carry out a thorough inspection. The following
illustrations show some of the techniques employed
in the fabrication of lead gutters and flats using
wood-covered rolls.

Setting out for welded joints
It will be seen that some careful setting out is
necessary when forming the details for welding. It
is recommended that the reader practise on a piece
of card, cutting with scissors and folding to make
the details shown. It is a practical operation and
worth knowing that card can be usefully employed
for making templates for repetitive work. The
following illustrations, which supplement those
in Book 1 are not exhaustive, but they show a
typical selection of components that can easily
be fabricated using the lead welding process.

Figure 10.35 illustrates how to set out and form
a drip in a box gutter so that all the welding can be

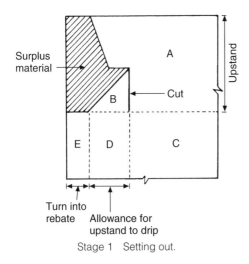

Surplus material

Cut

Upstand

A

B

E | D | C

Turn into rebate

Allowance for upstand to drip

Stage 1   Setting out.

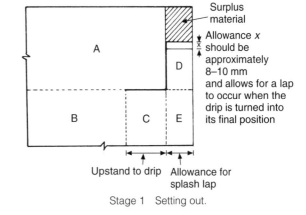

Surplus material

Allowance x should be approximately 8–10 mm and allows for a lap to occur when the drip is turned into its final position

A

D

B | C | E

Upstand to drip   Allowance for splash lap

Stage 1   Setting out.

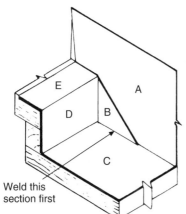

E

A

D

B

C

Weld this section first

Stage 2   The lead is turned and placed in position.

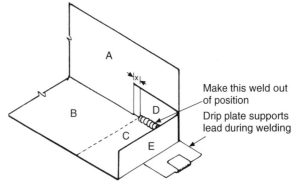

A

Make this weld out of position

Drip plate supports lead during welding

B

D

C

E

Stage 2   Turning the lead to make the first weld.

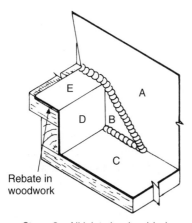

E

A

D

B

C

Rebate in woodwork

Stage 3   All joints lead welded.

(a) Welded drip undercloak

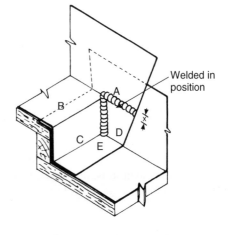

Welded in position

A

B

C   D

E

Stage 3   Welding completed.

(b) Welded drip overcloak

Fig. 10.35   Lead-welded drips in box gutters.

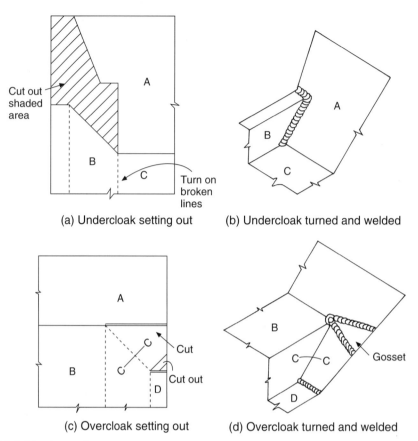

(a) Undercloak setting out

(b) Undercloak turned and welded

(c) Overcloak setting out

(d) Overcloak turned and welded

Fig. 10.36 Welded drips in tapering gutters.

done with the lead in position. Figure 10.36 shows a similar arrangement for a tapering gutter. In this case, due to possible fire risks, the undercloak would be better welded out of position. The method of fabricating lead rolls at an abutment is shown in Fig. 10.37; both the undercloak and overcloak can be welded out of position. As with drips, the undercloaks can often be bossed up as quickly as they can be welded. A disadvantage with welded undercloaks is that they tend to be bulky and can make it difficult to achieve a good finish on the overcloak.

A method of weathering rolls using lead welding, where they intersect with a drip, is shown in Fig. 10.38. The sequence of laying the lead is numbered 1–4. Both the undercloak (1) and the overcloak (2) can be bossed or welded using the technique shown in Fig. 10.37. Undercloak (3) is then laid and finally the overcloak (4). The detail for fabricating by welding ('x') is shown in Fig. 10.38(b); this is best done in position. The curved position can be welded by supporting it on a drip plate prior to turning it down so the two edges ('y') can be welded. As these are lap joints any fire risks are minimised.

The roll end at the eaves is not difficult to boss and, again, many plumbers prefer this method to welding. Figures 10.39(a) and (b) illustrate two methods commonly used where lead welding techniques are employed. Note the drip plate used to support the lead during the welding operation. As both of these methods employ the butt welding process there is a fire risk if they are welded in position. One method of overcoming this is to form

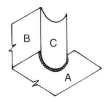

Note that the methods illustrated here are suitable only for small pieces of lead that can be welded out of position.

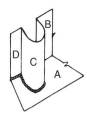

Stage 1 Illustrates position of lead during welding operation.

Finished job placed in position.

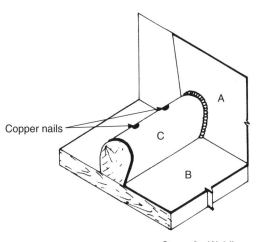

Copper nails

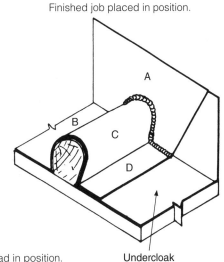

Undercloak

Stage 2 Welding completed with lead in position.

(a) Undercloaks

(b) Overcloaks

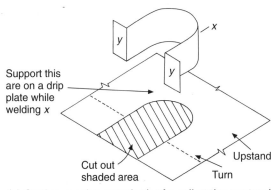

Support this are on a drip plate while welding x

Cut out shaded area

Upstand

Turn

(c) Setting out the overcloak of a roll at the upstand to a drip or abutment using welding techniques

See text for details of procedures.

Fig. 10.37 Lead welding wood-cored rolls

the roll end, then draw the bay forward and support the area of welding with an offcut of the wood roll. When the welding has been completed, the bay can be pushed back and dressed down.

Internal corners

These are time-consuming if bossed, and unless carried out with some understanding of the principles involved bossing can result in thinning of the lead; for this reason the use of welding is a better option. If the corner can be welded out of position the butt welding procedure can be used, but if this is not possible the recommended procedure is shown in Fig. 10.40. When the upstand is turned up, the area between the edge of the punched hole and intersection x is bossed up. This avoids an awkward starting point for the weld. The gusset is placed behind the upstand so that the 'overhand' method of welding can be employed.

Welded cleats

Where copper cleats or nails are welded to a section of lead, they must be coated with solder — the plumbers term for this is 'tinning'. Bear in mind that copper has a higher conductivity than lead and should be heated to near welding temperature

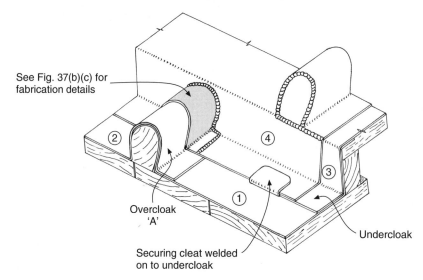

Fig. 10.38 Sequence of laying lead-welded overcloaks at a drip or abutment and the roll end at the eaves or drip edge.

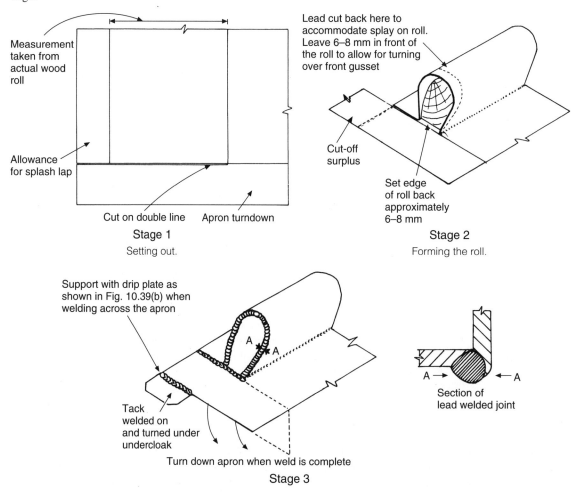

Fig. 10.39(a) Lead welding at roll ends at eaves.

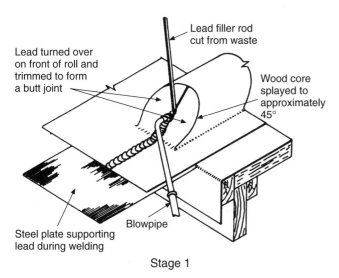

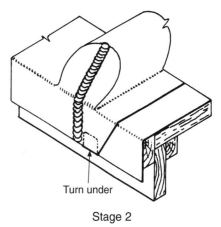

Stage 1

Stage 2

Welding completed and front edge turned down, the surplus lead being trimmed off leaving a cleat turned under the undercloak as shown.

Fig. 10.39(b) Lead welding roll ends at eaves.

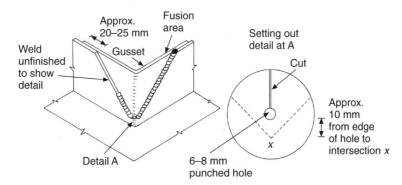

Fig. 10.40 Fabricating an internal corner.

before the weld is attempted. Some practice will be necessary before this operation can be carried out successfully.

Metal cladding of buildings with preformed panels

The term is generally understood to mean the covering of vertical surfaces of buildings as distinct from the weathering of sloping or flat surfaces. The technique has been used in Scandinavian countries for many years where the buildings are constructed of wood with a cladding of copper or aluminium for protection. Similar techniques have become popular in this country in recent years, not so much due to necessity, but to provide architectural features.

All the materials used for weathering buildings can be used for cladding, but one of the most common is lead, this being due to its good weathering qualities and to the ease with which it can be bossed or welded. Both copper and aluminium sheet are also used extensively for this purpose, and are often supplied in preformed panels which are usually fixed by specialists employed by the manufacturer. Lead cladding is more flexible because it can be fabricated, if necessary, on site by the plumber using traditional methods. Information about cladding with sheet lead is obtainable from the Lead Sheet Association who will advise on

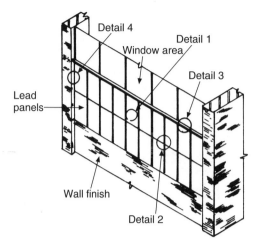

Detail 4
Window area
Detail 1
Detail 3
Lead panels
Wall finish
Detail 2

This illustrates part of a typical modern building and shows
how lead-weathered panels are used to cover areas
traditionally constructed with other materials.

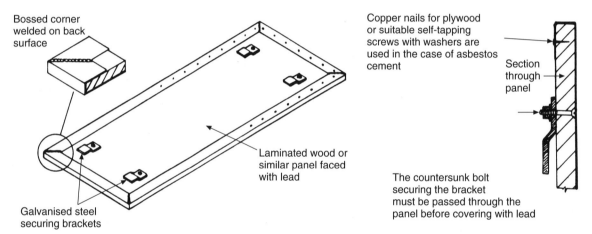

Bossed corner
welded on back
surface

Copper nails for plywood
or suitable self-tapping
screws with washers are
used in the case of asbestos
cement

Section
through
panel

Laminated wood or
similar panel faced
with lead

The countersunk bolt
securing the bracket
must be passed through the
panel before covering with lead

Galvanised steel
securing brackets

Detail 1 The fabrication of the panels

Note that provision is made only to secure the top half
of the lead covering to the panel, which allows for
thermal movement.

During fabrication the panel should be laid on a canvas
sheet to avoid scratching or indentation of the face side.

Fig. 10.41   Lead cladding for buildings.

specific techniques. Figure 10.41 shows some of
the details encountered in cladding with preformed
panels.

The panels are covered off site with lead sheet
as shown in the illustration. The lead is nailed
only on the top half of the panel which allows
the lead to expand freely; the maximum areas of

sheet are similar to those relating to dormer
cheeks.

The panels are hung on steel bars bolted to
timbers which are securely fixed to the fabric of
the building. These timbers are weathered with
a strip of lead having turned edges to prevent the
ingress of water behind the panels. The top of the

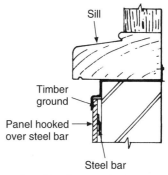

Detail 3 One method used to finish
the top of panels under a sill

An apron, fitted in a similar way as a dormer apron,
is turned over the top edge of the panels.

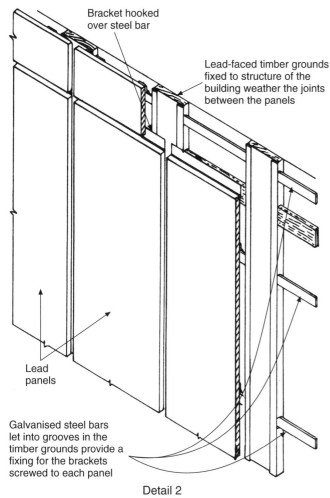

Detail 2

The lower edge of each panel is weathered with a strip of
lead welded on to the top of the panel immediately below.

Fig. 10.41 (*cont'd*)

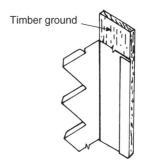

Detail 4 The way edges of the panels are
finished where they abut the brickwork

Angled flashing strips are used, one side being
turned into the brickwork joints, the other on to a
timber fixed to the face of the building.

panels can be weathered by means of a cover flashing or, as in the example shown, the window apron.

Patenation oil

Lead sheet forms its own protective patina when exposed to the atmosphere, but in damp conditions the formation of an uneven white carbonate will appear on newly fixed lead. As the protective coating forms the carbonate film disappears. Apart from not being pleasing to the eye, especially on external cladding, the white carbonate will be washed off by rain causing unsightly stains on the fabric of the building. To prevent this a coat of patenation oil can be applied, preferably at the end of each day's work. It should be applied with a soft cloth working from top to bottom of the work. One coat is usually sufficient but if the surface of the lead is marked by foot traffic during subsequent builders' work, a second coat may be applied after all other operations have been completed.

Inspection and maintenance of sheet weatherings

A well-designed and correctly fitted sheet metal roofing installation should have a long working life, and except in the event of a serious breakdown in the metal itself, should be as effective and long lasting as tiles or slates.

In cases where maintenance is necessary, in almost every case it is due to: faulty workmanship, incorrect thickness of metals used and poor design, a typical example of the latter being insufficient provision made for expansion occurring in hot weather, thereby causing fatigue cracking. Lead sheet is possibly more prone to 'creep' than other sheet materials due to its mass, but this can be minimised if it is well fixed and the maximum superficial areas in relation to its thickness have not been exceeded. Both creep and fatigue cracking are dealt with in Book One, Chapter 10.

Although regular inspection of roofs covered with any material is rare it is recommended, as a fault detected in its early stages is more easily corrected and damage to the substructure minimised.

Possibly the most common fault found is where a turn into the brick or stonework of the structure has pulled out of the joint. This is easily remedied by refixing, but the cause of the defect should be investigated and is usually due to one of the following:

(a) Insufficient turn in.
(b) In the case of cap flushing the maximum recommended length has been exceeded.
(c) The bottom edge of the flashing is not properly supported.

Where lead weatherings are fixed in exposed positions e.g. ridges and verges the effect of high winds can cause it to be torn from the decking. It should be noted, however, that this would certainly be due to insufficient or incorrect fixings.

During an inspection of lead any signs of cracking should be investigated. Such defects are almost always due to poor design and improper fixings. A repair may be effected by patching with wiping solder, or preferably by welding. It must be stressed, however, that in all such cases no amount of patching will permanently solve a problem caused by poor design. If it is possible to gain access to the area immediately below the roof, the underside of the decking and its supports can be examined for dampness or staining which will indicate any ingress of water. Any sign of corrosion should also be carefully investigated, paying special attention to the eaves of moss-covered tiled roofs where the acidic run-off may cause local pitting corrosion.

The only maintenance a well designed roof requires is periodic cleansing, especially if there are deciduous trees near by where the falling leaves could possibly lead to blocked gutters. If necessary any catchpits in gutters and hopper heads should be checked for cleanliness, also ensure the wire guards preventing debris entering the rainwater pipes are in good condition.

Further reading

The following are obtainable from HM Stationery offices:
BS 6915 Specification for design and construction of fully supported lead sheet roof and wall coverings. (Currently under revision.)
Lead sheet manuals & data sheets
The Lead Sheet Association, Hawkwell Business Centre, Maidstone Rd Pembury, Kent, TN2 4AM, Tel. 01892 822773.

Self-testing questions

1. From Table 10.2 calculate the total superficial area of one bay of a flat roof covered with code 6 lead sheet, allowing 200 mm for the under and overcloak for the drips, and 225 mm for the under and overcloak to the rolls.
2. Describe the underlying principles of fixing lead sheet to avoid fatigue cracking.
3. Sketch and describe the type of transverse expansion joint that must be used across a low-pitched lead covered roof with a fall of $2\frac{1}{2}°$.
4. (a) State the advantages of hollow roll work on a lead covered pitched roof.
   (b) Describe the procedure for laying such a roof and how the roll ends are finished at the eaves.

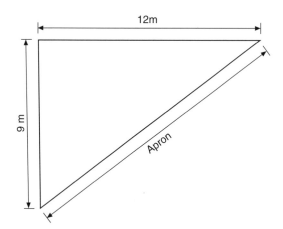

12m

9 m

Apron

Fig. 10.42

5. Name the type of gutter that would be used in the centre of a large, low-pitched lead roof.
6. (a) Describe the 3,4,5 rule and its use in building.
   (b) Calculate the length of metal apron flashing for a triangular asphalt roof shown as a plan drawing in Fig. 10.42.

7. Describe one situation where an L-shaped cleat would be employed.
8. (a) Make a sketch showing a torus roll.
   (b) Suggest a position where this detail can be used with advantage.
9. (a) List three types of dormer in common use.
   (b) Describe two methods of fixing and weathering the front edge of a dormer cheek on to the window frame.
10. (a) List the advantages of using lead-welding techniques for sheet roofing.
    (b) Set out on paper the folding and cutting lines for a lead-welded drip overcloak.
11. (a) State the type of lead-welded joint recommended for positional work on lead-covered roofs and indicate the reason for your choice.
    (b) Describe an alternative to a 90° vertical joint when welding lead, and explain its advantages.
12. List the precautions and methods you would use when welding positional details on a lead-covered roof with a timber substructure.

11 Electrical Systems

After completing this chapter the reader should be able to:

1. Recognise electrical supply systems in domestic properties.
2. Describe the components used to automatically cut off the supply in the event of failure of the conductors or appliances due to an electrical fault.
3. State the purpose of earthing and bonding.
4. Recognise potential dangers due to incorrect installation procedures in bathrooms, kitchens, and plumbing or heating appliances.
5. Understand the basic principle of power circuits.
6. Recognise the need for correct power ratings, fuses and cable sizes for plumbing appliances using mains electricity.
7. Describe the correct procedures for testing and isolating electrical supply systems.

The purpose of this chapter is to enable the student plumber to acquire a basic knowledge of electricity and its supply systems in order that electrically controlled and operated plumbing appliances may be installed correctly and safely. Even when electrical work is undertaken by a qualified electrician the plumber must be able to recognise the need for bonding appliances which can conduct electricity even when a supply is not connected directly to it.

Always remember that electricity is dangerous and is a potential killer; unlike water, gas or fire it cannot be seen or smelt, and the first indication of poor installation practice may be a severe shock which at best is unpleasant and painful and at worst fatal.

Regulations governing the installation of electrical supplies

In the interest of safety it is essential that electrical installations of any kind must be carried out by a competent person and that any appropriate regulations or recommendations are strictly adhered to.

IEE regulations

These are now incorporated in a British Standard which covers both the electrical wiring regulations and the Health and Safety Act, parts of which relate to the safety of electrical installation work. The wiring regulations are very comprehensive, state the recommended practice and are continually updated. Although they are not statutory, no electricity supplier would make a connection to any installation not meeting their requirements.

The Electricity Supply Regulations 1988

These are statutory, but relate mainly to the installation of supplies rather than work carried out in a building and the installation of appliances.

The Electrical Equipment Regulations

These are also statutory and specify the type and safety standards of equipment and appliances used by the consumer. If these items meet the requirements of British Standards it can be assumed the manufacturer has complied with these regulations.

The Electricity at Work Regulations 1989

These relate to the safe use of electrical installations and power tools in the workplace. Many of the requirements of this act are embodied in the Health and Safety Regulations and apply to both employers and employees. Other safety publications relating to specific items of equipment also quote these regulations where they apply.

Distribution of electricity

Electricity is best defined as an invisible source of energy which is conveyed from point to point via conductors. Apart from the various components of an appliance these conductors are usually insulated copper wires, copper being used because it is a good conductor and its malleability enables the cables to be manipulated very easily. Insulation of the conductors is necessary as electricity will always take the shortest possible route. If the 'phase', or live conductor comes into direct contact with the neutral or earth wires, it will result in a short circuit causing overloading, which in a correctly designed installation will result in a safety device, usually a fuse or miniature circuit breaker (m.c.b.), operating to cut off or isolate the supply.

Single-phase supplies

The term 'phase' relates to the current-carrying cable sometimes called 'live'. It should be noted that the wiring regulations also refer to the neutral as live, so one must be clear as to the term 'phase'. The supply of electricity to domestic properties normally has only one phase wire, hence the term 'single phase'. In industrial premises three-phase supplies are often used which produce higher voltages, so that more power is available for large machines. The normal voltage for single-phase supplies is 240 V, and for three-phase 415 V. It must be stressed that any work on three-phase supplies should only be carried out by a qualified electrician.

Types of supply

Direct current   In this case a flow of electricity takes place in one direction only. It should be noted that the flow is negative to positive, which is technically correct. However, for practical purposes, it is normally assumed that the flow takes place in the opposite direction, i.e. from positive to negative.

Direct current is not usually supplied for household use, unless it is provided by a private generating plant in areas where no main supply is available. Other examples include the use of dry batteries, e.g. for torches. It has very few applications in the type of electrical work undertaken by plumbers, unless they are fitting an electric bell system which, as a point of interest, was considered to be plumbers' work around the turn of the century.

Alternating current   This type of supply is normally provided by the electricity generating companies and, as its name implies, its direction of flow alternates continually in what are called cycles, see Fig. 11.1. The frequency, or number of times it changes direction, has been standardised in this country at 50 Hz (hertz) or cycles per second. Again, for all practical purposes, it should be assumed that the flow of current takes place from the live terminal through an appliance and returns via the neutral terminal.

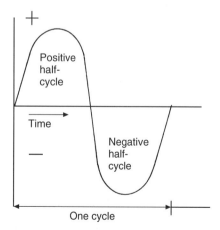

Fig. 11.1   Alternating current. Alternating current does not flow from the positive terminal to the negative terminal consistently, but alternates for half the cycle. Note that the term *hertz* is commonly used for *cycle*.

Electrical units

Voltage (symbol 'V')

The volt is the unit of electromotive force (e.m.f.) and just as in plumbing systems we know that the higher the head of water the greater will be the pressure, so with electricity the higher the voltage the greater will be its force to cause electrons to flow or drift along the conductor. An electric torch battery has a positive and negative terminal, and a potential difference is said to exist between them causing the electrons to flow from one to the other in a closed electric circuit. It should be noted that there is a drop in pressure called voltage drop as electricity flows around the circuit, and in long circuits this may be significant and require the use of a cable having a greater cross-section area. Although this is unlikely to have much effect on the type of work carried out by a plumber it is important that he is aware of this factor and pays attention to it.

Current (symbol 'I')

The ampere is the unit of current and is normally abbreviated to 'amp' and may be defined as the electrical unit of quantity and the amperage or volume of current passing through a conductor may be equated with the volume of water in litres passing through a pipe. Do not confuse the amp with the watt or kilowatt which are units of power. It is very important that a conductor or cable is of sufficient size to permit the current to pass through it without offering too much resistance, as this will lead to overheating and possible fire risk.

Resistance (symbol 'R')

The ohm (Ω) may be defined as the unit of resistance which opposes the flow of current. Resistance to current flow depends on the type of conductor used. Materials offering the least resistance to the current flow are termed good *conductors*, insulating materials are those that have great *resistance* to current flow. Most metals are good conductors. Good insulators include wood, glass, ceramics, paper, plastic and rubber. Providing the temperature does not change, the rate of drift of the electrons (the current) flowing through a wire is directly proportional to the potential difference (the voltage difference which makes the electrons move) between the two ends of a wire. This relationship is known as *Ohm's law*. A typical example of this law is to assume a current of 4 A is flowing through an electric element which is connected to a 240 V supply. When the supply voltage is halved, i.e. reduced to 120 V then the current is halved and becomes 2 A. In each case the ratio of the voltage to the current is the same, namely:

$$\frac{240}{4} = 60 \quad \text{and} \quad \frac{120}{2} = 60$$

and thus Ohm's law can be stated in the form:

$$\frac{\text{Voltage}}{\text{Current}} = \text{A constant, i.e. } \frac{V}{I} = R.$$

When the potential difference is measured in volts and the current is measured in amps, the constant, R, will be the resistance in ohms of the conductor. Thus, to find the resistance of a wire subjected to a mains supply of 240 V and a current of 6 A, since

$$\frac{V}{I} = R, \quad R = \frac{V}{I} = \frac{240}{6} = 40.$$

Thus the resistance of the wire is 40 Ω. Another example of the use of Ohm's law is to calculate a fuse rating for an appliance. In this case the formula must be rearranged thus, since

$$\frac{V}{I} = R, \text{ then } I = \frac{V}{R}.$$

Assuming a voltage of 240 V with an appliance having a resistance of 80 Ω then:

$$I = \frac{240}{80} \quad I = 3 \text{ A}$$

that is, the fuse rating will be 3 A. An easy way to remember Ohm's law, and to transpose the equation so that any one value can be found if the other two are known as shown in Fig. 11.2.

Power

Electrical energy is measured in kilowatts (1,000 W) and for metering and billing purposes a kilowatt-hour is referred to as one unit. Power in watts is the product of voltage × current. The cost of

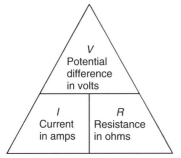

Cover the quantity to be found and the remaining two symbols will give the required formula

Ex 1 to find *V* multiply *I* by *R*
Ex 2 to find *R* divide *V* by *I*
Ex 3 to find *I* divide *V* by *R*

Fig. 11.2 Ohm's law equation.

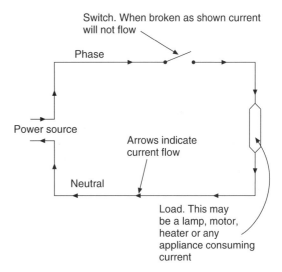

Fig. 11.3 Simple electric circuit.

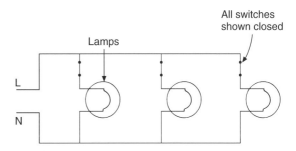

Fig. 11.4 Lamps wired into a parallel circuit. This system all the loads, in this case lamps, to be supplied by the full voltage and each can be independently switched.

operating electrical appliances can be calculated very simply and is shown as follows.

An electric water heater having a resistance of 40 Ω is supplied by a 240 V ring main. Determine its current rating and calculate its cost per hour, assuming the cost per unit is 7.5p. Since the amperage of the appliance must first be ascertained, Ohm's law equation is used in the following form:

$$I = \frac{V}{R}$$

$$\therefore I = \frac{240}{40} = 6\,\text{A}$$

As power in watts is the product of volts × amps: thus

$$240 \times 6 = 1,440$$

to convert this to kilowatts

$$\frac{1,440}{1,000} = 1.44\,\text{kW}.$$

This is approximately 1.5 kW and at 7.5p per unit the approximate cost of running this heater will be 11.25p per hour. It is appreciated that the electrical units and the calculation involved are unfamiliar to many plumbers and it is suggested that a little practice in them may be helpful. This can be done by using the information given by suppliers of electrical appliances of various types and also local electricity board shops and showrooms.

Simple electrical circuits
Before a flow of electricity can take place a complete circuit must be established from the positive terminal to the negative terminal as shown in Fig. 11.3. The consuming component, which may be a lamp, motor or heater, etc. is usually referred to as the 'load'. Any interruption by a break in the circuit, usually by operating a switch, and the current will cease to flow. From this it will be seen that a current will only flow through an unbroken circuit from positive to neutral. Figure 11.4 shows a parallel circuit with three lamps controlled by one switch. With this type of circuit each lamp takes the full voltage and power from the supply.

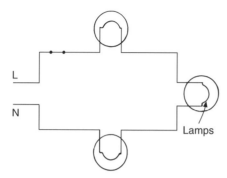

Fig. 11.5 Lamps wired in series. In this case the available voltage must be shared so that if 24 V are supplied, each lamp will only take 8 V. This system would not be suitable for lighting circuits as the illumination would be dimmed and independent switching would not be possible.

A series circuit is shown in Fig. 11.5. If three lamps are used again it will be seen that they give less illumination, due to the fact that, in effect, the current has to be shared between them. It is therefore unlikely that such a system would be used for lighting, but the term must be understood, as many control systems, especially those in boilers, employ components wired in series.

Electrical safety

Overcurrent

Overcurrent or overloading a conductor takes place where the current of electricity exceeds that which the conductor is safely capable of carrying. Just as a water pipe will only deliver a given quantity of water in litres for a given pipe size, so an electrical conductor of a given capacity can only carry a given current. If this is exceeded the conductor will become hot due to its resistance to the current flow. This could destroy the insulation, leading to the possibility of a fire which may be well established before the conductors make contact, causing the circuit protective device to operate. It must be quite clear that any cable selected for a specific task must be of adequate size for the current it is to carry.

Short-circuits

This is said to take place when, for various reasons, the neutral or earth cables are in direct contact with the phase. One reason for this could be careless

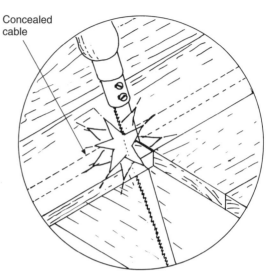

(a) Removing flooring using a padsaw. The use of a circular saw set for the correct depth would avoid most accidents of this type

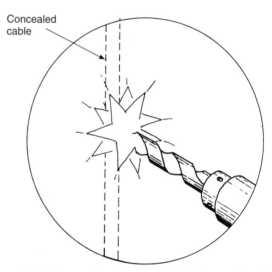

(b) Drilling walls can often result in the accidental damage to electrical cables. Avoid drilling near power sockets or switches and on a line above or below them

Fig. 11.6 Accidental damage to concealed cables. This usually results in short-circuiting the cables.

connection of a phase to the neutral or earth conductor or breakdown in the cable insulation due to ageing or damage. It can also happen accidentally, as shown in Fig. 11.6, during the course of one's work. It is worth mentioning that moderately priced instruments are available

enabling concealed pipes and cables to be
located before work is commenced.

Circuit protection devices

Isolation of a circuit in the event of overcurrent
or a dead short, is provided by fuse or miniature
circuit breaker (m.c.b.), both of which are designed
to automatically isolate or cut off the supply of
electricity.

Fuses A fuse consists of a single strand of wire
which is designed to overheat and melt under
overload conditions. There are two main types,
rewireable and cartridge fuses. The rewireable type
is simple and cheap. Should a fuse of this type blow
it can be replaced with wire obtained on a card
obtainable from most ironmongers or electrical
suppliers. The card will usually have fuse wire for
5, 10 and 15 A circuit control, the 5 A being the
smallest in cross-sectional area mostly used for
lighting circuits. It is important when replacing fuse
wires that the correct type is selected. An indication
of the circuit controlled by the fuse should be
provided on the cover of the consumer unit where
the fuses are located. The main disadvantage with
rewireable fuses is that they take at least three times
their rated value to operate. The alloy from which
the wire is made tends to deteriorate with age,
and if it melts due to overloading, white heat is
momentarily achieved, constituting a fire risk.
Cartridge fuses consist of a single wire completely
enclosed in a glass or ceramic tube, both ends of
which are sealed with a brass cap. The fuse rating
is usually marked quite clearly on the tube. In
comparison to fuse wire they are expensive to
replace, but they operate more quickly and there is
little fire risk.

Miniature circuit breakers These are basically
an automatic switch used instead of fuses in the
consumer unit to control various ring and lighting
circuits. While they are more expensive than fuses
they are considered safer — they do not present a
fire risk and can easily be reset when a fault is
corrected. A much heavier type of circuit breaker
called a residual current circuit breaker (r.c.c.b.), is
used in conjunction with earth electrode earthing
systems to isolate, not individual circuits, but the
whole installation. More details are given on these

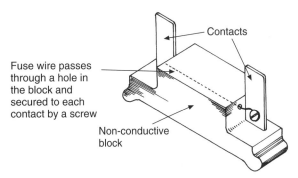

Fuses must always be correctly rated for the circuit or
appliance they protect.

(a) Renewable wire fuse. In the event of overload
current on the circuit the fuse wire will melt
automatically breaking the circuit

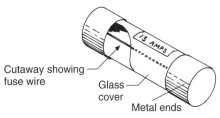

(b) Cartridge fuse. These work on the same
principle as a wire fuse but in this case it is
enclosed in a glass tube. They are designed to
protect against faults in individual appliances and
are used in fused plugs and switches

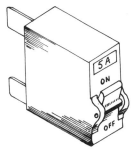

(c) Miniature circuit breaker. These are used in the
distribution panel instead of fuses. They are more
expensive, but much quicker acting. In the event of
circuit overload, a bimetallic strip operates a
solenoid which trips a switch to isolate the circuit.
They can then only be reset manually

Fig. 11.7 Circuit protective components.

components in the section 'Earthing and bonding'.
Figure 11.7 illustrates the main circuit protective
devices used in housing. Automatic isolating
switches are often incorporated in plumbing

equipment such as sink macerators to prevent overloading the motor in the event of jamming. They are used for the same reasons in many types of washing and dishwashing machines. They work on different principles to m.c.b.s. and should not be confused with them.

Earthing and bonding

The surface of our planet is referred to as the 'earth' and we use this as our reference, calling the earth 'zero potential'. As humans usually stand on the earth they are 'at earth' or 'zero potential'. If, however, they touch a metal object which has become electrically charged, they may receive an electric shock, as the current flows to earth through them.

In order to reduce the possibility of metal cases of electrical equipment becoming charged, due maybe to an internal fault, e.g. the breakdown of insulation on a line conductor, we 'earth' the metal case via a circuit protective conductor c.p.c. (the earth wire in the three-core cable feeding the equipment). An excessive current flowing to earth under fault conditions will cause the fuse feeding the equipment to blow. Providing the c.p.c. is of adequate size and has sound connections, there will be no substantial rise in voltage of the metal case and the fuse will blow in 5 seconds.

The c.p.c. of every final circuit in an installation is taken back to the earth terminal at the fuse board, and then to the main earth terminal at the incoming supply. Under fault conditions this current will flow back to the supply transformer, either through the ground or via the sheath of the supply cable. The neutral is connected to earth at the supply transformer — this provides a fault path back to the supply. Any metalwork which is earthed and associated with electrical equipment, is referred to as an 'exposed conductive part', examples being the metal case of a boiler, sink, water heater, etc.

Other items of metalwork could possibly become live due to a fault elsewhere and be connected by plumbing pipes, metal frame of the building, metal ducting of the air-conditioning system, etc. Now if under these conditions someone touches this metalwork, referred to as 'extraneous conductive parts', and are themselves earthed, they could get a fatal shock.

To prevent this difference of potential existing between the adjacent metalwork, we bond the items together — known as supplementary bonding in a bathroom and main bonding at the main service position. These bonding conductors must be of adequate size to allow the current to flow to earth without causing the metalwork to rise in potential above 50 V.

Earthing systems

It was common practice in days gone by to connect the main earth conductor to the water service, as being metal it provided a good conductor to earth. This is no longer permissible due to the fact that both water services and mains are, in many cases, made of polythene which is not a conductor of electricity.

Solid earthing   This is the TNS system (see Fig. 11.8). In most modern buildings where the electrical supply is underground the consumer's earth wire is connected to the armoured incoming cable. This armour not only protects the cable from accidental damage but also acts as an effective earth.

TT system   In some rural areas where the supply is taken from overhead cables an earthing rod is used. This is a copper or steel rod of approximately 12 mm in diameter which is driven into the ground so that any current leakage is dissipated and eventually finds its way back to the suppliers' neutral point. Bearing in mind that the path of a current to earth must have negligible resistance and that certain types of soil and the amount of moisture it contains will affect its conductivity, an automatic device for isolating the supply must be provided. This is called a residual current circuit breaker (r.c.c.b.) which is a heavy duty, very sensitive circuit breaker that will automatically shut off the supply in the event of any earth leakage. These devices work on the principle of a coil which operates in a magnetic circuit, which while the flow through the phase and neutral conductor is balanced, as under normal conditions, the coil will remain inactive. In the event of earth leakage this balance is lost causing the coil to react and automatically trip an integral switch to the off position which must be reset to restore the supply.

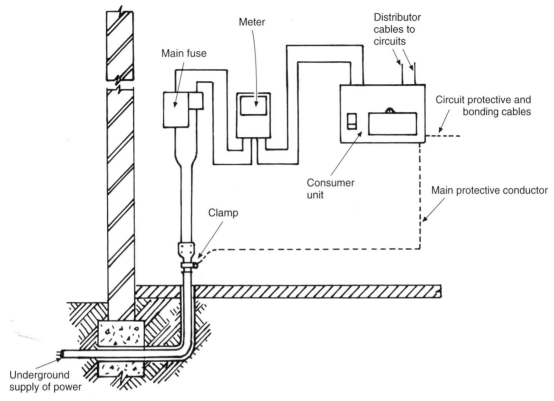

Fig. 11.8   TNS system of earthing. With this type of earthing arrangement the main earth cable is connected by means of a clamp to the metallic armoured sheath protecting the current-carrying cables.

If the r.c.c.b trips continuously it is indicative of a fault or short-circuit in the system. This earthing system is illustrated in Fig. 11.9.

Protective multiple earthing   This method of earthing (the TNCS system) shown in Fig. 11.10 uses the electrical supplier's neutral conductor as an earth, which in effect means that any current leakage goes straight back to the authority's neutral point. One of the advantages of this system is there is little resistance to the current leakage which means the safety cut-out devices will be rapidly activated.

Equipotential of phase and earth conductors   The word 'equipotential' in this context means that any bonding and earthing conductors must be of sufficient cross-sectional area in the event of a fault to carry away any flow of current with little or no resistance. If, for example, a short-circuit took place between a phase conductor fused for 13 A and an

earthing conductor capable of carrying only 10 A, it would be of little use, as the fuse would not melt and cut off the current.

Equipotential bonding   Any conductive material which is not a part of an electrical installation but which, due to a fault, can become energised is termed an *extraneous conductive part*. An example of this could occur where a fault has developed in an electric heater installed in a hot storage vessel causing all the metallic parts of the hot water system, and possibly the heating system as well, to become live. A similar situation could arise in connection with a metal boiler casing. From this it will be seen that although things like metal baths, sinks, taps and metal pipes have no direct contact with electrical supplies, due to a fault they may become energised and under these circumstances anyone touching them as shown in Fig. 11.11 could suffer an electric shock, possible fatal.

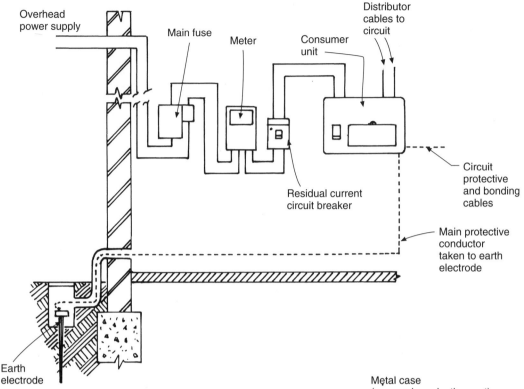

Fig. 11.9 TT system of earthing. This is used where the authorities' supply is taken from overhead cables which provide a phase and neutral supply only. Earthing in this case is by a metallic rod driven into the ground.

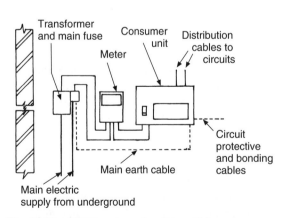

Fig. 11.10 TNCS system of earthing. This system differs from TNS only in that the main earth cable is connected to the neutral cable at the supply authorities' transformer.

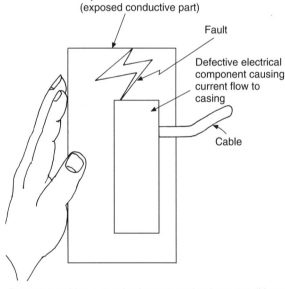

Persons touching a conductive part under these conditions could receive an electric shock, the severity of which depends upon the conductivity of the surface with which the person is in contact.

Fig. 11.11 Earthing extraneous conductive components. This illustration shows the necessity of earthing or bonding exposed conductive parts.

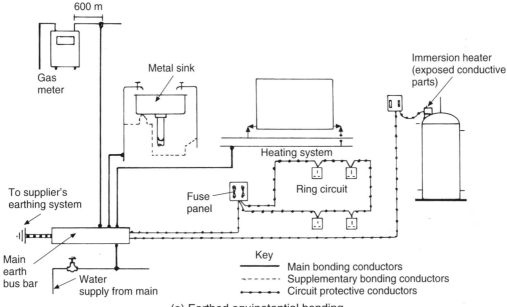

(a) Earthed equipotential bonding

The purpose of bonding metallic components is to ensure that if accidental contact is made between a part of the electrical installation through which an electrical leak is passing and other metallic or conductive components, they may become energised. Touching them could result in a severe electric shock. The illustration shows the bonding and earthing cables used to ensure any electrical leakage is safely conveyed to earth.

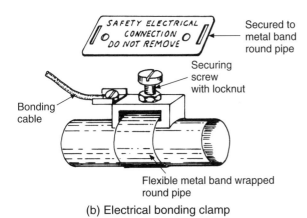

(b) Electrical bonding clamp

Fig. 11.12 Earthing and bonding.

It must also be remembered that water is a conductor of electricity due largely to impurities it may contain, so extra care needs to be taken when electrical connections are made to any equipment that is also in contact with water. See Fig. 11.12(a) which generally indicates the basic principles of equipotential bonding. A typical bonding clamp suitable for a bonding cable to a pipe is shown in

Fig. 11.12(b). If any clamps are removed during the course of repair or alteration work, they must be replaced securely on completion of the work. If any additions or modifications to existing work are carried out, a check should be made with a competent electrician as to whether any bonding is necessary.

Consumer unit This component is installed on a panel adjacent to the incoming electrical mains and provides the means of control and protection necessary to comply with the IEE regulations which are listed as follows:

(a) Protection against excess loading.
(b) Protection against earth leakage and fault conditions.
(c) Provision for isolating individual circuits and a main switch.

Protection against excess loading, earth leakage and fault conditions is provided by suitably rated fuses or m.c.b.s. Isolation of the supply is provided by a double-pole main switch.

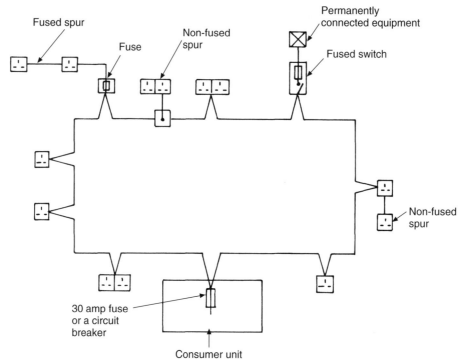

The main features of this type of ring main is that there is no limit to the number of socket outlets. The floor area must not exceed 100 m². The number of non-fused spurs must not exceed the total number of socket outlets or permanently connected equipment connected to the ring. The total number of fused spurs is unlimited. Permanently connected equipment must be protected against overload by either a fuse having a maximum rating of 13 amps or a suitable circuit breaker.

Fig. 11.13   Typical ring main circuit: note neutral conductors not shown.

Distribution of supplies in the property

The conductors to the various sub-circuits serving both power and lighting points are taken from the consumer unit. The term *power* here means the supplies to socket outlets and fixed appliances such as fires or hot water heaters. Each of these sub-circuits has its own fuse or circuit breaker. A 5 A fuse or equivalent m.c.b. is provided for lighting and 30 A for power circuits. It is not proposed to deal with lighting circuits here as this is normally the province of electricians, but it is important that the plumber is able to recognise power circuits and know the methods used to make additional connections to them.

Ring circuits   These are the most common form of power distribution in modern premises and there are slight variations in some types, mainly depending on the floor area covered and the current-carrying capacity of the main conductors. The one illustrated in Fig. 11.13 is possibly the most representative of those commonly used. The supply of power for domestic heating equipment is frequently taken from such a system. As its name implies it is simply cable laid out in the form of a ring which carries current to each socket outlet. There is no limit to the number of socket outlets on any one ring, but the floor area served must not exceed 100 m². This limit is to prevent the connection of too many appliances which could result in overloading. It is permissible to connect a spur to the ring main as shown, but the number of non-fused spurs on any one ring must not exceed the number of socket outlets and stationary apparatus connected directly in the ring. Figure 11.14 illustrates the wiring of a fused plug used in connection with a three-pin socket outlet in the

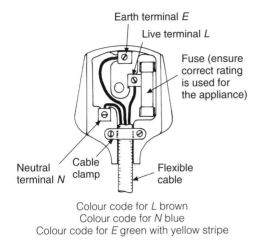

Colour code for *L* brown
Colour code for *N* blue
Colour code for *E* green with yellow stripe

Fig. 11.14 Three-pin plug used for connecting movable electrical equipment to a socket outlet.

ring. Another form of outlet is a fuse switch shown in Fig. 11.15. Figure 11.16 shows two approved methods of wiring an additional spur to an existing ring circuit. Non-permanent connections such as electric irons, radios and hair dryers, etc. are provided with a flexible cable and fused plug which is inserted into a socket outlet. Appliances which are permanently connected such as fires, boilers, water heaters, etc. are supplied with power via a fused switch which incorporates a fuse under a sealed cover. This cover, usually secured with a screw, is removable to allow access to the fuse. Whenever electrically controlled or operated plumbing equipment is used, or any electrical appliance for that matter, it is important that a

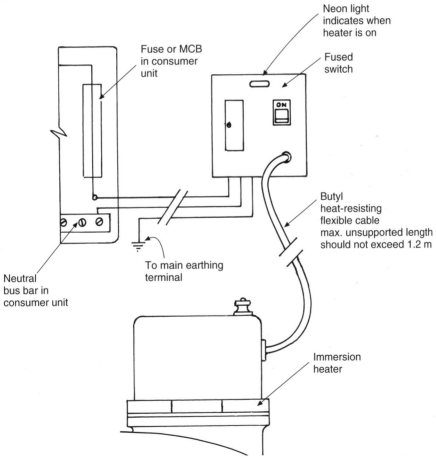

Fig. 11.15 Electrical supplies to water heaters. Because heaters of this type require a high amperage to avoid overloading the ring mains, they are supplied with power direct from the consumer unit having a separately fused circuit.

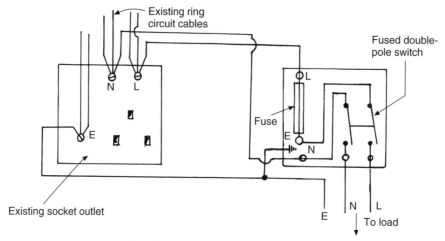

(a) Illustrates the cable connections from an existing socket outlet to a fused switch controlling an additional appliance. Additional socket outlets may be installed in a similar way.

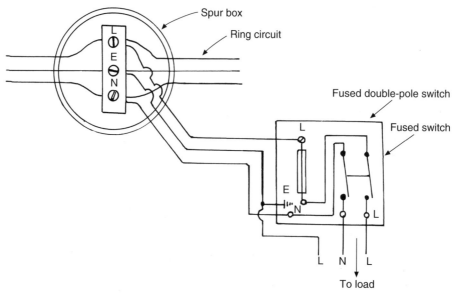

(b) Shows a method of making a direct connection to an existing ring main using a spur box

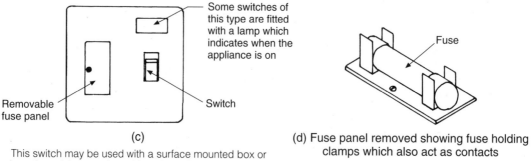

(c)

This switch may be used with a surface mounted box or where the cables are chased into the wall, a concealed box is used so the switch is flush with the surface of the wall.

(d) Fuse panel removed showing fuse holding clamps which also act as contacts

Fig. 11.16 Shows how additional connections may be made to existing ring mains.

fuse of the correct rating is fitted to avoid possible damage to the equipment or danger under fault conditions.

The principle of electrical ring circuits is based on the 'diversity' factor, that is to say an assumption is made as to the number of appliances in use and the power they consume at any one time. While it is most unlikely that the ring conductors will overload due to the use of electric kettles, fires, radios, etc. very high rated appliances such as immersion heaters and cookers might overload the circuit, causing the safety cut-out devices to be activated. For this reason such appliances are normally provided with their own separate circuits.

Circuits supplying electric immersion heaters
It is not recommended these heaters are connected to a ring main, due to their very high electrical loading. The types of heaters available are dealt with in the section on hot water supply (pp. 144–7). Figure 11.15 has illustrated the wiring from the consumer unit to the heater, and Fig. 11.17 shows how the element and thermostat are connected. It will be seen that the heater is supplied by a separately fused circuit taken from the consumer unit, and is usually controlled by a 20 A double-pole switch adjacent to the heater. If the immersion heater is situated in a bathroom the switch must be fitted externally.

All switches controlling electrical equipment in bathrooms must be fitted externally due to the

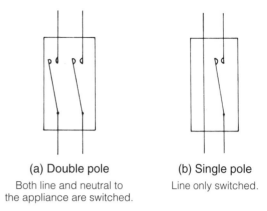

(a) Double pole
Both line and neutral to the appliance are switched.

(b) Single pole
Line only switched.

Fig. 11.18   Types of switches.

damp, humid conditions constituting a possible danger from electric shock. One exception is for switches at ceiling level, provided they are operated by a cord pull.

A double-pole switch differs from a single-pole switch in that both the phase and neutral conductors are switched see Fig. 11.18. A single-pole switch only isolates the phase conductor. A typical example of single-pole switches are those used for lighting circuits.

Current-carrying capacity of conductors (cables)

It is not always appreciated that wiring conductors of a given size through which a current of electricity is flowing has limitations on the capacity it can carry without an excessive amount of resistance. Failure to observe this fact will result in overheating, causing possible fire risk and destruction of the insulation.

Wiring conductors are measured by their cross-sectional area (c.s.a.) and typical examples are those used for lighting circuits which have a c.s.a. of 1 mm^2. Such circuits are normally protected by a 5 Amp safety device. Conductors used for ring circuits because of the possibility of heavy loading, have a c.s.a. of 2.5 mm^2 and are protected by a fuse of 30 Amp. It should be noted the two previous examples given relate to copper PVC sheathed wires. The other main factor to be considered is what is called voltage drop. Assuming a cable has been selected which is capable of carrying the required

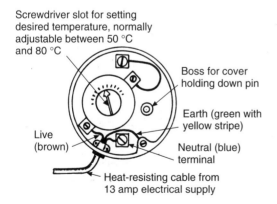

Screwdriver slot for setting desired temperature, normally adjustable between 50 °C and 80 °C

Boss for cover holding down pin

Earth (green with yellow stripe)

Live (brown)

Neutral (blue) terminal

Heat-resisting cable from 13 amp electrical supply

Fig. 11.17   View of electric immersion heater with the cover removed.

amperage but the cable lengths are very long, some account must be taken of this factor. Just as the fractional resistance of the pipe walls offer to a flow of water, so the voltage or pressure of electricity will be reduced progressively over long runs of wiring. This will have a similar effect as using wires of inadequate c.s.a., and while as a general rule cables of 1 mm² and 2.5 mm² c.s.a. are satisfactory for lighting and ring circuits. Cables for cookers and showers are usually 6 or 10 mm². In cases of doubt the IEE regulations should be consulted as they contain tables which show the voltage drop per amp per metre run. These regulations recommend the voltage drop in any final circuit should not be in excess of 4 per cent. To illustrate this in practical terms it could be said that if that supply at the inlet to a cable is 240 V the voltage at the furthermost outlet should not be less than 232 V.

Protection, insulation and fixing of cables

The methods and types of cables used vary considerably and depend mainly on the type of building, the purpose of the cable and the service conditions to which it will be subjected.

It is not proposed to deal with steel conduits and mineral insulated copper sheathed systems, as these are the subject of specialised work and do not normally fall into the province of work undertaken by the plumber in domestic premises.

Conduit

The main purpose of the conduit is to permit the cable to be withdrawn if necessary, and if made of metal it may also offer some protection to the cable from mechanical damage. Any mechanical conduit must be effectively earthed, and for this reason, PVC, being a non-conductor, is generally used except in industrial and commercial installations.

Oval conduit

Figure 11.19 illustrates this light conduit that is used for concealing cables in walls under the finished plaster. Its oval shape reduces the depth of chasing that is necessary, and if no power tools are available such as angle grinders, a bolster is probably the best hand tool to use to cut the chases, unless the wall is made of concrete or very hard brickwork.

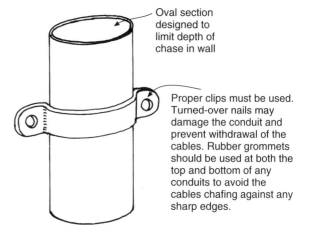

Oval section designed to limit depth of chase in wall

Proper clips must be used. Turned-over nails may damage the conduit and prevent withdrawal of the cables. Rubber grommets should be used at both the top and bottom of any conduits to avoid the cables chafing against any sharp edges.

Fig. 11.19 Light conduit.

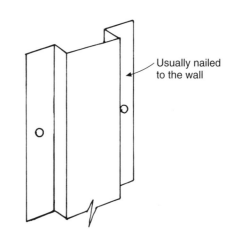

Usually nailed to the wall

Fig. 11.20 Channel section cable cove.

An alternative to oval conduit is the use of channel section PVC shown in Fig. 11.20. It normally makes cutting away unnecessary, as having little depth it can be covered by a normal thickness of plaster. Any concealed cable should always be run straight up or down when feeding any electrical components. Most building operatives are aware of this, therefore chances of piercing a cable with a nail or screw are minimised. Concealed cables running horizontally across a wall are not recommended. Where it is necessary to conceal cables run on the surface of a wall, a light type of PVC trunking is available, the details of which are

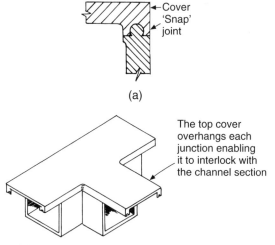

(a)

The top cover overhangs each junction enabling it to interlock with the channel section

(b) This system produces a complete range of components such as junctions and bends

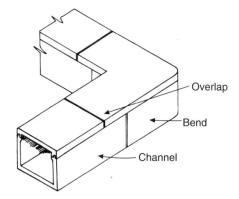

Overlap

Bend

Channel

(c) Showing how the sections overlap to ensure a rigid joint

Fig. 11.21 Light PVC trunking. The illustration shows one of many light trunking systems for the containment of cables on wall surfaces. It may be fixed on smooth surfaces by a self-adhesive backing strip or alternatively, traditional fixings such as screws. It is basically a channel. Which accommodates the cables, the top cover having a 'snap' closing arrangement shown at (a).

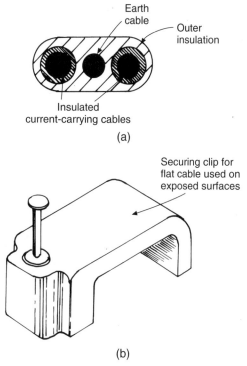

Earth cable

Outer insulation

Insulated current-carrying cables

(a)

Securing clip for flat cable used on exposed surfaces

(b)

Fig. 11.22 Section of flat section cable for general-purpose work. Cable-carrying capacities vary as to the length and temperatures to which it is subjected. Generally 2.5 CSA cable is suitable for ring circuits including any spurs. 1.5 is generally used for lighting circuits.

Cables

Flat section insulated sheathed cables
These cables are commonly used for domestic installations and are suitable for both surface and concealed work where they must be housed in the light types of conduit previously described. Figure 11.22 shows a typical section of this type of cable, the conductors of which are made of copper with the outer insulation being PVC. Due to the fact it is non-flammable and will resist attack by most acids and alkalis, and to some degree solvents, it is a very convenient material. It is manufactured with single, twin or twin and earth core for normal use. Unlike the phase and neutral cores which are insulated, the earth wire is bare except for the PVC outer sheath. For this reason when it is connected to any exposed earth the wire must be protected with

shown in Fig. 11.21. It is not suitable for situations where it could be damaged, but for domestic and light industrial work it does afford some degree of protection for cables and provides a very neat finish to exposed cables. Another advantage is it can be cut and fitted with tools normally carried by a plumber.

a corrosion-resistant and identifying sleeve which is available from electrical stores and is made with the correct earthing colours. The phase and neutral cables are normally red and black, red indicating phase. This type of cable can also be obtained with three insulated conductors and an earth cable. This is extremely useful for wiring components where two live cables are necessary, a typical example being room thermostats. It is not permissible to use the earth wire as a conductor at any time, even if the appliance is double insulated and no earth is necessary.

Bell wire
This term relates to cables used for electric bells which operate at very low voltages. It can in fact be used for any purpose where very low voltages are employed, but must not be used for 240 V, as it usually has only very light insulation.

Sheathed flexible cables
Sometimes called cords these are similar to the PVC sheathed cables previously mentioned, except for the fact they are circular in section and the conductors do not consist of one solid cable but are made up from many fine strands which make it very flexible and suitable for situations where trailing cables are used; typical examples are wiring to pumps and controls situated in a boiler casing. For normal use three-core cables are used with insulation colours of brown (phase), blue (neutral) and green with a yellow stripe (earth). Those having multiple cores employ a larger range of colours for identification purposes. The use of multicore cable avoids the unnecessary doubling up of standard three-core cables in control systems. Where any cables are used in situations where temperatures are higher than normal, i.e. cables connecting an electric immersion heater to a fused switch, butyl or ethylene-propylene insulation is used as these plastics are more temperature resistant than PVC. It is important, however, that no cable should be fixed to or adjacent to hot surfaces as no plastic material used for insulation is capable of withstanding prolonged intense heat without suffering damage.

Special care is necessary when installing cables in boiler casings. Figure 11.23 illustrates this type

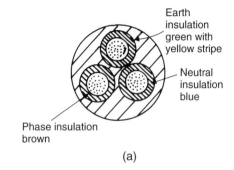

Earth insulation green with yellow stripe

Neutral insulation blue

Phase insulation brown

(a)

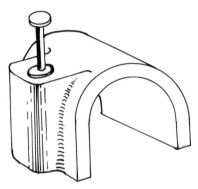

(b) Securing clip for flexible cables

Fig. 11.23 Flexible cables (cords).

of cable; Table 11.1 gives details of cable ratings and sizings.

Cable fixings and installation techniques
Because they are usually unsupported it is very important that flexible cables are firmly secured to the component without placing undue strain on the terminal connections. Figure 11.24 shows some of the methods of achieving this; Table 11.2 gives details of spacing of supports for cables.

To reduce the possibility of damage, runs of flexible cord should be as short as possible, and when unsupported should never exceed 1.2 m in length. Cables under suspended floors can be clipped to a joist where it runs in the same direction, but if not, a hole is bored through the joist at least 50 mm below its top edge through which the cable is threaded. This ensures they are unlikely to be penetrated by nails or screws when the finished flooring is fixed. In situations where it may be difficult to obtain

Table 11.1 Cable and flex sizes and ratings. Conductors in cable and flexes are described by their cross-sectional area, and will give a rough guide to their current-carrying capacity.

Cross-section (mm²)	Capacity (amps)	Domestic usages	Fuse* (amps)
Twin with c.p.c.			
1.0	11 A ⎫	Used for lighting circuits	5/6
1.5	14 A ⎭		10
2.5	18 A	Socket outlets, fixed equipment	15/20
4.0	25 A	Socket outlets, fixed equipment	15/20
6.0	32 A	Showers and cookers	30/32
10.0	43 A	Showers and cookers	45
Flexes			
0.5	3 A		2/3
0.75	6 A		5
1.0	10 A		—
1.25	13 A		13
1.5	16 A		—

* The relative fuse sizes to the above cable ratings give a 'rule of thumb' method of cable sizing.

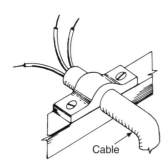

(a) Clamping arrangements used to secure cables in 13 amp plugs immersion heaters and pumps etc

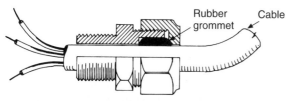

(b) This type of clamp is similar to a copper compression fitting. By tightening the nut the rubber grommet is squeezed on to the cable. It may be threaded directly into a component having a suitable thread or in some cases secured by a backnut (not shown)

Fig. 11.24 Security of cables at terminations.

Table 11.2 Details of cable supports (refer to Fig. 11.24(b)).

Overall Cable (mm)	Spacing of support for cables	
	Horizontal (mm)	Vertical (mm)
UP to 9	250	400
10–15	300	400
16–20	350	450

good surface fixings, the following alternatives may be used:

(a) A light batten screwed to the wall to provide a ground for the fixing clips.
(b) Light plastic trunking previously described.
(c) The use of quick-drying purpose-made adhesives which provide a secure fixing to the cable throughout its length.

When stripping the insulation from cables prior to fixing to terminals always check that the insulation and the conductors are undamaged. Special care is necessary when stripping flexible cords as the fine wires forming the conductors are easily cut and damaged.

It is recommended that correctly adjusted, cable-stripping pliers are used and the setting

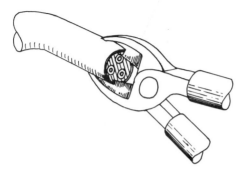

In all cases the following points must be observed:
(a) Care must be taken not to damage the conducting cable or the remaining insulation.
(b) Remove only sufficient insulation necessary to make a satisfactory connection to the terminal.

Fig. 11.25 Cable stripping. Side-cutting pliers may be used to cut the outer insulating sleeve of flexible cord prior to peeling it back to expose the insulated cables. Stripping with a knife is possible, but the cable insulation may be damaged using this method unless extreme care is applied.

is checked on a scrap of wire before use. Figure 11.25 shows how side-cutting pliers are used to strip the outer cable insulation.

Termination of conductors to components
Cables are connected to components by means of terminals which vary depending on the type of component, the c.s.a. of the cables, and whether it is used in an industrial or domestic environment. It is important, however, that any connections are secure, that no bare cable is exposed outside the connecting point and the cables are not subject to stress. Figure 11.26 shows a variety of terminals used for terminating cables to electrical components. Where flexible cables are used, the strands of wire should be twisted together to avoid loose strands protruding from the terminal.

Safe isolation
Prior to any work being carried out on existing installations, or when servicing electrically controlled plumbing equipment, it is essential to isolate the supply. In domestic properties any switches should be in the off position, with the fuses removed that control the circuit on which one is working. Fuses should be stored in a safe place

until the work is completed. Special care must be employed if work is carried out in industrial premises. The first step is to inform a responsible person that the electrical supply will be shut off prior to isolating the supply. A prominent notice should be displayed on the switch or fuse box warning personnel that electrical work is in the process of being carried out and the power must not be switched on. If it is possible to lock the fuse box do so, and keep the key along with any fuses in your toolbox or overall pocket. Finally a test must be made to check the system is isolated by using an approved test lamp shown in Fig. 11.27 or a voltmeter. Circuit testing screwdrivers are not always reliable.

Inspecting and testing electrical installations

Any work carried out on an electrical installation must be tested to comply with the recommendations of the current edition of the *Wiring Regulations*, which covers all the requirements for both visual inspections and tests carried out using special instruments. It is recommended that more detailed information is obtained on the subject from publications listed in the Further reading section of this chapter. Only those tests applying to electrical work carried out by a plumber are dealt with here, and it is important they are performed in a logical sequence. Testing procedures shown do not include three-phase installation.

Visual inspection
The main points to consider here are:

(a) All connections to terminals are properly secured.
(b) Cables are correctly supported and fixed having, where necessary, suitable protection against heat and mechanical damage.
(c) Where applicable, electrical components, switches, etc. should be clearly marked for easy identification.
(d) All protective devices such as fuses must be verified as to their correct rating.
(e) Checks should be made to ensure that all earthing and bonding cables are fitted and checked for security.

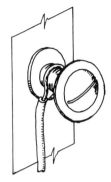

(a) Type of terminal common to domestic heating controls

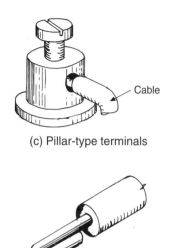

(c) Pillar-type terminals

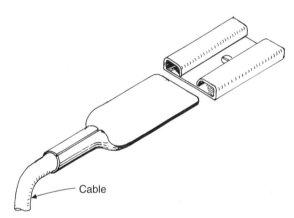

(d) Illustrates the method of doubling the conductor to ensure maximum security

Terminal screws

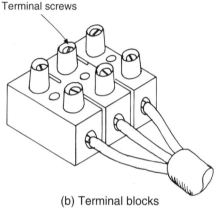

(b) Terminal blocks

These are basically lengths of brass tube encased in a plastic moulding. The cables are secured in the insulated tube by the terminal screws as shown. These blocks are obtained in lengths containing 7 or 10 connections which may be easily cut with a sharp knife depending on the number required. They are made in three sizes to accommodate a range of cables having differing cross-sectional areas. Terminal blocks should always be housed in a suitable box and not simply wrapped in insulating tape.

(e) Spade connection commonly used for extra low voltage work, e.g. 12–24 V

Fig. 11.26 Terminations. This term is used to denote the connections of cables to components.

Installation tests

Testing complete electrical installations is not usually carried out by a plumber. It is, however, necessary to be aware of the associated testing procedures. A multifunction test meter is the most suitable for the kind of work a plumber is likely to undertake, and they test voltage resistance and current over a wide range of applications. Such an instrument is shown in Fig. 11.28. They are very sensitive and care must be taken when they are

handled; they must be carefully stored when not in use. They are definitely not the type of tool to be thrown into a tool bag. It should be noted that these instruments can be seriously damaged if used incorrectly, and it is important that instructions for their use are read and understood. The following tests must be carried out in the sequence shown. The reader should note that an instrument called a 'Mega' is necessary for 'insulation testing'.

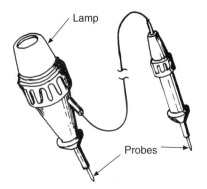

Fig. 11.27 Approved test lamp. Principal use is to check whether the phase conductor of a circuit is energised. To use, the probes are held against the phase and neutral terminals of a switch or appliance, if the lamp lights the phase of obviously live.

Continuity of conductors

This test is carried out to ensure that all conductors or wires are not broken and any connection such as terminals are secure. Figure 11.29 illustrates the method of carrying out this test. A slight resistance may be noted during this test, depending on the length of cable, but if it does not exceed more than a few ohms, the test results are acceptable.

Insulation resistance test

This test is made to ensure the insulation of both phase and neutral conductors are intact, and capable of preventing fault conditions (short-circuiting) between the current-carrying conductors and earth. If, for example, a cable has been subject to abrasion during installation, or inadvertently cut in such a

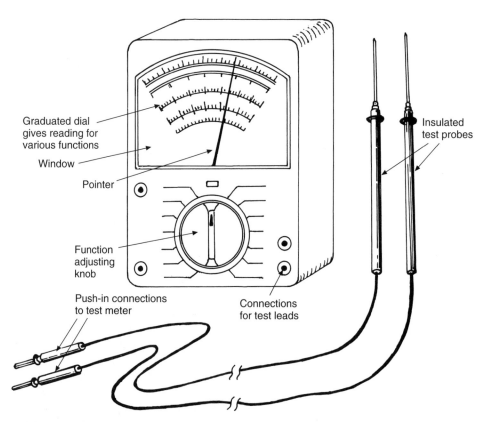

Fig. 11.28 Multi-function test meter. This type of meter is suitable for most of the test functions required for the installation and testing of heating controls and equipment. They are not suitable for all the test procedure described, some of which require an instrument called a 'mega'.

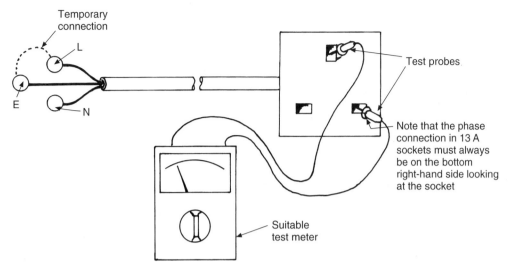

Fig. 11.29 Testing the continuity of conductors. Note that the main supply is isolated while carrying out this test. This test is conducted to ensure that all the current conductors are sound, e.g. unbroken and any connections are secure. The illustration shows the phase cable, connected to a socket outlet being tested. In this case the earth cable is used as part of the circuit and is temporarily connected to the phase terminal in the distribution panel. This test can also be carried out using a battery and bulb.

way that the insulation is damaged, this test will detect it. Prior to conducting the test, any pilot or indicator lamps must be disconnected to avoid inaccurate readings being shown. Any components that may be damaged by the high test voltage employed should be temporarily disconnected. This includes most of the components used in heating control systems. The tests must be conducted with all fuses in place, with switches and circuit breakers closed. Where it is not practical to disconnect current-using equipment, the local switches controlling them must be in the open position. The tests are carried out using an insulation tester with a scale reading in ohms. The illustration in Fig. 11.30 shows how an insulation test is conducted on an electric immersion heater circuit.

Polarity testing
The test must be carried out to ensure that all circuit protection devices, for example fuses and switches, are connected to the phase conductor only. Figure 11.31 illustrates the possible result

of reversed polarity, and this test is carried out to ensure a dangerous mistake such as this does not occur. Testing may be carried out as shown in Fig. 11.32 using a test meter, but it can be done using an electric bell or test lamp powered by a battery.

Ancillary components

The following relates to some of the common components used with electrical equipment common to plumbing installations. The object is to enable the reader to identify them and understand their working principles.

Capacitors
The plumber will be familiar with this component where it is fitted to the motor of the pumps used for domestic heating. In this situation its purpose is to provide the starting torque for the rotor on the type of electric motor used. This electric motor will not start on the single-phase current, and to overcome this a means must be introduced to give the starting

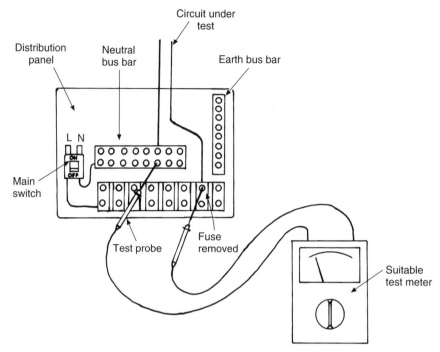

Fig. 11.30   Insulation resistance test. The illustration shows this test between the phase and neutral cables. The circuit protective cable (earth) must also be checked in the same way. Between the phase and earth terminals for mains supply installations the test voltage should not exceed 500 V. The minimum insulation resistance should not be less than 0.5 MΩ.

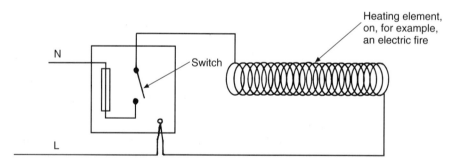

Fig. 11.31   Illustrates the dangerous effects of incorrect polarity. This switch has been incorrectly wired so that although it is open and the circuit is switched off and the element does not glow it is still live and anyone touching it could be subjected to a fatal electric shock. The fuse is also unlikely to function should a fault occur.

torque. This is effected by the capacitor in the way illustrated in Fig. 11.33. The solid line shows the a.c. flow from a single-phase supply. The broken line shows the effect of introducing a capacitor. Figure 11.33 shows a diagrammatic layout of a capacitor start type electric motor. When the motor is energised both the starting and running windings cause the rotor to revolve. At a predetermined speed the centrifugal switch will break the supply to the starter winding as once rotating, its motor will enable the running winding to supply the necessary torque or momentum.

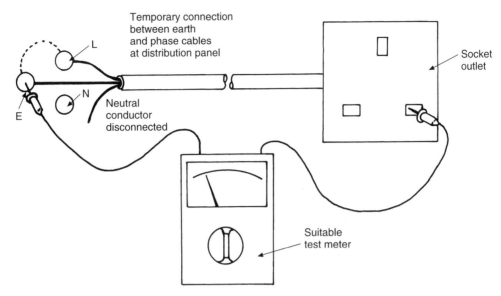

Temporary connection between earth and phase cables at distribution panel

L

N

E

Neutral conductor disconnected

Socket outlet

Suitable test meter

Fig. 11.32 Polarity testing. This test ensures the correct polarity of the circuit throughout. It will be seen that this and the continuity test is very similar and in some cases is carried out simultaneously. It is essential that all fuses, circuit breakers and switches in the phase cable are in position and in the closed position prior to conducting this test.

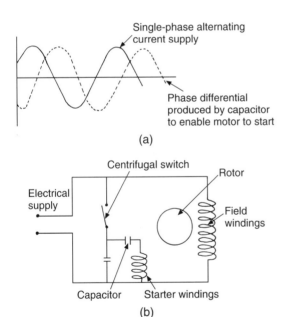

Single-phase alternating current supply

Phase differential produced by capacitor to enable motor to start

(a)

Centrifugal switch

Rotor

Electrical supply

Field windings

Capacitor   Starter windings

(b)

Fig. 11.33 Capacitors — how a capacitor is necessary to start a single-phase electric motor. An electric motor will not start on single-phase current. A capacitor is normally used on small motors used for heating pumps, washing machines, etc. to provide a secondary phase to start the motor. When the rotor is turning the integral centrifugal switch contacts open closing off the current to the starter windings.

One of the most common causes of domestic heating pump failure is a faulty capacitor, usually indicated by the humming sound from the pump and no rotation of the motor. Care must be exercised when handling capacitors as they store an electrical charge, and even when isolated they are capable of delivering an unpleasant electrical shock. The terminals of a capacitor should be earthed to remove the charge when it is disconnected for maintenance purposes.

Electromagnets

These consist of a wire wound on to an iron core and when the coil is energised, the resulting magnetic field in the core is similar to that of the familiar permanent magnet. A simple experiment may be carried out by winding a length of fine insulated wire on to a wire nail about 75 mm long and connecting the two ends of the wire to a torch battery. The result will be a fairly powerful magnet. Electromagnets have many applications, one of the most common being used in an electric bell. Yet another is to operate a relay, which is basically an electric switch, when the coil of the relay is energised. This opens or closes another switch to

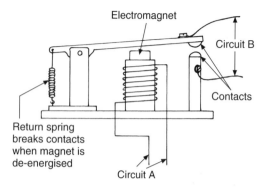

Fig. 11.34 Application of electromagnet. Electromagnet acting as a relay, when circuit A is energised it makes the contacts of circuit B. This principle has many applications in the automatic functions of gas- and oil-fired equipment.

control a separate circuit to that which operates the relay. In this way a low-voltage circuit can be used to control a higher voltage. Figure 11.34 shows a simple relay used in this way. Relays are widely used in automatic control systems and may be obtained with various contacts which can be either opened or closed by the single operation of a relay. Relays are also used in safety devices, when the magnetic force resulting from excessive current can be made to overcome the force of a retaining spring, resulting in the isolation of the voltage source. This forms the basis of many of the automatic cut-out devices on plumbing equipment.

Solenoids are another application of electromagnets. Figures 2.31, 2.35 and 2.40 show typical examples of their use.

Transformers
Electricity is produced in power stations and distributed by the national grid at high voltages, which must be reduced to 240 V for domestic supplies. It is quite often necessary to reduce this voltage still further in the interest of safety, the reduction to 110 V for power tools is a common example. Many components in heating equipment used by the plumber require an increase or reduction of mains voltage. In other cases it is increased to produce a spark for ignition devices. The component used for reducing or increasing an a.c. voltage is known as a step-down or step-up transformer. A transformer works on the principle

of electromagnetic induction and consists in its basic form of two independent windings magnetically interlinked by an iron or iron alloy frame or core. When the current flows through one of the windings, called the *primary winding*, the magnetic field *induces* a voltage into the other winding, called the *secondary winding*. If there are more turns of wire on the secondary winding than on the primary, the voltage across the secondary will be proportionately greater. If the secondary has fewer turns of wire than the primary then the converse is true. The two coils are placed in close proximity. Figure 11.35 illustrates in diagrammatic form a simple transformer.

Regulations relating to electrical appliances in bathrooms

Amendment No. 3 to BS 7671, requirements for electrical installations, is now mandatory. The amendment covers section 601 on locations containing a bath or shower. Water, being a conductor of electricity, poses a real danger when it comes into contact with electrical appliances and switches, and previous legislation has been stringent. However, the new regulations are very specific. Bath and shower rooms are divided into 4 zones: 0, 1, 2, and 3. Figure 11.36 shows these zones in connection with a shower. The lower the numeral, the greater will be the possibility of a dangerous situation. Thus zone '0' relates to a situation where the appliance is likely to be completely submerged, which is unusual with building services. The following is a guide to the types of controls and switches which may be used in the various zones. There are, however, certain exceptions where they are incorporated into the equipment designed for use in a specific zone. Normally the following applies:

Zone 0.   No switch or controls normally permissible.
Zone 1.   Only switches of up to 12 V d.c. or 30 V 'ripple-free' d.c. may be fitted, with all safety sources being installed outside of zones 0–1–2. The term ripple-free relates to certain types of rectifying equipment which are integral to the appliance.

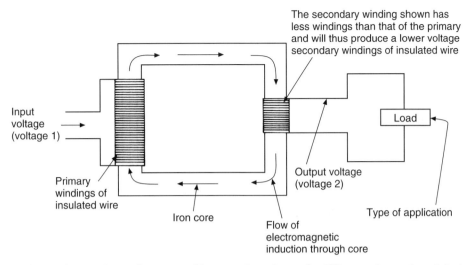

The secondary winding shown has
less windings than that of the primary
and will thus produce a lower voltage
secondary windings of insulated wire

Input
voltage
(voltage 1)

Load

Primary
windings of
insulated wire

Iron core

Output voltage
(voltage 2)

Flow of
electromagnetic
induction through core

Type of application

Fig. 11.35 Basic transformer. A transformer provides an outlet voltage of a different value to that of the input. They can be used to step up or increase the voltage to provide the spark for ignition devices, or step down where the output or secondary voltage may be less than that of the primary.

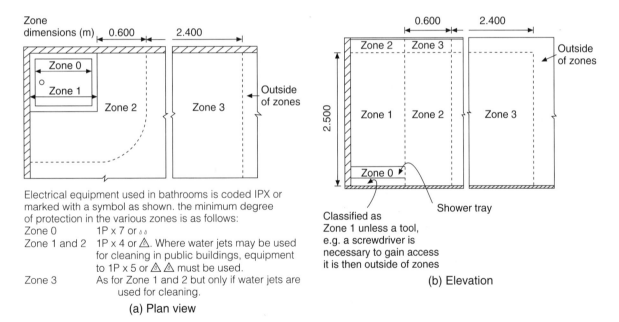

Zone
dimensions (m)

0.600

2.400

Zone 0

Zone 1

Zone 2

Zone 3

Outside
of zones

Electrical equipment used in bathrooms is coded IPX or
marked with a symbol as shown. the minimum degree
of protection in the various zones is as follows:

Zone 0   1P x 7 or ◊◊

Zone 1 and 2 1P x 4 or △. Where water jets may be used
for cleaning in public buildings, equipment
to 1P x 5 or △ △ must be used.

Zone 3   As for Zone 1 and 2 but only if water jets are
used for cleaning.

(a) Plan view

0.600

2.400

Zone 2

Zone 3

Outside
of zones

Zone 1

Zone 2

Zone 3

2.500

Zone 0

Shower tray

Classified as
Zone 1 unless a tool,
e.g. a screwdriver is
necessary to gain access
it is then outside of zones

(b) Elevation

Fig. 11.36 Electrical switches and controls in shower rooms.

Zone 2.   The recommendations are the same
as for zone 1 except for showering
supply outlets complying with BS
EN 60742.

Zone 3.   As for zones 1 and 2 and additionally
low-voltage socket outlets incorporating a

safety isolating transformer complying to
BS 3535.

The following fixed electrical current-using
equipment may be used in zone 1, providing it meets
the relevant standards and is suitable for the zone.

(a) A water heater, e.g. electric shower units.
(b) A shower pump.
(c) Other equipment which meets the requirements of IEE Regulations 412-06 (certain types of extractor fan may be suitable).
(d) Low-voltage current-using equipment.

All the foregoing are suitable for inclusion in zone 2, as are protection lighting, extractor fans, heaters, and whirlpool bath equipment. Insulated pull-cord switches complying with BS 3676 are suitable for use in zones 1, 2 and 3.

All the appliances listed in zones 1 and 2 may be used in zone 3, which also permits equipment other than fixed current devices which meet the requirements of IEE regulations 412-06. They must be protected by a suitable residual current device affording automatic disconnection of the supply, and earth equipotential bonding. Where possible they should be used in a non-conducting location e.g. out of reach of splashing or accidental contact with water.

Further reading

IEE Regulations
Details of publications relating to electrical work can be obtained from
The Institute of Electrical Engineers, Savoy Place, London WC2R 0BL.
MK Electrics Ltd, Marketing Services Department, Shrubbery Road, Edmonton, London N9 0OB.

Electrical testing equipment
AVO International, Archcliffe Rd, Dover, Kent, CT17 9EN, Tel. 01304 502100.

Self-testing questions

1. Name the regulations to which all electrical installation work must comply.
2. Identify the difference between direct and alternating currents.
3. Identify the electrical terms used to define pressure of current flow, quantity of current and resistance to flow.
4. Assuming an appliance is supplied with a voltage of 240 and has a resistance of 60 Ω, state the amperage of a commercial cartridge fuse that can safely be used.
5. State the effect on a cable subject to overcurrent.
6. State the reasons for earthing electrical systems and explain the three main methods used.
7. Define the term 'equipotential bonding' and explain why it is necessary.
8. State the reason why high rating appliances such as electric water heaters and cookers are not recommended to be connected to a ring circuit.
9. Specify the type of insulation necessary for flexible cables where higher than normal temperatures may be encountered, e.g. adjacent to boilers or water heaters.
10. List the procedures taken to ensure safe electrical isolation when rewiring an electric immersion heater from an existing fused switch.
11. List the visual checks that must be made on an electrical installation prior to commissioning.
12. (a) State the purpose of conducting a polarity test after having installed a fused switch from an existing supply.
    (b) Describe the possible effect of incorrect polarity.

Appendix A: Assignments

Dormer window weathering assignment

A dormer window is to be weathered in lead sheet, the relevant details of which are shown in Fig. A.1.

1. State the recommended codes of lead sheet for weathering:
   (a) The apron.
   (b) The cheeks.
   (c) The top.
2. (a) From the measurement shown in Fig. A.1(b) specify the width of the lead sheet required to form the apron.
   (b) Set out half full size on paper the cuts and folds necessary to form the apron using lead welding techniques round the corner post and on to the roof.
   (c) Cut and fold the paper, make and insert the necessary gussets securing them by masking or adhesive tape, and check the measurements and roof pitch against the details shown.
3. (a) Name three intermediate fixings that could be used to secure the dormer cheeks.
   (b) Make a sketch showing two methods of securing and finishing the front edge of the cheeks to the corner posts.
   (c) Specify the type of nails and cleats used for fixing lead sheet work.
4. To prevent high winds lifting the lead, the drip edges are shown secured by bale tacks welded to the edge. An alternative method of finishing this detail is the use of a torus roll. Make a sectional sketch through a torus roll including the lead weatherings.
5. Show by means of a sketch a suitable method for fabrication by lead, welding the roll overcloak at the junction with the roof. The weld is to be made out of position.

6. From the measurements shown in Fig. A.1 calculate the total area of lead required to weather the dormer top, making an allowance of 85 mm for the undercloak and 175 mm for the overcloak. Make an allowance of $12\frac{1}{2}$ per cent for waste.

Hot and cold water supply assignment

The water service in a domestic property, which has been unoccupied during the winter months, is to be commissioned, and the estate agent responsible for the property has also asked for a general report on the hot and cold water services. Compile a report listing the remedial action necessary to correct the faults and defects you have found in the system.

1. Although the water was shut off and the pipework drained when the building became unoccupied, it is noted that some of the pipework is below the level of the drain cock and may have suffered frost damage. As much of the work is concealed below the floor level, describe how it could be tested for soundness prior to admitting water to the system.
2. The original cold water storage cistern has at some stage been replaced with two smaller ones. Figure A.2 illustrates these and the feed and expansion cistern. State the reasons why the cisterns as shown contravene the Water Regulations.
3. The cold water supply to a shower in the ground floor cloakroom is found to be connected directly to the cold water service pipe, and the water temperature at the shower rose is found to be very difficult to adjust. It is also noticed there is a discharge from the overflow of the cisterns in the roof space when the shower is operating. State on your report

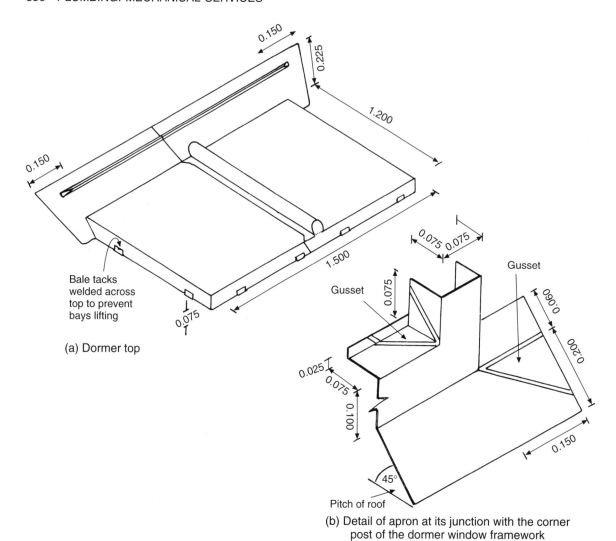

0.150

0.225

1.200

0.150

0.075 0.075

Gusset

Bale tacks
welded across
top to prevent
bays lifting

Gusset

0.075

0.060

0.200

1.500

0.075

(a) Dormer top

0.025

0.075

0.100

0.150

45°

Pitch of roof

(b) Detail of apron at its junction with the corner
post of the dormer window framework

Fig. A.1 Weathering dormer windows.

the cause of these problems and how they
should be rectified.

4. It is also noticed that when the cold tap on the
bath is turned off suddenly, a loud noise is
heard throughout the entire cold water system
indicating it is subject to water hammer.
Despite changing the washer and reducing the
flow rate at the stopcock, the noise persists. List
two possible components that could be used to
solve this problem.

5. It is noted that a large quantity of water has
to be drawn off from the hot supply to the

washbasin in the downstairs toilet before
water at a usable temperature is delivered. This
is due to a very long draw-off. Describe a less
wasteful method of providing hot water if an
electrical power supply is available.

Oil-fired heating assignment

A sealed system of heating is to be installed in the
office suite of a medium-sized industrial building.
The system is to be heated using an atomising oil
fired boiler.

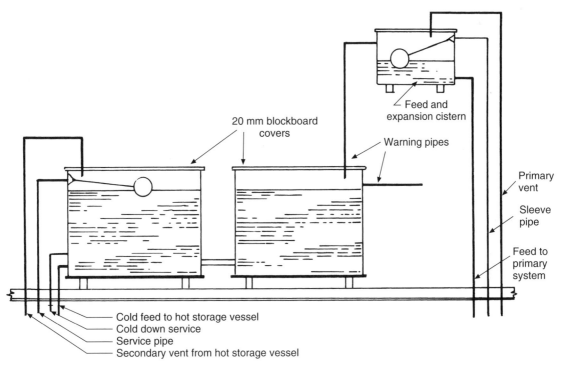

Fig. A.2 Illustrates cold water storage cisterns and pipework in roof space. Insulation not shown.

1. (a) State the recommended maximum working temperature of the heating system.
   (b) In the event of thermostat failure, what provision must be made to ensure the safety of the installation?
2. (a) Specify two types of heat emitters that would be suitable for a sealed heating system and explain the reasons for your selection.
   (b) Specify two types of pipework systems that would be suitable for the heat emitters you have selected.
3. (a) Explain how the system should be filled from the main cold water service to comply with the Water Regulations.
   (b) Describe an alternative method of filling the system.
4. From the illustration shown in Fig. A.3, list the components A to J and describe their function.
5. Describe the procedures taken when commissioning the atomising burner, including all the tests necessary to ensure its efficient operation.
6. While conducting the tests it is found the chimney is subject to an excessive updraught. Specify a suitable component that must be fitted in the flue to overcome this problem and explain its working principles.
7. The owner of the premises has indicated that he would like to enter into a maintenance agreement. Prepare a schedule indicating all the items that must be checked during an annual service on the installation.

Sanitation and drainage assignment

A three-storey office building is to be extended, necessitating an increase in the sanitary accommodation. Due to problems involved in connecting the new sanitary pipework to that in existence, it has been decided to install an additional discharge pipe serving a new sanitary annexe on the first floor. A 6 m length of drain will

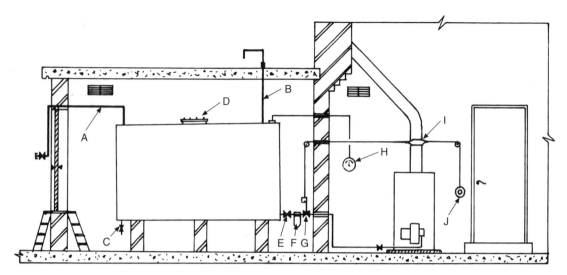

Fig. A.3 Oil storage and boiler installation.

also be necessary from the stack to the existing PVC drain. The connection at the drain will be made by constructing an additional inspection chamber.

1. (a) If the invert of the existing drain is 1.2 m below ground level, and the gradient is 1 in 40, what will be the invert level at the upper end.
   (b) Describe a suitable method of checking the invert on a short length of drain.
   (c) Describe the type of bedding necessary for flexible underground drain pipes to comply with the Building Regulations.
2. (a) Explain how you would connect the new branch into the existing drain to cause as little inconvenience as possible to the occupants of the existing building.
   (b) Select a suitable component for making the actual connection of the new branch drain to the existing main drain.
   (c) Describe an approved method of water testing the underground drain on completion.
3. The first and second floors of the extension are constructed of reinforced concrete, with holes left at the points where the main discharge stack passes through them.

(a) State how the gap where the discharge pipe passes through the floor is sealed to prevent the spread of fire in the event of an emergency.
   (b) State the recommendations relating to the main discharge stack where (i) it terminates as a vent to the atmosphere, and (ii) the type of bend used at the base of the stack.
4. (a) Explain why an excessive number of bends on a branch discharge pipe may cause siphonage of the appliance traps.
   (b) In a situation where the maximum recommended length of a basin waste is exceeded, describe three acceptable methods of overcoming the possible siphonage of the trap seals.
5. An automatic flushing cistern is fitted serving three bowl urinals in the male toilet.
   (a) State two methods of controlling the frequency of flushing which complies with the Water Regulations.
   (b) List the advantages and disadvantages of bowl urinals.
6. Describe the approved method of conducting soundness and performance tests on the completed sanitary pipework installation.

Appendix B: Colour Coding of Pipelines

Table B.1 Basic identifying colours

Pipe contents	Colour	BS colours ref.
Water	Green	5-65
Steam	Silver grey	9-099
Combustible oil	Brown	2-030
Gases not air	Yellow ochre	368
Acids/alkalis	Violet	797
Air	Light blue	8-088
Other fluids	Black	9-105
Electrical	Orange	557

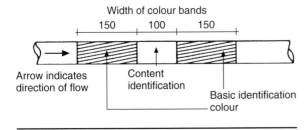

Width of colour bands

Arrow indicates direction of flow

Content identification

Basic identification colour

Table B.2

Pipe contents	Basic colour	Content colour
Drinking water	Green	Blue
Cooling water	Green	White
Cold down services	Green	White blue white
Hot water services	Green	White red white
Boiler feed water	Green	Red white red
Fire services	Green	Red
Condensate water	Green	Red green red
Chilled water	Green	White green white
Central htg < 100 °C	Green	Blue red blue
Central htg > 100 °C	Green	Red blue red
Drainage		Black
Compressed air		Light blue
Natural gas		Yellow
Town gas		Yellow green yellow

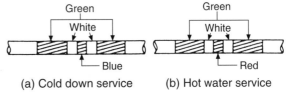

Green

White

Blue

Red

(a) Cold down service   (b) Hot water service

Appendix C: BS 1192: Piped and Ducted Services

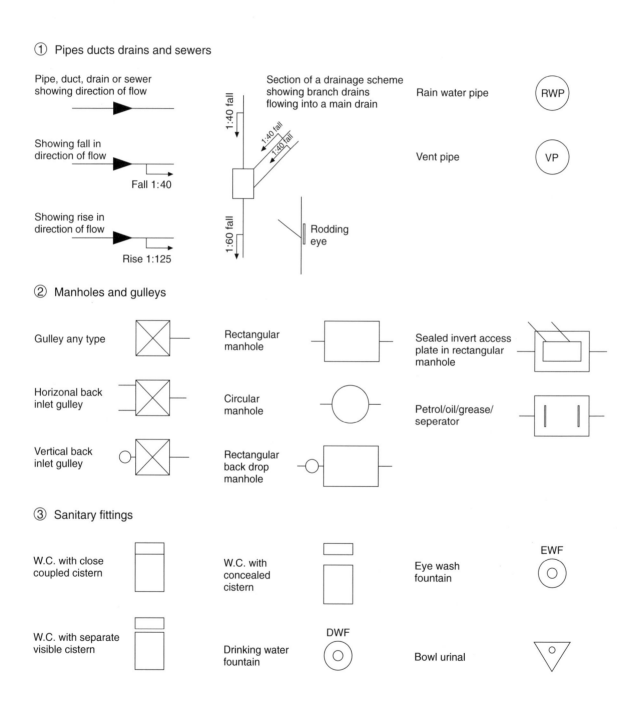

① Pipes ducts drains and sewers

Pipe, duct, drain or sewer showing direction of flow

Showing fall in direction of flow

Fall 1:40

Showing rise in direction of flow

Rise 1:125

Section of a drainage scheme showing branch drains flowing into a main drain

1:40 fall

1:40 fall

1:40 fall

1:60 fall

Rodding eye

Rain water pipe

RWP

Vent pipe

VP

② Manholes and gulleys

Gulley any type

Horizonal back inlet gulley

Vertical back inlet gulley

Rectangular manhole

Circular manhole

Rectangular back drop manhole

Sealed invert access plate in rectangular manhole

Petrol/oil/grease/ seperator

③ Sanitary fittings

W.C. with close coupled cistern

W.C. with separate visible cistern

W.C. with concealed cistern

Drinking water fountain

DWF

Eye wash fountain

EWF

Bowl urinal

③ Sanitary fittings (*cont'd*)

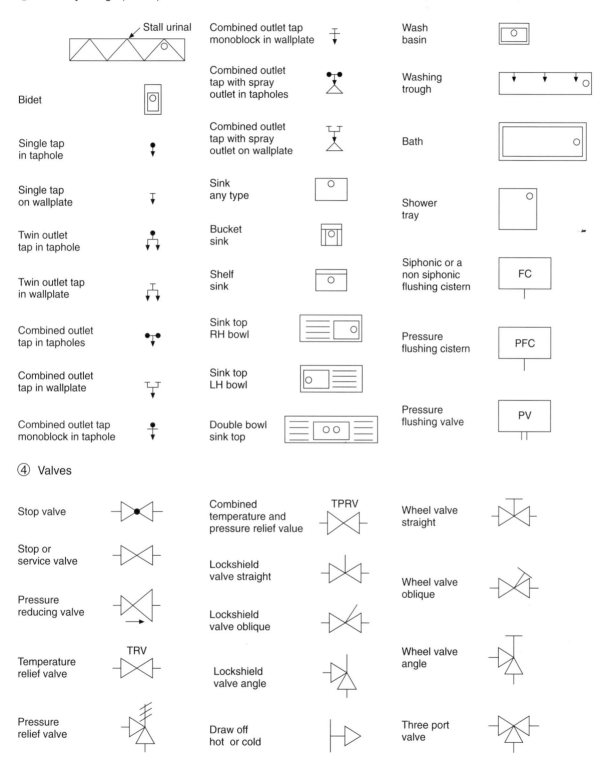

④ Valves

④ Valves (*cont'd*)

Four port valve

Motorised
two port valve

Regulating
valve

Drain tap

Drain tap
alternative symbol

Double
check valve

Single
check valve

Float operated
valve

Pipe interupter
with permanent
atmospheric vent

Pipe interupter with
atmospheric vent
and moving element

Anti vacuum
valve

⑤ Pipe fittings

Tun dish with air gap

Union joint

Flanged end

Flanged joint

Bonded joint

Anchor point

Pipe guide

sliding coupling

Flexible coupling

Capped end

Steam trap
seperator

Condensate
release trap

Open vent

Valved vent

Bottle vent

Automatic
air vent

Gas meter

Water meter

Capped end

Blanking plate

Spray outlet

Venturi

Prssure
tapping

Test point

Test/sensor
pocket

Line strainer

Alternative strainer

⑥ Pipework components

Radiator		Forced draught convector		Pump, any type	
Heated towel rail		Radient strip		Expansion vessel	
Natural draught convector		Unit heater		Alternative expansion vessel	

⑦ Ductwork

Air flow direction inward		Test hole		Inspection plate face view	
Air flow direction outward				Edge view	
Turning valves		Damper sectional		Grill/diffuser face view	
Flexible duct/connector		Damper elevation		Edge view	

Adapted from BS 1192 Part 3: 1987. This standard has been withdrawn and is replaced by BS EN ISO 3766: 1999, BS EN ISO 7518: 1999 and BS EN ISO 11091: 1999. British Standards can be obtained from the British Standards Institution. (Tel +44(0)2089969001).

Appendix D: National Vocational Qualifications

NVQs in mechanical engineering services — plumbing (currently under review)

National Vocational Qualifications (NVQs) are a new system of certification which have been introduced in England and Wales. They have replaced the long-established method of assessing awards by external examinations, which has been accepted by industry for many years, in recognition of the achievement of a craft qualification. NVQs are based on continuous assessment rather than examinations and, like their Scottish equivalent, Scottish Vocational Qualifications (SVQs), reflect standards that have been agreed by the plumbing industry. They have been developed jointly by the British Plumbing Employers Council, plumbing employers, organisations and the Joint Industrial Board (JIB) for the Plumbing Industry.

In England and Wales two levels of NVQ in plumbing have been introduced: levels 2 and 3 in Mechanical Engineering Services — Plumbing. Level 2 replaces the craft certificate and level 3 the advanced craft certificate which were awarded by the City and Guilds of London Institute.

Structure

Each level has a specified number of units of competence, which are divided into elements of competence. Each element of competence has performance criteria and range statements. The element defines the standards which have to be met and the range defines the circumstances in which the standard must be applied.

Assessment

The syllabus (which summarises what has to be assessed) is dictated by the element of competence, the performance criteria and the range statements. Generally, the assessment for each unit comprises:

- practical observations
- oral questions
- written questions

Written evidence is required for the practical observations.

There is no need to assess all elements of a unit at once. Each element can be assessed separately. When a candidate has satisfied all the performance criteria for each element in a unit the candidate will be credited with that unit. The candidate must meet all performance criteria successfully for each element and satisfy the knowledge requirements of the element.

There is no requirement for units to be completed in a particular order. Candidates can complete separate elements in several units before completing any one unit. Units and elements can also be assessed in any order. In addition, units do not have to be completed within a certain time.

Work-based assignments

An integral part of the course is the continuous work-based assessment of the candidate's work which can be carried out in the workplace or at an appropriate centre. Assessments are made on a range of practical activities and on documentary evidence provided by the candidates to prove their competence. The evidence may be in the form of photographs or written statements by employers or clients. Where assessment in the workplace is not possible, it can be carried out through simulated activities and tasks in an approved centre.

Activity record

An activity record is a written description of the activities that have been undertaken by the

candidate during the completion of a unit.
The record lists details of the following:

- Activities undertaken, including production techniques used accompanied by appropriate sketches.
- Any problems or difficulties encountered and how they were resolved.
- Knowledge of what was done and of any relevant legislation and company policies that applied to the activities.
- How the job was planned, including any liaison with other personnel, co-contractors and customers.
- All safety precautions taken.
- Materials and equipment used.

Related knowledge and understanding

A candidate's performance may be demonstrated directly. However, in many cases satisfactory performance alone is not enough and the candidate must show an understanding of the task being done. Standard assessments have been prepared by the awarding body to allow candidates to show that they have this knowledge and understanding.

The assessments comprise a series of questions which require a short written answer or a sketch. Sometimes this will need to be supplemented by oral questions, which will be asked by the assessor. Centres may sometimes provide their own assessments.

Accreditation of prior learning (APL)

Candidates can use evidence from work or other activities undertaken before starting the NVQ in the assessment of a unit. The evidence must satisfy the performance criteria and range statements in the same way as evidence gathered while working towards the NVQ.

Portfolio of evidence

The 'portfolio of evidence' is the documentation which the candidate submits to the assessor for assessment of an element or unit. This will usually be a file or folder in which the candidate keeps all their evidence. It should be emphasised that it is the candidate's responsibility to keep the portfolio and add the appropriate evidence.

The portfolio should include a completed record of assessment and the supporting evidence. The evidence can be anything that illustrates the candidate's competence. Forms of evidence can include authenticated photographs, job cards, time sheets, appraisals from line managers or supervisors, testimonials, video and audio tapes and computer disks.

The portfolio must be accessible to the reader. It should be divided into sections which relate to the units and elements of the NVQ. The information must be coherent and include a contents page, dates on all entries and titles and headings to describe the contents. All related entries should be cross-referenced, indicating the relevant units and elements of the standards.

Sources of information

Further information on NVQs and SVQs can be obtained from the following organisations.

British Plumbing Employers Council,
14–15 Ensign Business Centre,
Westwood Business Park,
Westwood Way,
Coventry CV4 8JA
Tel. 01203 470626.

Scottish Vocational Educational Council (ScotVEC),
Hannover House,
Douglas Street,
Glasgow
Tel. 0141 248 7900.

Appendix E: Study Guide for NVQ Units

This appendix shows how the underpinning knowledge given in this volume of *Plumbing: Mechanical Services* relates to the NVQ units. It follows the **main** performance requirements in the 'type of evidence' sections of the unit specification, indicating the pages or chapters in Book 2 that contain the relevant underpinning knowledge; Book 1 of *Plumbing: Mechanical Services* is referenced where appropriate. The supplementary evidence requirements are not covered.

The requirements are listed unit by unit and element by element. The number of each performance requirement matches the number in the unit specifications.

The reader should use this information as a guide only, for details relating to specified components and systems reference should be made to relevant chapters and the index.

Unit P2/1 Install and test the components of the system

Element P2/1.1 Interpret the installation requirements of the components of the system

1. Hot and cold water, hot water heating
   Chapters 3, 4 and 5 and *Book 1*
2. Unvented and non-storage hot water systems
   Pages 122–30
3. Above ground discharge pipework and sanitation systems   Pages 249–54 and *Book 1*
4. Below ground drainage system
   Pages 266, 284–7
5. Gas supply   Chapter 2
6. Electrical systems   Chapter 11
7. Mains services   *Book 1*
8. Oil supply   Chapter 6

Element P2/1.2 Prepare sites for installation and testing

1. Checks, health and safety with respect to:
   (a) access equipment   *Book 1*
   (b) excavations   *Book 1*
   (c) hazardous conditions   *Book 1*
2. Input services   *Not covered*
3. Schedules specifications   *Not covered*
4. Checks on site conditions   *Not covered*

Element P2/1.3 Fabricate, position and fix components

Assessment method I
Bending of materials:

* copper   *Book 1*
* low carbon steel   *Book 1*

Assessment method II
Jointing of materials:

* cast iron   *Book 1*
* copper   *Book 1*
* low carbon steel   *Book 1*
* pressure pipe (plastic)   *Book 1*
* soil and waste systems (plastic)   *Book 1*

Assessment method III

1. Hot and cold water, hot water heating
   Chapters 3, 4 and 5 and *Book 1*
2. Unvented and non-storage hot water systems
   Pages 122–30, 142–4, 233
3. Above ground discharge pipework and sanitation systems
   Pages 255–8, 260 and *Book 1*
4. Below ground drainage system
   Pages 255–61, 268, 275, 277–8

5. Gas supply   Pages 40, 43–9, 51–4
6. Electrical systems   Pages 334, 340, 345

Element P2/1.4 Connect and test components

Assessment method I

1. Connection to incoming service:
   (a) cold water   *Book 1*
   (b) gas   Pages 31–3 and *Book 1*
   (c) soil system to drain termination   *Book 1*
2. Connection of new pipework into an existing gas supply   Pages 31–3

Assessment method II

1. Water soundness   Page 114
2. Gas soundness:
   (a) existing systems   Page 31
   (b) system after connection   Page 33
   (c) new installation   Page 32
3. Soil and waste soundness   Pages 261–2
4. Drainage   Pages 295–8
5. Flue:
   (a) draught   Page 39
   (b) soundness   Page 40
6. Electrical:
   (a) earth continuity   Pages 347–9
   (b) polarity   Page 349
7. Mains water pressure   Page 115
8. Water flow rate   Page 115

Unit P2/2 Commission and decommission systems

Element P2/2.1 Carry out checks prior to performance testing

1. Hot and cold water, hot water heating
     Chapters 3 and 4 and *Book 1*
2. Unvented and non-storage hot water systems
     Pages 123–4
3. Above ground discharge pipework and sanitation systems   Pages 261–3 and *Book 1*
4. Below ground drainage system   Pages 295–8
5. Gas supply   Pages 32–3
6. Electrical systems   Page 347

Element P2/2.2 Monitor and compare the dynamic performance

1. Hot and cold water, hot water heating
     Chapters 3, 4 and 5 and *Book 1*
2. Cold water   Chapter 3 and *Book 1*
3. Above ground discharge pipework and sanitation systems   Chapter 8 and *Book 1*
4. Below ground drainage system   Chapter 9
5. Gas supply   Chapter 2
6. Electrical systems   Chapter 11

Element P2/2.3 Decommission systems

1. Hot and cold water, hot water heating
     Page 116 and *Book 1*
2. Unvented and non-storage hot water systems
     Page 152
3. Above ground discharge pipework and sanitation systems   Page 299 and *Book 1*
4. Below ground drainage system   Page 299
5. Gas supply   Page 69
6. Electrical systems   Page 346

Unit P2/3 Maintain the effective operation of systems

Element P2/3.1 Routinely maintain system components

1. Hot and cold water, hot water heating   Page 77
2. Unvented hot water systems   Pages 129–30
3. Non-storage hot water systems
     Refer to individual manufacturer's maintenance instructions

Components:

- Stop valve   *Book 1*
- Gate valve   *Book 1*
- Ball valve   *Book 1*
- Radiator valve (thermostatic)   Page 175
- Float operated valves (including diaphragm type)   Pages 77–8 and *Book 1*
- Pumps   Pages 77–140
- Motorised valves   Page 164
- Pressure relief valves   Pages 126, 170
- Temperature relief valves   Page 128
- Vacuum relief valves   Pages 95–6, 125

- Pressure reducing valves   Page 84
- Mixing valves   Pages 227, 230–1, 255
- Pressure/storage vessels   Pages 79–81, 191–4

Element P2/3.2 Diagnose and rectify the cause of faults

1. Insufficient or no water supply   *Book 1*
2. Air locks   *Book 1*
3. Noise in system
          Pages 77, 79, 80–4, 189 and *Book 1*
4. Component failure
          Refer to individual manufacturer's maintenance instructions
5. Blockage   *Refer to all related chapters*
6. Leakage   *Refer to all related chapters*
7. Corrosion of components
          Pages 118–19, 189, 266 and *Book 1*
8. Loss of trap seal   *Book 1*

Unit P2/4 Maintain the safe working environment

Element P2/4.1 Monitor and maintain one's own health and safety

Assessment method I

1. Checks for suitability:
   (a) step ladders   *Book 1*
   (b) trestle scaffold   *Book 1*
   (c) ladders   *Book 1*
   (d) roof ladders   *Book 1*
   (e) scaffolds   *Book 1*
   (f) mobile towers   *Book 1*
2. Lifting equipment checked   *Book 1*
3. Appropriate lifting techniques used   *Book 1*
4. Large heavy object is lifted   *Book 1*
5. Checks carried out on:
   (a) hand tools   *Book 1*
   (b) immediate work area   *Book 1*
6. Appropriate precautions:
   (a) combustible/noxious/explosive/dangerous
       gases   Pages 1–3
   (b) hazardous materials
       Book 1, also *refer to COSHH Regulations*
7. Personal protective equipment
       Page 2 and *Book 1*

Assessment method II
Accident procedures:

- Authorised person   *Book 1*
- Details entered in accident book   *Book 1*
- Appropriate forms completed   *Book 1*

Assessment method III
Unsafe working practices:

- Authorised person   *Book 1*
- Appropriate forms completed   *Book 1*

Element P2/4.2 Contribute to the limitation of damage

1. First aid/emergency procedures
   (a) minor cuts
       Refer to an approved first aid manual
   (b) minor burns
       Refer to an approved first aid manual
   (c) electric shock   *Book 1*
   (d) shock
       Refer to an approved first aid manual
2. Assistance   *Book 1*
3. Warning/alerting people   *Book 2*
4. Damage to premises   *Not covered*

Element P2/4.3 Contribute to the limitations of damage
This element is beyond the scope of these books.

Element P2/4.4 Agree and maintain a safe environment

1. Hazards identified   *Book 1*
2. Hazards made safe   *Book 1*

Unit P2/5 Maintain effective working relationships

Element P2/5.1 Establish and develop professional relationships with customers and co-contractors
This element is beyond the scope of these books.

Element P2/5.2 Establish and maintain professional relationships with authorised site visitors
This element is beyond the scope of these books.

Element P2/5.3 Maintain effective working relationships with colleagues
This element is beyond the scope of these books.

Unit P2/6 Contribute to quality development and improvement

Element P2/6.1 Promote the organisation's/ industry's image
This element is beyond the scope of these books.

Element P2/6.2 Encourage energy efficiency
Refer to chapters 4 and 5, also *Book 1*.

Unit P2/7 Fabricate, install and check sheet weathering systems components

Element P2/7.1 Interpret the installation requirements
Sheet lead installations:

1. Chimney penetration through pitched roof
   Book 1
2. Pitched roof to vertical wall   *Book 1*
3. Weathering to a dormer   Pages 310–16
4. Weathering to a roof penetration   *Book 1*
5. Weathering to a canopy (flat)   *Book 1*
6. Weathering to a bay top (flat):
   Page 306 and *Book 1*
   (a) system material, material specification, etc.
   Book 1
   (b) schedule of materials and components
   Book 1

Element P2/7.2 Prepare sites for fabricating, installing and checking

1. Checks: health and safety, etc.   *Book 1*
2. Surface preparation   *Book 1*
3. Temporary protection of building and contents
   Not covered

Element P2/7.3 Fabricate, position and fix components

1. Marking out sheetwork
   Pages 318–22 and *Book 1*
2. Preparatory work   *Book 1*

3. Laying inodorous/proprietary underlays   *Book 1*
4. Fabricating:
   (a) internal corners by bossing   *Book 1*
   (b) external corners by bossing   *Book 1*
   (c) internal corners by lead welding
   Pages 321–3
   (d) external corners by lead welding
   Pages 9, 17, 319–20
5. Fixing sheet lead to vertical brickwork   *Book 1*

Unit P2/8 Maintain sheet weathering systems

Element P2/8.1 Interpret the requirements for routine maintenance
Routine maintenance requirements:

1. Keeping free of debris   Page 326
2. Checking for defects   Pages 7, 8, 326
3. Checking fixings   Page 326

Schedule of maintenance requirements   Page 326

Element P2/8.2 Maintain and rectify faults in sheet weathering systems   Page 326

Unit P3/1 Design systems to meet customer's requirements

Element P3/1.1 Identify and establish customer's requirements
This element is beyond the scope of these books.

Element P3/1.2 Design and agree systems to meet customer's needs

1. The designed system and layout complies with relevant statutory requirements for gas, water, electricity, health and safety, building control and planning regulations.
   Pages 24, 26, 71, 86–7, 91–100, 151, 176, 265–6
2. The agreed option complies with British, European and International standards and codes of practice recommended for industry use.   *Not covered*
3. The agreed option takes account of site type, site conditions and site features.   *Not covered*

4. The agreed option meets customer requirements for costs and is compatible with the organisation's pricing policy.   *Not covered*
5. The agreed option matches the customer's stated requirements for performance and aesthetics, and accommodates the customer's constraints.   Pages 121–3, 130
6. System requirements are presented in a manner that enables customer agreements.
                                    Not covered
7. Systems designs are presented in an appropriate format.   *Book 1* and Appendix C

Unit P3/2 Specify and monitor programmes for installing and commissioning systems

Element P3/2.1 Design and agree systems to meet customer's needs
This element is beyond the scope of these books.

Element P3/2.2 Negotiate and agree terms and conditions for implementing systems
This element is beyond the scope of these books.

Unit P3/3 Install and test the components of the system

Element P3/3.1 Interpret the installation requirements of the components of the system

1. Installation requirements in respect of systems type, function and performance are identified completely and accurately, using drawings, specifications, manufacturers' instructions provided or from verbal instructions.
    Book 1 and Appendix C. See also relevant B.S.
2. Installation requirements in respect of site types, site conditions, site features and existing main services are identified completely and accurately.   Appendix B
3. Proposed system and layout complies with statutory requirements for gas, water, electricity, health and safety, building control and planning regulations.   *All systems shown comply with these requirements.*
                      See also individual "Regulations"
4. Components, materials and equipment required for the installation are accurately quantified and recorded.   *Not covered*

5. Where there is a variation between the job specification/instructions and the customer's/ contractor's stated requirements, the appropriate action is taken before commencing installation.
                                    Not covered

Element P3/3.2 Prepare sites for installation and testing

1. Site layout, condition and structure is safe for the work to proceed.   *Book 1*
2. Site location and condition is suited to the operational requirements of the components being installed or tested.   Pages 1, 2, 3
3. Site is accessible and free from obstruction for the delivery and storage of materials and other resources necessary for the work to proceed.
                                    Not covered
4. Input services are located and correctly identified and proved suitable for intended purposes.   *Book 1*
5. Where input services prove unsuitable, deficiencies are recorded accurately and promptly and appropriate action taken.   *Not covered*
6. Specified plans, materials, components and equipment are available at the site according to schedule and stored in a safe and secure manner.   *Not covered*
7. In cases where plans, materials, components and equipment are not available, the problem is accurately recorded and appropriate personnel are informed.   *Not covered*
8. Potential disruption to the normal activities of the customer is identified and reported to the customer or the customer's nominated representative in sufficient time for disruption to be minimised.   *Not covered*
9. A commencement, completion date and schedule is confirmed with customers and communicated to suppliers where appropriate.
                                    Not covered
10. Information is passed onto the customer or the customer's nominated representative.
                                    Not covered
11. Disturbance, damage to the customer's site/ fabric/structure is minimised.   *Not covered*
12. Storage of materials, components and equipment conforms to specifications and/or manufacturers' recommendations.   *Book 1*

Element P3/3.3 Fabricate, position and fix components

1. Components to be positioned and fixed are undamaged, of specified type and quantity and are fit for their intended purpose.
   Refer to all related chapters
2. Components are fabricated, positioned and fixed according to the installation specification and schedule using safe and approved methods that meet statutory requirements for gas, water, electricity, health and safety, building control and plumbing regulations.
   Refer to all related chapters
3. The fabrication, positioning and fixing of components complies with British, European and International standards and codes of practice recommended for industry use.
   Refer to all related chapters
4. Where non-specified materials are used, these are of the required type and quantity and are fit for their intended purpose.
   Not covered
5. Where fabrication, positioning and/or fixing cannot be achieved in accordance with the installation specification and schedule, an approved alternative solution is agreed with the customer.
   Not covered
6. Input services are located and correctly identified and proved suitable for their intended purpose. *Refer to all related chapters*
7. Where input services prove unsuitable, deficiencies are recorded accurately and promptly and appropriate action is taken.
   Not covered
8. Existing site/property defects and potential dangers are accurately recorded and reported to the appropriate person before the work begins.
   Not covered
9. Where defects and potential dangers affect the achievements of the installation specification to schedule, appropriate alternative action is taken.
   Not covered

Element P3/3.4 Connect and test components

1. Service connections are in accordance with installation specification, manufacturers' recommendations, statutory requirements for gas, water, electricity, health and safety, building control and plumbing regulations.
   Refer to lists of manufacturers, relevant Codes of Practice and the Building Regulations
2. Connecting and testing of components complies with British, European and International standards and codes of practice recommended for industry use. *Refer to related chapters*
3. Input services are located and correctly identified and proved suitable for their intended purpose. *Refer to related chapters*
4. Where input services prove unsuitable, appropriate action is taken. *Not covered*
5. Connecting and pre-commissioning testing is conducted using safe and approved methods in the required sequence.
   Refer to all related chapters
6. Where defects are identified, appropriate action is taken. *Refer to all related chapters*

Unit P3/4 Commission and decommission system

Element P3/4.1 Carry out checks prior to performance testing

1. The system is clean, flushed and charged for its dynamic operation using methods and agents particular to one system.
   Pages 32, 103, 156
2. Mechanical components function according to manufacturers' specifications.
   Tests to be carried out on site or during practical activities
3. Components meet the design specification for type and are positioned, fixed and connected as specified.
   Refer to individual manufacturer's maintenance instructions and all related chapters
4. The system is leak free at specified pressures.
   Pages 26, 29, 114–15, 186–7, 261, 295–7
5. Electrical power, wiring and control systems are safe for loading. Pages 341, 345, 347–51
6. Required tests are complete according to specified statutory regulations for procedure and outcomes including test certificates.
   Tests to be carried out on site or during practical activities

7. Required information for the safe effective operation of the system is complete, accurate, legible and positioned appropriately for access.

Refer to individual manufacturer's maintenance instructions

8. Identified faults in the static system are rectified prior to dynamic operation.

Faults to be identified during practical activities

Element P3/4.2 Monitor and compare the dynamic performance of the system with design specification and statutory requirements

Work to be carried out on site or during simulated activities

Element P3/4.3 Decommission systems

1. Relevant persons are notified in advance of the decommissioning of systems and appropriate authorisation or agreement is obtained.

Not covered

2. Site conditions, structures/features are safe for the work to proceed. *Not covered*

3. Systems are isolated from power/supply sources using safe and approved procedures.

Pages 68, 115, 299

4. Where necessary, systems are emptied using safe and approved methods.

Pages 68, 115, 299

5. Where contents are to be recovered for use or disposal, safe and approved recovery methods are used. *Not covered*

6. Where spillage/leakage occurs, prompt appropriate action is taken and relevant documentation is complete, accurate and available to an authorised person.

Work to be carried out on site or during simulated activity

7. Systems are verified empty using approved tests. *Work to be carried out on site or during simulated activity*

8. Recovered contents are labelled accurately and legibly for re-use or disposal.

Work to be carried out on site or during simulated activity

9. Systems are designated safe for access for further work and authorised persons are notified promptly when decommissioning is completed.

Work to be carried out on site or during simulated activity

Unit P3/5 Maintain the effective operation of systems

Element P3/5.1 Plan and schedule for planned maintenance of systems

1. Maintenance requirements are planned and recorded using manufacturers' recommendations and design specifications to meet customers' requirements, organisations' requirements and codes of practice.

Pages 67–9, 77, 92–4, 211–15, 235 and refer to Codes of Practice and individual manufacturer's maintenance instructions

2. Maintenance plans conform to statutory regulations, codes of practice, organisational requirements and comply with customers' requirements.

Pages 67–9, 77, 92–4, 211–15, 235 and refer to Codes of Practice and individual manufacturer's maintenance instructions

3. Maintenance schedules identify correct location, specified frequency, notified dates and times and records of maintenance schedules are complete, accurate, legible and available to users.

Pages 67–9, 77, 92–4, 211–15, 235 and refer to Codes of Practice and individual manufacturer's maintenance instructions

4. Materials, equipment and tools are available for the implementation of the maintenance plan.

Chapter 1 and Pages 187, 188 and Book 1 Chapter 2

Element P3/5.2 Routinely maintain system components to sustain effective system performance

1. Components are routinely maintained according to plans and schedules or instructions.

Pages 67, 110–14, 194, 227, 238, 276, 326 and refer to Codes of Practice and individual manufacturer's maintenance instructions

2. Maintenance routines comply with manufacturers' recommendations (or other authorised recommendations) for required tests

and checks for correct operation and are carried out using approved methods.

> Pages 67, 110–14, 194, 227, 238, 276, 326
> and *refer to Codes of Practice and individual
> manufacturer's maintenance instructions*

3. Where components are not meeting performance expectation, these are cleaned or repaired or replaced with components of equal technical capability, using approved and safe methods after isolation from the power/supply source.

> Pages 67, 110–14, 194, 227, 238, 276, 326
> and *refer to Codes of Practice and individual
> manufacturer's maintenance instructions*

4. Components are cleaned using appropriate cleaning agents and equipment.   Page 262
5. Where cleaning agents are hazardous, appropriate protective equipment is used.

> Page 262

6. Records of routine maintenance are complete, accurate, legible, in the required format and available to authorised persons.   *Not covered*

Element P3/5.3 Diagnose and rectify the cause of faults to restore effective system performance

1. Faults once reported are responded to promptly in accordance with organisational procedures.
> *Not covered*
2. Where it is safe to proceed, system faults are located and identified using approved or appropriate diagnostic methods.
> Pages 68–9, also *refer to all related chapters
> and individual manufacturer's instructions*
3. Diagnostic and rectifying actions minimise risk to individuals and the environment.
> Page 3, also *refer to all related chapters
> and individual manufacturer's instructions*
4. Rectifying action is carried out in agreement with customers.   *Refer to all related chapters
> and individual manufacturer's instructions*
5. Rectifications restore effective system performance.   *Refer to all related chapters
> and individual manufacturer's instructions*
6. Diagnosing and rectifying action is recorded and reported in accordance with statutory regulations, codes of practice and organisational requirements and comply with manufacturers' warranties where applicable.   *Not covered*

7. Where fault rectification is not possible, or where only partial or temporary system performance is achieved, this is reported to the customer and authorised persons accurately and promptly and the system is left safe.

> Page 21, see *Gas Regulations*

Unit P3/6 Maintain the safe working environment

*For more detailed information refer to the
Construction (Health and Welfare) Regulations*

*Element P3/6.1 Monitor and maintain one's own
health and safety, and contribute to the maintenance
of the health and safety of others*

1. Current regulations, recommendations and guidelines for health and safety protection of self and others are followed.   *Book 1*
2. The immediate working area (including all equipment, fixtures, fittings, materials and components within the job holder's responsibility) is free from hazards.   *Book 1*
3. Accidents are reported promptly to authorised persons and recorded accurately, completely and legibly in the approved format.   *Book 1*
4. Suppliers' and manufacturers' instructions relating to the safety of all equipment, fixtures, fittings, materials and components are followed.
> *Refer to individual manufacturer's instructions*
5. Approved/safe methods and techniques are used when lifting.   *Book 1*
6. Personal safety equipment is maintained, worn and used as appropriate to the nature of the work situation.   *Book 1*
7. Where unsafe work practices are identified, prompt remedial action is taken and reported.
> *Book 1*

*Element P3/6.2 Contribute to the limitation of
damage to persons and premises in the event of
accident and emergency*

1. Injuries to others resulting from accidents and emergencies receive prompt and appropriate assistance.   *Book 1*
2. Damage to property is identified and specified, and appropriate action is taken to minimise further damage   *Not covered*

3. Visitors are alerted to potential hazards.

Not covered

4. Agreed procedures in the event of an emergency are followed where appropriate.

Book 1

Element P3/6.3 Contribute to the limitation of damage to persons or property in the event of fire, explosion or toxic atmospheres

1. Professional emergency services are summoned immediately in the event of a fire/explosion or symptoms of a fire/explosion. *Not covered*

2. Alarm/alert/evacuation/disaster systems and safety systems are activated promptly

Not covered

3. Containable fires are extinguished using appropriate equipment, only in circumstances where a delay in evacuation does not increase risk to the job holder and others. *Book 1*

4. In the event of fire alarm/evacuation/ emergency warnings, work is suspended immediately, equipment is isolated from the power/fuel source, if safe to do so, and job holders proceed to a safe place in accordance with assembly procedures by approved (or quickest safe alternative) routes.

Not covered

5. In the event of toxic atmosphere occurring, appropriate action is taken promptly. *Book 1*

Element P3/6.4 Agree and maintain a safe environment with co-contractors

1. Systems for safeguarding persons are agreed and disseminated to workers. *Book 1*

2. Agreed systems conform to Health and Safety Act. *Book 1*

3. Responsibility and procedures for monitoring the agreed systems are effective. *Book 1*

4. Identified hazards in the working environment are made safe and/or, where appropriate, notified to relevant and authorised persons promptly. *Book 1*

Unit P3/7 Maintain effective working relationships

Element P3/7.1 Establish and develop professional relationships with customers and co-contractors
This element is beyond the scope of these books.

Element P3/7.2 Establish and maintain professional relationships with authorised site visitors
This element is beyond the scope of these books.

Element P3/7.3 Maintain effective working relationships with colleagues
This element is beyond the scope of these books.

Unit P3/8 Contribute to quality development and improvement of products and services

Element P3/8.1 Promote the organisation's/industry's image to existing and potential customers and co-contractors
This element is beyond the scope of these books.

Element P3/8.2 Identify the organisation's/industry's image to existing and potential customers and co-contractors
This element is beyond the scope of these books.

Element P3/8.3 Encourage energy efficiency
This element is beyond the scope of these books.

Index